Oxidative Stress in Applied Basic Research and Clinical Practice

Series Editor

Narasimham Parinandi, Dept Internal Med, Pulmonatry Division
Ohio State University
Columbus, USA

The current definition of oxidative stress (OS) is an abnormal level of reactive oxygen-nitrogen species leading to cellular injury, which can be prevented by antioxidant (AOX) intake providing a therapeutic advantage. When this delicate balance is perturbed, homeostasis is compromised and disease may develop. Extensive research has shown a strong correlation with these events. Springer has been a leader in publishing books related to oxidative stress since 1998. The goal of the Oxidative Stress in Basic Research and Clinical Practice series is to present data showing relationships between OS and biomarkers, as well as results from clinical trials illustrating the efficacy of AOX supplementation. Each volume covers a medical specialty, or major disease known to be associated with an imbalance between free radical production and AOX defense. The expectation of the series is to critically evaluate the potential of AOXs in therapeutic intervention using simple and sophisticated methodology that could be adapted to point-of-care testing for diagnosis, intervention strategies and prognosis. This approach of combining methodology, technology and epidemiology will benefit medical practitioners and may subsequently improve health care and reduce medical cost. Editors anticipate updating their respective volumes as frequently as warranted by emerging new studies.

Sakima A. Smith • Shane S. Scott
Editors

Myocardial Oxidative Stress in Anti-Cancer Therapies

Springer

Editors
Sakima A. Smith
Dorothy M. Davis Heart and Lung Research Institute
The Ohio State University College of Medicine
Columbus, OH, USA

Shane S. Scott
Dorothy M. Davis Heart and Lung Research Institute
The Ohio State University College of Medicine
Columbus, OH, USA

The Ohio State Medical Scientist Training Program (MSTP)
Columbus, OH, USA

ISSN 2197-7224 ISSN 2197-7232 (electronic)
Oxidative Stress in Applied Basic Research and Clinical Practice
ISBN 978-3-031-97516-5 ISBN 978-3-031-97517-2 (eBook)
https://doi.org/10.1007/978-3-031-97517-2

This Springer imprint is published by the registered company Springer Nature Switzerland AG
The registered company address is: Gewerbestrasse 11, 6330 Cham, Switzerland

Preface

Cardiovascular disease (CVD) accounts for significant global morbidity, and mortality. While novel anticancer therapies have improved survival for many cancer patients, several of these drugs impart serious short- and long-term cardiotoxicity. In fact, cardiovascular disease is the leading cause of death among cancer patients and survivors, not cancer. Myocardial oxidative stress resulting from an imbalance between reactive oxygen species and endogenous antioxidants is the crux of anticancer drug-induced cardiac dysfunction and heart failure. In addition, patients' comorbid CVD risk factors also seem to correlate with increased risk of cardiotoxicity. Worthy of note, among women treated with anthracyclines, anthracylcline-induced cardiomyopathy disproportionally affects black women and other marginalized groups.

At this time, development of pharmacotherapies against myocardial oxidative stress-mediated cancer therapies is limited, as successful animal modeling to evaluate anticancer drug-related cardiotoxicity does not always translate clinically. Moreover, early proof-of-concept clinical trials evaluating cardiotoxicity, prevention, and treatment are lacking.

This Special Issue aims to summarize the present knowledge of oxidative stress in anticancer drug-induced cardiotoxicity including cellular and molecular mechanisms, prevention and treatment strategies, drug research, and population experiments. Chapters also enumerate the racial and gender disparities in anticancer therapy-related cardiotoxicity. This book should serve as source of knowledge and guidance for research, clinical prevention, and treatment of cardiotoxic anticancer drugs for cardiologists, cardio-oncologists, and heart failure researchers.

Risk factors Associated with Oxidative Stress-Mediated Cardiac Dysfunction

Several factors preclude and augment the development of myocardial oxidative stress and subsequent cardiac dysfunction. In the chapter titled "Risk factors Associated with Anthracycline Induced Cardiac Dysfunction," Ailemen *et al* delineate the critical risk factors associated with anthracycline-induced cardiotoxicity (AIC). While anthracyclines have established themselves as a fundamental component in multiple cytotoxic anticancer regimens and have contributed to the reduction of cancer mortality. Despite their efficacy in cancer treatment, anthracyclines are correlated with considerable cardiac dysfunction, affecting patients with varying levels of severity. Ailemen *et al* systematically identifies modifiable and non-modifiable risk factors for AIC which includes cumulative drug dose, smoking history, alcohol use, obesity, sex, race, and prior cardiovascular history. The authors also illuminate genes, and single nucleotide polymorphisms that confer increased AIC. Furthermore, this chapter underscores the importance of identifying high-risk patients and implementing targeted cardioprotective measures prior to anthracycline treatment for preemptive management of cardiotoxicity induced by these life-saving anticancer therapies.

Racial and Gender Disparities in Anticancer Drug-Related Cardiotoxicity

Recent studies have determined the presence of disparate risk for development of cardiovascular disease and mortality both during and after anticancer therapies based on sex, race, and membership of other underrepresented groups. In the chapter titled "Multilevel Drivers of Disparities in Cardio-Oncology," McCrohan *et al* provide in-depth elucidation of the current state of cardio-oncology regarding these disparities. The authors review current data on cardio-toxicity disparities, specifically by sex, focusing on imbalances in the risk of adverse cardiac events related to cancer treatment in women with a variety of cancer diagnoses, such as lymphoma, non-small cell lung cancer, breast cancer, and esophageal cancer. Evidence for health inequities by ethnicity, race, and membership in other underrepresented groups, such as lesbian, gay, bisexual, transgender, queer, intersex, and/or asexual individuals, and pediatric patients who have received radiotherapy and targeted or immune-based therapies are also described. Further, both non-biologic and biologic sources for disparate cardiovascular risk and outcomes are highlighted. Non-biologic contributors encompass social determinants of health, lack of clinical trial representation for underrepresented groups, and differential environments. Biologically, studies have pinpointed epigenetic stressors, genomic markers, and inflammation pathways as contributors to cardiotoxicity risk. The authors also emphasize areas for further exploration toward addressing disparities in the field of cardio-oncology.

Cardiac Cell Signaling Mechanisms in Myocardial Oxidative Stress

Owing to the high myocardial oxygen demand and delicate redox balance, anticancer therapy-induced myocardial oxidative stress is a major mechanism of interest. In the chapter titled "Cardiotoxicity, Cancer Therapies, and Cardiac Cell Signaling Mechanisms," Desai and Baliga provide an overarching review of the cardiac cell signaling mechanisms altered by prolific cardiotoxic anticancer therapies in the chapter. The authors describe the intricate relationship between cancer therapies and their cardiotoxicity, detailing the causative alterations in critical cardiac cell signaling pathways, including oxidative stress, mitochondrial dysfunction, inflammation, apoptosis, and autophagy dysregulation. Together these disruptions compromise the structural and functional integrity of cardiomyocytes, leading to arrhythmias, cardiomyopathy, and heart failure. The discussion provides an overview of the cardiac signaling pathways altered by anthracyclines, trastuzumab, and tyrosine kinase inhibitors, which are widely used in cancer treatment but are associated with both distinct and overlapping mechanisms of cardiac injury. Understanding the molecular mechanisms underlying these adverse cardiac effects is essential for developing effective strategies to mitigate cardiotoxicity without compromising the efficacy of cancer therapies.

Phytochemicals in Cardioprotection and Cardiotoxicity

Phytochemicals have garnered increasing attention as therapeutic agents due to their structural diversity, bioavailability, and low toxicity compared to synthetic drugs. Phytochemicals are particularly compelling targets for cardiovascular research and clinical management due to their ability to modulate multiple molecular pathways, including oxidative stress reduction, inflammation suppression, and mitochondrial stabilization. In the chapter on "Phytochemicals in Cardioprotection and Cardiotoxicity," Scott *et al* explore the multifaceted roles of phytochemicals in addressing cardiovascular diseases (CVDs) and cardiotoxicity, which pose significant health challenges. The authors provide a comprehensive review of several classes of phytochemicals, including polyphenols, flavonoids, alkaloids, terpenes, terpenoids, and quinones. Key compounds from each class and their mechanisms of action in mitigating myocardial oxidative stress—from in vitro and in vivo animal studies are discussed.

The chapter incorporates clinical trial evidence validating the cardioprotective and therapeutic roles of several of these compounds. Where possible, the authors also provide current evidence supporting the emerging role of several phytochemicals in mitigating cardiotoxicity associated with chemotherapeutic agents, such as doxorubicin. This chapter advances the cardiovascular pharmacology discourse on the potential of phytochemicals to fill the gaps where many synthetic drugs fail to attenuate myocardial oxidative stress in CVD and cardiotoxicity and offer insights that will guide both researchers and clinicians in advancing therapeutic strategies.

Detection, Prevention, and Management of Anthracycline-Induced Cardiotoxicity

Cardiotoxicity can manifest both as an acute phenomenon or late sequalae. As cancer survivorship has increased, the long-term irreversible cardiovascular effects have become more evident. As such, several groups have called for the development of guidelines for the prevention and management of AIC. In the chapter titled "Detection, Prevention, and Management of Anthracycline-Induced Cardiotoxicity," Schmidt *et al* highlight current understanding in the literature of tools for detection, prevention, and management of myocardial damage caused by anthracyclines. The chapter begins by explaining the underlying mechanisms of anthracycline-induced cardiotoxicity, such as myocardial oxidative stress, as well as emerging cellular pathways, including ferroptosis and necroptosis. It highlights the latest diagnostic tools, such as strain imaging and biomarkers, that allow for earlier and more accurate detection of subclinical myocardial injury. The discussion continues to elaborate on current treatment options, including FDA-approved medications like dexrazoxane, and other conventional cardioprotective therapies like β-blockers, dietary/nutritional intervention, and exercise-based programs. Moreover, the chapter illuminates promising new research on treatments targeting the ferroptosis and necroptosis pathways. The examination and management of AIC is critical, because it addresses the necessary need to protect the myocardium from anticancer agents, without compromising their efficacy against cancer targets.

Current Pharmacologic Prevention Treatments Against Oxidative Stress-Induced Cardiotoxicity

Oxidative stress is arguably the most common overlapping cellular signaling mechanism underlying cardiotoxicity for several anticancer therapies. As such, pharmacotherapies targeting these signaling cascades go a long way to help reduce myocardial oxidative stress, improve the safety of anticancer therapies, and reduce patient morbidity and mortality. In the chapter titled "Pharmacotherapies in Prevention of Myocardial Oxidative Stress Mediated by Anthracyclines, Taxanes, and Alkylating Agents," Osuji and colleagues explore current preclinical and clinical evidence-based therapies being used to combat cardiotoxicity with a focus on nonspecific agents such as anthracyclines, taxanes, and alkylating agents. For newer targeted therapies, the discussion continues with the chapter titled "Optimizing Cardiovascular Safety: Cardiotoxicity Prevention in Tyrosine Kinase Inhibitor,

Monoclonal Antibody, Proteasome Inhibitor, and 5-Fluorouracil Anticancer Therapies." Wright *et al* consolidate and compare multiple mechanisms driving cardiac oxidative stress caused with tyrosine kinase inhibitors, monoclonal antibodies, 5-fluorouracil, and proteasome inhibitors, and outline the current understanding of the pharmacologic treatment strategies for oxidative stress induced by these novel anticancer chemotherapies.

The Future of Cardio-oncology

Cardiotoxicity is a significant concern for cancer patients, oncologists, and cardiologists, as it presents with variable risk depending on age, type of anticancer therapy, and cancer type. Significant strides made in the field of cardio-oncology have largely focused on understanding the mechanistic underpinning of cardiotoxicity and developing cardioprotective strategies against anthracyclines. However, gaps exist in the availability of cardioprotective therapies for newer agents, such as immune checkpoint inhibitors (ICIs). Scott *et al* briefly review the history of field in the chapter on "The Future of Cardio-oncology." The authors discuss the role of cardiac imaging, emerging challenges of inflammatory cardiotoxicity syndromes stemming from ICIs and CAR-T cell therapies and discuss current cardioprotective landmark randomized clinical trials. The discussion continues to illuminate current gaps in care, health disparities, use of human inducible pluripotent stem cell cardiomyocytes, and the potential of artificial intelligence to augment risk stratification. Additionally, the authors detail the need for continued research, further advancement of diagnostic tools, and more diverse randomized clinical trials for the mitigation of long-term cardiovascular complications stemming from cancer therapies.

Columbus, OH, USA

Sakima A. Smith
Shane S. Scott

Acknowledgments

We are extremely grateful to the Scott family, Candice Bariffe, James W. Scott, Jaconda Jackson-Scott, and Smith family, Ahmad Smith, and Renae Smith for their support, love, and encouragement in successful completion of this book.

We express our deepest gratitude to our contributors of chapters and the reviewers who have made this book possible.

We extend our special thanks to Merry Stuber and Amrita Unnikrishnan for their diligence, unwavering support, and patience during the editorial process.

Sakima A. Smith
Shane S. Scott

Contents

Part I Epidemiology and Cardiac Signaling

Risk Factors Associated with Anthracycline-Induced Cardiotoxicity ... 3
Israel Ailemen, Shane S. Scott, Majd Alfayez, Joel Ferrall, Anna Matzko, Ashley N. Greenlee, and Sakima A. Smith

Multilevel Drivers of Disparities in Cardio-Oncology ... 21
Megan McCrohan, Tyren Busby, Dionisia Quiroga, Mary Branch, Arnethea L. Sutton, Sherry-Ann Brown, Sarju Ganatra, Sanam Ghazi, Onaopepo Kola-Kehinde, Mussammat Ferdousi, Karen P. Williams, Eric D. Miller, and Daniel Addison

Cardiotoxicity, Cancer Therapies, and Cardiac Cell Signaling ... 33
Virang Desai and Ragavendra R. Baliga

Phytochemicals in Cardioprotection and Cardiotoxicity ... 51
Shane S. Scott, Cora Albers, Maryam Iqbal, Israel Ailemen, Sakima A. Smith, and Narasimham L. Parinandi

Part II Current Trends in Detection and Management

Detection, Prevention, and Management of Anthracycline-Induced Cardiotoxicity ... 91
Jonathan Schmidt, Shane S. Scott, Majd Alfayez, Maryam Iqbal, Joel Ferrall, Anna Matzko, Ashley N. Greenlee, and Sakima A. Smith

Pharmacotherapies in Prevention of Myocardial Oxidative Stress Mediated by Anthracyclines, Taxanes, and Alkylating Agents ... 105
Evans Osuji, Jonathan Wright, Tanvi Brar, Shane S. Scott, and Sakima A. Smith

Optimizing Cardiovascular Safety: Cardiotoxicity Prevention in Tyrosine Kinase Inhibitor, Monoclonal Antibody, Proteasome Inhibitor, and 5-Fluorouracil Anticancer Therapies 157
Jonathan Wright, Evans Osuji, Tanvi Brar, Shane S. Scott, and Sakima A. Smith

The Future of Cardio-Oncology 193
Shane S. Scott, Cora Albers, Ashley N. Greenlee, Thai H. Ho, Daniel Addison, and Sakima A. Smith

Index 207

Authors Biography

Shane S. Scott, PhD, MS is a MD/PhD candidate, and aspiring cardiothoracic surgeon-scientist at The Ohio State University College of Medicine. He received Bachelor of Science degrees with Honors in chemistry and biochemistry from Brandeis University in 2013. He received a Master of Science degree in Medical Sciences from Boston University College of Medicine in 2018. He recently completed his PhD at The Ohio State University.

During his graduate training Dr. Scott's research was twofold. On the one hand, he explored role of tyrosine kinase inhibitors in cardiac arrhythmia and heart failure. On the other, he explored the mechanism(s) by which the absence of cardiac βII-spectrin promotes the development of arrhythmias, heart failure, and premature death. He is currently a third-year medical student and hopes to pursue residency in the field of cardiothoracic surgery.

He has published several peer-reviewed articles and contributed several book chapters. The *Myocardial Oxidative Stress in Anti-cancer Therapies and Cardiovascular Disease* represents his first volume as the lead co-editor.

Sakima A. Smith, MD, MPH is currently an Associate Professor of medicine with tenure at The Ohio State University. He obtained his medical degree from Temple University School of Medicine in 2006. He completed his internal medicine residency at Barnes-Jewish Hospital/Washington University in 2009 and completed his general cardiology and heart failure/transplant fellowships at Ohio State. He also completed a research fellowship and Masters in Public Health from Ohio State in 2014. He was awarded the James V. Warren Fellowship Research Award in 2014, the Denman Distinguished Undergraduate Research Mentor award in 2017, and Distinguished Researcher of the Year Award from Landacre Research Honor Society in 2019. He established an independent research lab with funding from the Robert Wood Johnson Faculty Development Grant, a NIH K08 Grant, and a NIH R01 Grant. He has current R01 funding as PI, is Co-I on several grants, and is the co-PI on a novel T32 postdoctoral fellowship grant. His research program spans basic cardiovascular research to population health, diversity, and health outcomes. His current R01 funded work aims to expand the field of cardio-oncology, with a focus on understanding how tyrosine kinase inhibitors and other anticancer agents contribute to cardiotoxicity. His seminal research using both animal and human models has demonstrated that TKIs can induce adverse cardiac events through multiple mechanisms.

Part I
Epidemiology and Cardiac Signaling

Risk Factors Associated with Anthracycline-Induced Cardiotoxicity

Israel Ailemen, Shane S. Scott, Majd Alfayez, Joel Ferrall, Anna Matzko, Ashley N. Greenlee, and Sakima A. Smith

Abbreviations

AIC	Anthracycline-induced cardiotoxicity
HTN	Hypertension
CHF	Congestive heart failure
CDC	Center for Disease Control and Prevention
NHBs	Non-Hispanic Blacks
LVEF	Left ventricular ejection fraction
WHO	World Health Organization
CAD	Coronary artery disease
DM	Diabetes mellitus
GFR	Glomerular filtration rate

I. Ailemen · M. Alfayez · A. Matzko · A. N. Greenlee · S. A. Smith
Dorothy M. Davis Heart and Lung Research Institute, The Ohio State University College of Medicine, Columbus, OH, USA
e-mail: israel.ailemen@osumc.edu; majdalfayez1126@gmail.com; ashley.greenlee@osumc.edu; sakima.smith@osumc.edu

S. S. Scott (✉)
Dorothy M. Davis Heart and Lung Research Institute, The Ohio State University College of Medicine, Columbus, OH, USA

The Ohio State Medical Scientist Training Program (MSTP), Columbus, OH, USA
e-mail: shane.scott@osumc.edu

J. Ferrall
Keck School of Medicine, University of Southern California, Los Angeles, CA, USA
e-mail: jferrall@usc.edu

S. A. Smith, S. S. Scott (eds.), *Myocardial Oxidative Stress in Anti-Cancer Therapies*, Oxidative Stress in Applied Basic Research and Clinical Practice,
https://doi.org/10.1007/978-3-031-97517-2_1

Introduction

Therapeutic advancements have reduced the long-term cancer mortality rates for many solid and blood cancers. Anthracyclines emerged as a cornerstone for the treatment of several cytotoxic anticancer regimens and have reduced overall mortality of breast cancer, sarcoma, lymphoma, and some leukemias malignancies [1]. However, anthracycline-induced cardiotoxicity (AIC) has emerged as an unwanted cardiotoxic effect [2]. Specific research has been developed to explore the risk factors that confer increased development of AIC in patients. Both nonmodifiable and modifiable risk factors have been identified in the preclinical and clinical cohort and case studies.

A significant amount of research has demonstrated a dose-related increased risk for AIC—and represents an important modifiable risk factor. Other risk factors include smoking, alcohol use, nutrition, and obesity [3–5]. Nonmodifiable risk factors include extremes of age (age >65 years or <4 years), female gender, and African American race [6]. Available studies also show that concomitant cardiovascular disturbances are associated with AIC. Prior history of hypertension (HTN), diabetes, renal failure, prior mediastinal radiation therapy, and concomitant treatment with other cytotoxic agents including cyclophosphamide, trastuzumab, or paclitaxel have also been recognized [6, 7]. Recently, the presence of prior cardiac disease including left ventricular dysfunction and genetic mutations has also been reported. Socioeconomic demographics and disparities have also demonstrated increased risk and role in predisposition to AIC [8].

Owing to the irreversibility of AIC, the development of targeted research and interventions should be aimed at minimizing its impact by risk-stratifying patients based on risk factors. In this chapter, we will discuss the modifiable and nonmodifiable risk factors that contribute to cardiotoxicity in patients who receive anthracycline therapy. In addition, the current mechanisms underlying increased AIC will also be explored (Table 1).

Table 1 Risk factors associated with anthracycline-induced cardiotoxicity

Nonmodifiable risk factors
Female sex
Age
Race: Non-Hispanic Blacks (NHB), Hispanic, and Asian
Modifiable risk factors
Smoking
Alcohol
Obesity
Cardiovascular risk factors
Hypertension
Diabetes
Renal failure

Risk Factors

There are two general categories that create an increased risk of AIC: modifiable and nonmodifiable factors. Modifiable factors include both medical and nonmedical risk factors. Medical factors include pre-existing cardiovascular conditions (HTN, prior ventricular dysfunction or myocardial infarction), smoking habits, renal disease, previous or concurrent chemotherapies, chest radiotherapy, and cumulative anthracycline dose. Nonmodifiable risk factors include demographic and lifestyle factors such as age, sex, and race [6]. This division of risk is arbitrary, and many of these factors are multifactorial since most are medical and social.

Nonmodifiable Risk Factors

Sex

Female sex has been associated with an increased risk of developing AIC [5, 9]. This is likely due to the fact that men more efficiently clear, and metabolize DOXO, as evidenced by increased serum metabolite (doxorubicinol) in males, indicating a higher metabolic rate [5, 10]. Alternatively, it has been hypothesized that estrogen is a key distinguishing factor by serving in a cardioprotective role [5]. The authors hypothesize that the sexual dimorphism of some proteins related to mitochondrial oxidative stress may prove to be a driving mechanism for the increased risk of AIC, especially in prepubertal and postmenopausal female populations who lack the cardioprotective effects of estrogen [5]. With this in mind, it is conceivable that both female sex and age are risk factors for increased risk of AIC.

Age

Age is the most studied risk factor for cancer, and cancer can be considered an age-related disease [11]. Owing to this, patient age at the time of anthracycline treatment appears to be a significant determinant of AIC development, particularly the extremes of age (age <4 years or >65 years). While extensive research is currently being performed to uncover the mechanisms behind this phenomenon, the true explanation is likely multimodal.

In the United States, an estimated greater than 50% of childhood cancer survivors have been treated with anthracyclines and thus are at risk of developing cardiotoxicity [12]. Patients less than 18 years of age are unlikely to present with pre-existing cardiovascular conditions before anthracycline treatment. However, childhood cancer survivors treated with anthracyclines face a significantly elevated risk of developing AIC and AIC-associated CHF, with these survivors having a risk 5–15 times higher than their peers without anthracycline exposure [5, 7, 13].

Analogous to older patients, the exact reason for this disparity is unclear; however, some inferences can be made. There is evidence to suggest that the pharmacokinetics of anthracyclines are different in this population. In a 2011 study, researchers discovered that absorption, plasma protein binding, metabolism, and excretion rates of various medications were reduced in this population, while the volume of distribution was greater compared to their adult counterparts [10]. Although this study did not specifically examine the pharmacokinetics of anthracyclines, it is speculated that altered tissue distribution of anthracyclines could lead to increased drug retention and exposure, thereby increasing the risk of developing AIC. Further research in this area holds promise for exploring this possibility in future studies.

For older patients, several factors could contribute to their elevated risk. As the prevalence of cardiovascular diseases escalates with age (Table 2), older individuals generally have a higher predisposition to suffer from cardiovascular dysfunction in the setting of anthracycline treatment [4]. Consistent with this, a systematic review examining baseline blood pressure and the development of AIC demonstrated that the risk of AIC is exacerbated in the presence of pre-existing hypertension [5]. Older patients are also significantly more likely to develop AIC-associated congestive heart failure (CHF). This increase in CHF in this population may be explained by the diminished reserve capacity and adaptative ability of the aging heart [13, 14]. Additionally, decreased renal function, which increases with age, may result in poor

Table 2 Prevalence of AIC risk factors in US adult population

Comorbid disease	Age, years	Prevalence (%)	References
CHF	60–70	4.93% (95% CI: 3.58–6.76)	[16]
	70–79	6.96% (95% CI: 5.36–8.98)	[16]
	≥80	10.14% (95% CI: 7.83–13.04)	[16]
HTN	>60	60%	[17]
	≥80	74%	[17]
CAD	40–59	~40%	[18, 19]
	60–79	~75%	[18, 19]
	≥80	~86%	[18, 19]
HLD	40–49	15.1%	[20]
	≥60	11.7%	[20]
PAD	40–49	1.43%	[21]
	≥70	16.62%	[21]
Chronic renal insufficiency[a]	65–79	21.7%	[22]
	≥80	51.1%	[22]
T2DM	50–64	13.3% vs 16.3%[b]	[23]
	≥65	20.1%[c]	[23]

[a]GFR eGFR 15–59 ml/min/1.73m^2

[b]comparing metropolitan and nonmetropolitan areas

[c]both metropolitan and nonmetropolitan areas

Abbreviations: CAD coronary artery disease, *CI* confidence interval, *CHF* congested heart failure, *HLD* hyperlipidemia, *HTN* hypertension, *PAD* peripheral artery disease, *T2DM* type II diabetes mellitus anthracycline blood clearance, increased anthracycline circulation, and cardiac exposure [15]. Together, these data suggest that age-related changes in renal function promote AIC-related cardiovascular dysfunction. Nonetheless, further exploration is required to validate the higher incidence of AIC in this population.

Race

Race, despite being a social construct, has emerged as an important consideration for the development of AIC. All cancer survivors are at increased risk for cardiovascular disease; however, the risk is higher for marginalized racial and ethnic minorities. While research and data are limited regarding AIC as a variable of patient race, a large cohort study indicated that non-Hispanic black (NHB), Hispanic, and Asian groups were more likely to suffer AIC-associated heart failure than non-Hispanic White patient groups with similar treatment conditions and control for cardiovascular risk factors [24]. The exact source of this disparity is still under contention; however, racial disparities in access to quality medical care may be just one cause for this difference in the group [25]. Additionally, these groups could also contain varying economic classes which would also greatly affect the quality of care that a patient receives and could be a key area of future research.

Disparities

Disparities in AIC exist between sex, race, and age groups [7]. In this section, we wish to highlight the disparities in the occurrence and treatment of AIC and theorize on the underlying, "fundamental causes" [15]. Marginalized and minority populations experience lower rates of oncologic and cardiovascular care from screening to survivorship [26]. As such, oncologic treatment-induced cardiotoxicity may exacerbate these health disparities. For example, Black female breast cancer survivors have an increased incidence of cardiotoxicity morbidity and mortality compared with their White counterparts [8]. It is important to recognize that risk factors are often defined retrospectively with significant limitations, and while generalizable, they are still limited. In many studies of AIC, the authors mention "African American ancestry" as a significant risk factor. Yet, there is likely no true ancestral difference, but rather historical and ongoing structural violence that manifests as health disparities. It is well known that there are racialized disparities within oncology as well as cardiology [5, 14, 27]. This creates an area of intersection in cardio-oncology where significant disparities persist. In a retrospective study, Hasan and colleagues observed a threefold increased risk of developing AIC among a 100-patient African American cohort compared to an established reference cohort [3]. Lotrionte et al. also showed that African American ethnicity is an independent, significant risk factor for the development of AIC [28].

Modifiable Risk Factor

Cumulative Anthracycline Dose

Cumulative dose of anthracyclines is a well-documented risk factor for AIC [29–31]. For doxorubicin, the most commonly used anthracycline, the risk of left ventricular dysfunction and development of heart failure increases significantly with higher cumulative doses. At a cumulative dose of 400 mg/m^2, the incidence of heart failure is approximately 5%, 26% at 550 mg/m^2, and 48% at 700 mg/m^2 [30]. Therefore, it is recommended that lifetime cumulative dose of doxorubicin does not exceed 400–450 mg/m^2 [30]. In comparison, anthracyclines, including epirubicin, have different dose thresholds for cardiotoxicity due to variations in pharmacokinetics and potency. For epirubicin, cumulative doses exceeding 900 mg/m^2, significantly increase risk of AIC [31]. In older patients, the cumulative anthracycline dose required for AIC is lower [32]. Notably, Hoff et al. demonstrated higher incidence of AIC in patients >60 years compared to younger patients [32]. At cumulative doses of 250 mg/m^2, anthracycline-induced cardiomyopathy occurred at rate of 1.5% (40–59-years-olds) compared to 2.4% (age ≥60 years) while at 400 mg/m^2: 2.3% vs 4.6%, and at 600 mg/m^2: 14.9% vs 22.4% [32]. Comorbid infection [29, 33] and renal impairment/disease may further complicate anthracycline administration and worsen AIC. Although anthracyclines are primarily metabolized in the liver [34], severe renal dysfunction can alter drug clearance and exacerbate toxicity. Impaired renal function can increase systemic exposure to anthracyclines by affecting drug-metabolizing enzymes and transporters, leading to prolonged drug half-life and higher cardiac risk. Consequently, patients with renal dysfunction require careful dose adjustments and close monitoring to reduce the risk of AIC.

Modifiable Lifestyle Risk Factors

Several lifestyle risk factors including smoking, alcohol consumption, and stress have been associated with increases in cardiovascular disease. Emerging work also suggests that these risk factors are associated with increased incidence of AIC.

Smoking

Smoking is among the most well-researched lifestyle risk factors and has been established as a source of cardiovascular oxidative stress, inflammation, and endothelial damage [4]. Among patients who develop AIC, there is conflicting evidence regarding the association between AIC and smoking. In a recent systematic review involving 9 studies and 5862 patients, Qiu et al. found no significant association between smoking and anthracycline-associated cardiac events [5]. However, in a separate systematic review and meta-analysis involving 11 studies and 2192 patients, researchers demonstrated an association between AIC and smoking utilizing the Framingham risk scores [35]. Additionally, in a single-center study seeking to examine the long-term effects of standard chemotherapy on myocardial function in asymptomatic breast cancer survivors who received anthracycline treatment with and without adjuvant trastuzumab and control group, Ho et al. concluded that cigarette smoke was a negative predictor of longitudinal strain only in the chemotherapy group [36].

Numerous studies exploring the mechanistic link between these factors are sparse, and the underlying mechanism remains partially understood. Patients with a history of smoking and complications associated with chronic smoking have induced a "*double hit*" effect as many of the diseases overlap with AIC [13]. Another mechanism may involve increased concentrations of reactive oxygen species via increased TRPC3-Nox2 complexes, which can be induced by doxorubicin and smoking [37, 38]. Nishiyama et al. showed that these TRPC3-Nox2 complexes are associated with cardiac atrophy in AIC and that *ibudilast*, an anti-asthmatic drug, can prevent both doxorubicin and passive smoking cytotoxicity [38]. Additionally, it may be likely that a larger proportion of those with a history of smoking would receive anthracycline-based therapies as a link has been established between an increased risk of cancers that are typically treated with anthracyclines, like breast and lung cancer, and smoking practices [14]. It may also be likely that smoking inadvertently increases the risk of AIC by prolonging the duration of anthracycline-based chemotherapies. In vitro testing has established that increased nicotine concentrations can weaken the cytotoxic effects of anthracyclines, as a result reducing the tumor response to treatment [27]. This reduction in tumor response, as a result of a substance that is found in the majority of tobacco and smoking products, may lead to extended treatment with anthracyclines or an increased number of anthracycline treatment cycles, which in turn would increase the risk of AIC. This relationship has yet to be established, and further research is required to examine the effect of nicotine on anthracyclines in vivo and with clinically relevant concentrations.

Alcohol Use

Excessive alcohol consumption is another common cardiovascular disease risk factor and has also been linked to increased myocardial oxidative stress, mitochondrial dysfunction, accelerated protein degradation, hypertension, cardiomyopathy, and even arrhythmias [4, 15]. Among patients with AIC, Reinbolt et al. demonstrated increased risk (OR = 2.39) for AIC development in patients consuming greater than eight alcoholic drinks per week compared to controls, whereas patients with moderate consumption (<8 drinks per week) had a significantly lower risk of cardiotoxicity (OR = 0.49) [15]. Despite this compelling data, more research is required to establish alcohol use as a risk factor for AIC.

Obesity

Overweight and obese patients, defined by the CDC as having a BMI above 25 and 30, respectively, account for over 40% of Americans and are an independent risk of developing cancers and cardiovascular diseases [4]. In a meta-analysis by Qiu et al. examining 7488 patients receiving anthracyclines without trastuzumab, obesity was demonstrated as a risk factor for developing AIC (OR:1.72), while being overweight was not associated with a significant increase in risk (OR:1.08) [5]. In another study examining the impact of obesity and dyslipidemia among patients treated with anthracyclines and trastuzumab, a significant association was found between obesity and cardiotoxicity; however, dyslipidemia did not present as a significant risk for developing cardiotoxicity [39]. In a separate study, the association of patient body mass index (BMI) and cardiotoxicity by using data from 26 French cancer centers revealed that obesity was independently associated with cardiotoxicity (OR: 3.02, 95% CI) defined as a reduction in left ventricular ejection fraction [LVEF] > 10 percentage points from baseline to LVEF <50% [40]. A systematic review and meta-analysis by Guenancia et al. including 15 studies and 8745 breast cancer patients treated with either anthracyclines or sequential anthracyclines and trastuzumab revealed that both obesity and overweight were risk factors for cardiotoxicity (OR: 1.47, 1.15 95% CI, respectively) [41]. While the exact mechanism underlying increased cardiotoxicity risk in obese patients is not fully understood, obesity has also been linked to cardiovascular conditions like hypertension which have shown evidence to exacerbate AIC and have been proven to be a risk factor exasperating AIC [5, 15]. Another possible explanation for the increased risk of AIC in obese cancer patients is anthracycline pharmacokinetics. Furlanetto et al. demonstrated that in patients with BMI >30, increased dosages given to obese patients to account for their weight resulted in a higher risk of developing severe toxicities albeit no effect on survival rates [42].

Nutrition

Patient nutritional status may also serve as a risk factor for AIC. Initial research suggesting a role for nutritional diet in cardiotoxicity centered on alterations in anthracyclines pharmacokinetics. For example, protein malnutrition in rats resulted in a significant decrease in anthracycline elimination, which prolonged cardiac exposure to anthracyclines [43]. This occurrence may be of increased significance among cancer patients experiencing malnutrition during chemotherapy and is of significant concern in approximately 75% of younger patients according to WHO. The relationship between anthracycline exposure time and incidence of AIC has already been established, and the documented increase in exposure time due to malnutrition may be significant to clinicians and future research. Furthermore, dose adjustments may be required for nutritionally deficient patients. The mechanism that causes this increase in exposure and decrease in elimination has not been studied extensively.

Table 3 Natural compounds that may be cardioprotective in AIC

Natural compounds	References
Curcumin	[45]
Resveratrol	[46]
Quercetin	[47]
Omega-3 fatty acids	[48]
Naringenin	[49]

There has been a significant push in food science to study natural compounds present in common foods and traditional medicines for their potential therapeutic effects. To date, websites have been developed to summarize the research into foods associated with AIC, such as Foods for Breast Cancer [44]. Many of the studied components are antioxidants with established cardioprotective effects against the development of AIC in animal models or cell cultures. Many of these antioxidants target and attenuate critical AIC mechanisms including reduction in ROS generation, lipid peroxidation, ferroptosis, inflammation, and apoptosis. While there are several eloquent reviews describing these drugs, they are quickly losing their comprehensiveness due to the rapid pace of research generated. Table 3 summarizes some of the drugs or compounds currently being employed in clinical trials and/or synergistically promote anthracycline cancer death while preserving cardiac function.

Medical Risk Factors

In identifying high-risk patients, several clinical trials and retrospective studies have explored risk predictors/factors for AIC [5, 9, 50]. Major high-risk medical factors for AIC include cumulative anthracycline dosage and concomitant chemotherapy involving alkylating or antimicrotubular agents, immunotherapies, or previous radiation therapy with documented effects on the heart, which are among the most significant risk factors for AIC [50]. Interestingly, several studies document that the cumulative dose of anthracycline showed the strongest association with AIC [50, 51]. However, in this section, we will focus on the most prevalent medical risk factors including preexisting cardiovascular dysfunction, diabetes, and renal disease.

Cardiovascular Risk Factors

Pre-existing and comorbid cardiovascular disease (CVD), especially in the older adult population, are associated with increased AIC incidence [52, 53] and are summarized in Table 2. Recent work implicates several myocardial diseases augment the risk of AIC, including heart failure, asymptomatic LV dysfunction (LVEF <50%), evidence of coronary artery disease (CAD), moderate and severe valvular heart disease with left ventricular hypertrophy or left ventricular impairment, hypertension with cardiac remodeling, hypertrophic cardiomyopathy, dilated cardiomyopathy, restrictive cardiomyopathy, cardiac sarcoidosis with myocardial involvement, and significant cardiac arrhythmias [50]. While data linking hyperlipemia to AIC is sparse, one study demonstrated that higher concentrations of cardioprotective HDL-cholesterol was associated with lower risk of anthracycline-induced subclinical cardiotoxicity in patients with diffuse large B cell lymphoma [54]. Furthermore, given the bidirectional relationship between cancer and CVD [55], it is conceivable that cancer and anticancer therapies beget cardiac dysfunction. Notwithstanding, further studies evaluating the impact and sufficient management of arterial hypertension and CAD are needed to avoid permanent cardiac damage.

Hypertension

Hypertension is a well-documented risk factor for AIC development [9, 51, 56]. An early retrospective study by Ryberg et al. showed that independent of cumulative dose, patients with cardiac risk factors, including HTN, diabetes, and obesity, had a three times greater incidence of cardiac toxicity dose [51]. Pre-existing arterial hypertension has a very important role in the development of AIC in breast cancer patients treated with doxorubicin [9]. For example, in a study by Jin et al., baseline risk factors, including hypertension, hyperlipidemia, diabetes, and obesity, were significantly associated with an increased risk of cardiac events among 55,708 breast cancer patients treated with anthracyclines [35]. Similarly, Vaitiekus and colleagues reported a significant association between pre-existing arterial hypertension and left ventricular systolic dysfunction following doxorubicin-treated breast cancer patients ($P < 0.004$) [9]. In another study, Zhang et al. reported that breast cancer patients who suffered from HTN and CAD prior to anthracycline treatment were more likely to develop AIC ($p < 0.0001$) [50]. In a meta-analysis, Qui et al. demonstrated that in a cohort of 7, 488 patients receiving anthracycline chemotherapy, pre-existing HTN was associated with increased risk of AIC [5]. Further, the OR for developing AIC was (OR: 1.99; 95% CI: 1.43–2.76) [5]. Experimental studies in animals confirm that hypertension may be a significant factor predisposing anthracycline-based chemotherapy cardiotoxicity [9].

Diabetes

The evidence suggesting DM is a risk factor for AIC development is mixed [9, 51]. Ryberg et al. were among the first to show that diabetic patients treated with the anthracycline epirubicin were associated with AIC [51]. In one study, Vaitiekus et al. showed that DM was not associated with left ventricular systolic dysfunction caused by doxorubicin therapy [9]. On the other hand, Zhang et al. reported that diabetic breast cancer patients treated with anthracycline therapies were more likely to develop AIC [50]. In a recent meta-analysis with close to 7500 breast cancer patients, it was shown that the odds of developing AIC are greater in patients treated with anthracyclines (OR: 1.74; 95% CI: 1.11–2.74) [5]. Due to these discrepancies, further studies are required to demonstrate a clear association between DM and AIC development.

Renal Failure

While further studies are required to elucidate the mechanism by which prior renal dysfunction directly or indirectly increases myocardial toxicity of cancer therapies, there is a growing body of research supporting renal insufficiency as an additional risk factor and must be considered by cardio-oncologist prior to anthracyclines treatment. Furthermore, given the increased incidence of comorbid chronic renal insufficiency in the older adult population [29, 57], special considerations should be made in this population. Worthy of note, anthracyclines alone do not seem to promote nephrotoxicity [58]. Early studies by Russo et al. reported that breast cancer patients treated with doxorubicin had lower glomerular filtration rate (GFR), and increased incidence of cardiotoxicity [58]. Interestingly, GFR lower than 78 ml/min/1.73 m^2 was the strongest predictor of cardiotoxicity independent of doxorubicin treatment, as well as reduced LVEF [58], suggesting that poor renal clearance increases dose-dependent cardiac exposure. While the clinical evidence for this is sparse, volume dysfunction overload in the setting of poor kidney function also promotes cardiac injury [58]. Furthermore, renal failure and/or electrolyte imbalances such as hypokalemia and hypomagnesemia may also worsen AIC when used in combination with platinum-based chemotherapies including cisplatin [53, 59, 60].

Genetic Risk Factors

Clinical and treatment-related risk factors do not adequately explain individual and population susceptibility to chemotherapy-related cardiac dysfunction [61]. There has been a growing body of research that illuminates genetic risks for AIC, with several single nucleotide polymorphisms (SNPs), and genetic variants identified, although the research to date is not comprehensive. Research by Aminkeng et al. suggests genetic testing recommendations for the following SNPs, *RARG* rs2229774 (topoisomerase-2β expression), *SLC28A3* rs7853758 (drug transport), and *UGT1A6* rs17863783 (drug metabolism), which have the most consistent evidence for increased AIC risk [62]. In a separate study exploring genetic risk factors for AIC, Linschoten and colleagues also suggest the addition of other variants including CELF4 rs1786814 (sarcomere structure and function) and 1 intergenic variant (rs28714259) to this list [61]. The CELF4 protein is associated with TNNT2 pre-mRNA splicing, the gene for cardiac troponin T, and a biomarker of myocardial injury [63]. CELF4 rs1786814 showed a gene–environment interaction and increased doses of anthracyclines resulting in AIC [63]. For patients exposed to greater than 300 mg/m^2 of anthracyclines, the rs1786814 GG genotype conferred a 10.2-fold (95% CI, 3.8- to 27.3-fold; $P < 0.001$) increased risk of chemotherapy-related cardiomyopathy [63].

SLC28A3 rs7853758 was seen as a minor, protective allele, found more commonly in the control population of its foundational study [64]. Importantly, the authors recognize that the association studies were done in pediatric patient populations, and there is a need to repeat the research in adult cancer patients [62]. Recently, Siemens and colleagues found that the SLC28A3 variant imparts a protective role in patients receiving high cumulative anthracycline doses (>250 mg/m^2), while RARG and UGT1A6 were associated with an increased risk of AIC at low anthracycline does (<150 mg/m^2) in a pediatric population [65].

In a cohort of Korean breast cancer patients, Park et al. performed a genome-wide association study (GWAS), which demonstrated finding SHISA3 rs17530621, MPP4 rs11894115, RPL7 rs58328254, and PRUNE2 rs117299725 significant [66]. However, following adjustments for obesity and trastuzumab, the significance was lost [66], suggesting integration of genetics and clinical characteristics may be used to enhance the ability to predict the risk for anthracycline-related cardiomyopathy. In a recent study of 176 women with breast cancer, Kopeva et al. found that p53 (rs1042522), NOS3 (rs1799983), GPX1 (rs1050450), and NADPH oxidase (rs4673) were associated with higher risks of developing AIC [67].

In a machine learning study, 31 genes were identified as variants when comparing 183 patients receiving anthracyclines compared to control patients (106 total) [68]. Of these genes, which were predominantly associated with the PI3K/AKT/mTOR pathway, and p53 pathway genes, PI3KR2 and ZNF827 were chosen for further validation and analysis [68]. Both of these gene variants were seen in the control population, suggesting that they may serve a cardioprotective role [68]. PI3KR2 and ZNF827 are associated with autophagy, important in cellular pathways that are altered in the AIC [68]. Metformin is a ZNF827 inhibitor via inhibition of TR4 recruitment of ZNF827 and was found to be a cardioprotective [68].

There is also evidence that genetic variants within the following genes convey some risk of developing AIC with more validating research required: ABCC1, ABCC2, ABCC5, ABCB1, ABCB4, CBR3, RAC2, NCF4, CYBA, GSTP1, CAT, SULT2B1, POR, HAS3, SLC22A7, SCL22A17, HFE, and NOS3 [69]. Table 4 summarizes a list of the most significant genes and is well reviewed by Smita Bhatia et al. and Folefac Aminkeng and colleagues [62, 69]. In contrast to data from pediatric cancer survivors [70], validation studies by Reinbolt and colleagues did not demonstrate any significant correlation between the genetic variation of CBR1 and CBR3 in the development of AIC in an adult female breast cancer population [15]. Despite this, further testing and validation of these genetic variants with associated AIC risk are required, and the use of human induced pluripotent stem cells has been identified as a promising avenue [71].

Table 4 Genes commonly associated with increased AIC risk

Single nucleotide polymorphisms (SNPs)
RARG rs2229775
SLC28A3 rs7853758
UGT1A6 rs17863783
Genetic variant
rs28714259
CELF4 rs1786815
SHISA3 rs17530621
ABCC1

Abbreviations: ABCC1 ATP-binding cassette subfamily C member 1, *CELF4* CUGBP Elav-like family member 4, *RARG* retinoic acid receptor gamma, *SHISA3* Shisa family member 3, *SLC28A3* solute carrier family 28 member 3, *UGT1A6* UDP-glucuronosyltransferase 1A6

Conclusions

Anthracyclines remain a critical class of chemotherapeutic agents in cancer treatment but are associated with significant cardiotoxic effects, which pose a major clinical challenge. This review highlights key risk factors, including obesity, smoking, alcohol consumption, hypertension, and genetic predispositions, which contribute to the cardiotoxicity of anthracyclines. Patients presenting with these risk factors require close monitoring and the implementation of targeted protective strategies both during and after anthracycline therapy to mitigate adverse outcomes. Future research requires focus on combining genetic, biomarker screening, and clinical risk factors to identify patients most at risk for AIC. This approach could facilitate the creation of personalized monitoring and intervention strategies. Furthermore, studies should assess the effects of customized lifestyle interventions—including dietary changes, exercise programs, and smoking cessation—on reducing cardiotoxicity during and after treatment. Worthy of note, such studies face limitations, including the heterogeneity of risk factors, insufficient data on genetic predispositions, and the lack of representation from diverse populations. By tackling these challenges and following the proposed future directions, the clinical management of anthracycline-induced cardiotoxicity can be significantly enhanced.

Disclosures The authors have no relevant affiliations or financial involvement to disclose.

Funding This work was supported by National Institute of Health grant, T32HL134616 to SSS.

References

1. Smith LA, Cornelius VR, Plummer CJ, Levitt G, Verrill M, Canney P, et al. Cardiotoxicity of anthracycline agents for the treatment of cancer: systematic review and meta-analysis of randomised controlled trials. BMC Cancer. 2010;10:337.
2. Sawyer DB. Anthracyclines and heart failure. N Engl J Med. 2013;368:1154–6.
3. Piano MR. Alcohol's effects on the cardiovascular system. Alcohol Res Curr Rev. 2017;38:219–41.
4. Caliri AW, Tommasi S, Besaratinia A. Relationships among smoking, oxidative stress, inflammation, macromolecular damage, and cancer. Mutat Res Mutat Res. 2021;787:108365.
5. Qiu S, Zhou T, Qiu B, Zhang Y, Zhou Y, Yu H, et al. Risk factors for Anthracycline-induced cardiotoxicity. Front Cardiovasc Med. 2021;8:736854.
6. Volkova M, Russell R. Anthracycline cardiotoxicity: prevalence, pathogenesis and treatment. Curr Cardiol Rev. 2011;7:214–20.
7. Philip LJ, Findlay SG, Gill JH. Baseline blood pressure and development of cardiotoxicity in patients treated with anthracyclines: a systematic review. Int J Cardiol Cardiovasc Risk Prev. 2022;15:200153.
8. Ohman RE, Yang EH, Abel ML. Inequity in cardio-oncology: identifying disparities in cardiotoxicity and links to cardiac and cancer outcomes. J Am Heart Assoc. 2021;10:e023852.
9. Vaitiekus D, Muckiene G, Vaitiekiene A, Maciuliene D, Vaiciuliene D, Ambrazeviciute G, et al. Impact of arterial hypertension on Doxorubicin-based chemotherapy-induced subclinical cardiac damage in breast cancer patients. Cardiovasc Toxicol. 2020;20:321–7.
10. CDC. Adult Obesity Facts [Internet]. Obesity. 2024 [cited 2024 Sep 9]. Available from: https://www.cdc.gov/obesity/php/data-research/adult-obesity-facts.html
11. White MC, Holman DM, Boehm JE, Peipins LA, Grossman M, Henley SJ. Age and cancer risk: a potentially modifiable relationship. Am J Prev Med. 2014;46:S7–15.
12. Toro-Salazar OH, Hor KN, O'Loughlin M, Burke G, Stainsby JA, Gillan E, et al. Defining late onset occult asymptomatic cardiotoxicity in childhood cancer survivors exposed to anthracycline therapy: a cardiac magnetic resonance imaging study. J Cardiovasc Magn Reson. 2013;15:P163.
13. Zhou B, Carrillo-Larco RM, Danaei G, Riley LM, Paciorek CJ, Stevens GA, et al. Worldwide trends in hypertension prevalence and progress in treatment and control from 1990 to 2019: a pooled analysis of 1201 population-representative studies with 104 million participants. Lancet. 2021;398:957–80.
14. Jacob L, Freyn M, Kalder M, Dinas K, Kostev K. Impact of tobacco smoking on the risk of developing 25 different cancers in the UK: a retrospective study of 422,010 patients followed for up to 30 years. Oncotarget. 2018;9:17420–9.
15. Reinbolt RE, Patel R, Pan X, Timmers CD, Pilarski R, Shapiro CL, et al. Risk factors for anthracycline-associated cardiotoxicity. Support Care Cancer. 2016;24:2173–80.
16. Bozkurt B, Ahmad T, Alexander KM, Baker WL, Bosak K, Breathett K, et al. Heart failure epidemiology and outcomes statistics: a report of the Heart Failure Society of America. J Card Fail. 2023;29:1412–51.
17. Oliveros E, Patel H, Kyung S, Fugar S, Goldberg A, Madan N, et al. Hypertension in older adults: assessment, management, and challenges. Clin Cardiol. 2020;43:99–107.
18. Rodgers JL, Jones J, Bolleddu SI, Vanthenapalli S, Rodgers LE, Shah K, et al. Cardiovascular risks associated with gender and aging. J Cardiovasc Dev Dis. 2019;6:19.
19. Yazdanyar A, Newman AB. The burden of cardiovascular disease in the elderly: morbidity, mortality, and costs. Clin Geriatr Med. 2009;25(563–77):vii.
20. MacDonald T. Prevalence of high cholesterol among the adult U.S. population: NHANES 2013–2018. Curr Dev Nutr. 2022;6:925.
21. Eraso LH, Fukaya E, Mohler ER, Xie D, Sha D, Berger JS. Peripheral arterial disease, prevalence and cumulative risk factor profile analysis. Eur J Prev Cardiol. 2014;21:704–11.

22. Murphy D, McCulloch CE, Lin F, Banerjee T, Bragg-Gresham JL, Eberhardt MS, et al. Trends in prevalence of chronic kidney disease in the United States. Ann Intern Med. 2016;165:473–81.
23. *QuickStats:* percentage* of adults aged ≥18 years with diagnosed diabetes,† by urbanization level§ and age group — National Health Interview Survey, United States, 2022. MMWR Morb Mortal Wkly Rep. 2024;73:49.
24. Zhang L, Song J, Clark R, Mangoni A, Goldberg Y, Slipczuk L, et al. Abstract 13090: racial and ethnic differences in Anthracycline Cardiotoxicity. Circulation [Internet]. 2021 [cited 2024 Sep 9];144. Available from: https://www.ahajournals.org/doi/10.1161/circ.144.suppl_1.13090.
25. Bulatao RA, Anderson NB, National Research Council (US) Panel on Race E. Health Care. Underst Racial Ethn Differ Health Late Life Res Agenda [Internet]. National Academies Press (US); 2004 [cited 2024 Sep 9]. Available from: https://www.ncbi.nlm.nih.gov/books/NBK24693/.
26. Bikomeye JC, Awoyinka I, Kwarteng JL, Beyer AM, Rine S, Beyer KMM. Disparities in cardiovascular disease-related outcomes among cancer survivors in the United States: a systematic review of the literature. Heart Lung Circ. 2024;33:576–604.
27. Zhou Y, Gu X, Ashayeri E, Zhang R, Sridhar R. Nicotine decreases the cytotoxicity of doxorubicin towards MCF-7 and KB-3.1 human cancer cells in culture. J Natl Med Assoc. 2007;99:319–27.
28. Lotrionte M, Biondi-Zoccai G, Abbate A, Lanzetta G, D'Ascenzo F, Malavasi V, et al. Review and meta-analysis of incidence and clinical predictors of anthracycline cardiotoxicity. Am J Cardiol. 2013;112:1980–4.
29. Neuendorff NR, Loh KP, Mims AS, Christofyllakis K, Soo W-K, Bölükbasi B, et al. Anthracycline-related cardiotoxicity in older patients with acute myeloid leukemia: a Young SIOG review paper. Blood Adv. 2020;4:762–75.
30. Swain SM, Whaley FS, Ewer MS. Congestive heart failure in patients treated with doxorubicin: a retrospective analysis of three trials. Cancer. 2003;97:2869–79.
31. Ryberg M, Nielsen D, Skovsgaard T, Hansen J, Jensen BV, Dombernowsky P. Epirubicin cardiotoxicity: an analysis of 469 patients with metastatic breast cancer. J Clin Oncol. 1998;16:3502–8.
32. Von Hoff DD, Layard MW, Basa P, Davis HL, Von Hoff AL, Rozencweig M, et al. Risk factors for doxorubicin-induced congestive heart failure. Ann Intern Med. 1979;91:710–7.
33. Getz KD, Sung L, Ky B, Gerbing RB, Leger KJ, Leahy AB, et al. Occurrence of treatment-related cardiotoxicity and its impact on outcomes among children treated in the AAML0531 clinical trial: a report from the Children's Oncology Group. J Clin Oncol Off J Am Soc Clin Oncol. 2019;37:12–21.
34. Le Bot MA, Glaise D, Kernaleguen D, Ratanasavanh D, Carlhant D, Riche C, et al. Metabolism of doxorubicin, daunorubicin and epirubicin in human and rat hepatoma cells. Pharmacol Res. 1991;24:243–52.
35. Jin H, Xu J, Sui Z, Wang L. Risk factors from Framingham risk score for anthracyclines cardiotoxicity in breast cancer: a systematic review and meta-analysis. Front Cardiovasc Med. 2023;10:1101585.
36. Ho E, Brown A, Barrett P, Morgan RB, King G, Kennedy MJ, et al. Subclinical anthracycline- and trastuzumab-induced cardiotoxicity in the long-term follow-up of asymptomatic breast cancer survivors: a speckle tracking echocardiographic study. Heart Br Card Soc. 2010;96:701–7.
37. Shimauchi T, Numaga-Tomita T, Ito T, Nishimura A, Matsukane R, Oda S, et al. TRPC3-Nox2 complex mediates doxorubicin-induced myocardial atrophy. JCI Insight. 2017;2(e93358):93358.
38. Nishiyama K, Numaga-Tomita T, Fujimoto Y, Tanaka T, Toyama C, Nishimura A, et al. Ibudilast attenuates doxorubicin-induced cytotoxicity by suppressing formation of TRPC3 channel and NADPH oxidase 2 protein complexes. Br J Pharmacol. 2019;176:3723–38.

39. Maaroufi A, Bendahou H, Benmalek R, Abouriche A, Belkouchia S, Ammouri Z, et al. Impact of dyslipidemia and obesity on the incidence of cardiotoxicity in breast cancer patients treated with anthracyclines and Trastuzumab. Arch Cardiovasc Dis Suppl. 2021;13:209.
40. Kaboré EG, Guenancia C, Vaz-Luis I, Di Meglio A, Pistilli B, Coutant C, et al. Association of body mass index and cardiotoxicity related to anthracyclines and trastuzumab in early breast cancer: French CANTO cohort study. PLoS Med. 2019;16:e1002989.
41. Guenancia C, Lefebvre A, Cardinale D, Yu AF, Ladoire S, Ghiringhelli F, et al. Obesity as a risk factor for Anthracyclines and Trastuzumab cardiotoxicity in breast cancer: a systematic review and meta-analysis. J Clin Oncol. 2016;34:3157–65.
42. Furlanetto J, Eiermann W, Marmé F, Reimer T, Reinisch M, Schmatloch S, et al. Higher rate of severe toxicities in obese patients receiving dose-dense (dd) chemotherapy according to unadjusted body surface area: results of the prospectively randomized GAIN study. Ann Oncol. 2016;27:2053–9.
43. El-Demerdash E, Ali AA, El-Taher DEM, Hamada FMA. Effect of low-protein diet on anthracycline pharmacokinetics and cardiotoxicity. J Pharm Pharmacol. 2012;64:344–52.
44. Cobos Á, Díaz O. "Superfoods": reliability of the information for consumers available on the web. Foods Basel Switz. 2023;12:546.
45. Sarawi WS, Alhusaini AM, Fadda LM, Alomar HA, Albaker AB, Aljrboa AS, et al. Nano-Curcumin prevents cardiac injury, oxidative stress and inflammation, and modulates TLR4/NF-κB and MAPK signaling in copper sulfate-intoxicated rats. Antioxid Basel Switz. 2021;10:1414.
46. Dolinsky VW, Dyck JRB. Calorie restriction and resveratrol in cardiovascular health and disease. Biochim Biophys Acta. 2011;1812:1477–89.
47. Sharma A, Kashyap D, Sak K, Tuli HS, Sharma AK. Therapeutic charm of quercetin and its derivatives: a review of research and patents. Pharm Pat Anal. 2018;7:15–32.
48. Kromhout D, Yasuda S, Geleijnse JM, Shimokawa H. Fish oil and omega-3 fatty acids in cardiovascular disease: do they really work? Eur Heart J. 2012;33:436–43.
49. Kwatra M, Kumar V, Jangra A, Mishra M, Ahmed S, Ghosh P, et al. Ameliorative effect of naringin against doxorubicin-induced acute cardiac toxicity in rats. Pharm Biol. 2016;54:637–47.
50. Zhang M, Yang H, Xu C, Jin F, Zheng A. Risk factors for Anthracycline-induced cardiotoxicity in breast cancer treatment: a meta-analysis. Front Oncol. 2022;12:899782.
51. Ryberg M, Nielsen D, Cortese G, Nielsen G, Skovsgaard T, Andersen PK. New insight into Epirubicin cardiac toxicity: competing risks analysis of 1097 breast cancer patients. JNCI J Natl Cancer Inst. 2008;100:1058–67.
52. Baldassarre LA, Yang EH, Cheng RK, DeCara JM, Dent S, Liu JE, et al. Cardiovascular Care of the Oncology Patient during COVID-19: an expert consensus document from the ACC Cardio-Oncology and Imaging Councils. J Natl Cancer Inst. 2021;113:513–22.
53. Bovelli D, Plataniotis G, Roila F. ESMO Guidelines Working Group. Cardiotoxicity of chemotherapeutic agents and radiotherapy-related heart disease: ESMO clinical practice guidelines. Ann Oncol. 2010;21(Suppl 5):v277–82.
54. Ou W, Jiang T, Zhang N, Lu K, Weng Y, Zhou X, et al. Role of HDL cholesterol in anthracycline-induced subclinical cardiotoxicity: a prospective observational study in patients with diffuse large B-cell lymphoma treated with R-CHOP. BMJ Open. 2024;14:e074541.
55. Bertero E, Ameri P, Maack C. Bidirectional relationship between cancer and heart failure: old and new issues in cardio-oncology. Card Fail Rev. 2019;5:106–11.
56. Cai F, Luis M, Lin X, Wang M, Cai L, Cen C, et al. Anthracycline-induced cardiotoxicity in the chemotherapy treatment of breast cancer: Preventive strategies and treatment (Review). Mol Clin Oncol [Internet]. 2019 [cited 2024 Dec 29]; Available from: http://www.spandidos-publications.com/10.3892/mco.2019.1854
57. Mills KT, Xu Y, Zhang W, Bundy JD, Chen C-S, Kelly TN, et al. A systematic analysis of worldwide population-based data on the global burden of chronic kidney disease in 2010. Kidney Int. 2015;88:950–7.

58. Russo G, Cioffi G, Di Lenarda A, Tuccia F, Bovelli D, Di Tano G, et al. Role of renal function on the development of cardiotoxicity associated with trastuzumab-based adjuvant chemotherapy for early breast cancer. Intern Emerg Med. 2012;7:439–46.
59. Rachma B, Savitri M, Sutanto H. Cardiotoxicity in platinum-based chemotherapy: mechanisms, manifestations, and management. Cancer Pathog Ther. 2024;3:101.
60. Csapo M, Lazar L. Chemotherapy-induced cardiotoxicity: pathophysiology and prevention. Clujul Med. 1957;2014(87):135–42.
61. Linschoten M, Teske AJ, Cramer MJ, van der Wall E, Asselbergs FW. Chemotherapy-related cardiac dysfunction: a systematic review of genetic variants modulating individual risk. Circ Genomic Precis Med. 2018;11:e001753.
62. Aminkeng F, Ross CJD, Rassekh SR, Hwang S, Rieder MJ, Bhavsar AP, et al. Recommendations for genetic testing to reduce the incidence of anthracycline-induced cardiotoxicity. Br J Clin Pharmacol. 2016;82:683–95.
63. Wang X, Sun C-L, Quiñones-Lombraña A, Singh P, Landier W, Hageman L, et al. CELF4 variant and Anthracycline-related cardiomyopathy: a Children's Oncology Group Genome-Wide Association study. J Clin Oncol. 2016;34:863–70.
64. Bernstein D, Burridge P. Patient-specific Pluripotent stem cells in Doxorubicin cardiotoxicity: a new window into personalized medicine. Prog Pediatr Cardiol. 2014;37:23–7.
65. Siemens A, Rassekh SR, Ross CJD, Carleton BC. Development of a dose-adjusted polygenic risk model for Anthracycline-induced cardiotoxicity. Ther Drug Monit. 2023;45:337–44.
66. Park B, Sim SH, Lee KS, Kim HJ, Park IH. Genome-wide association study of genetic variants related to anthracycline-induced cardiotoxicity in early breast cancer. Cancer Sci. 2020;111:2579–87.
67. Kopeva KV, Grakova EV, Shilov SN, Berezikova EN, Popova AA, Neupokoeva MN, et al. Anthracycline-induced cardiotoxicity in women without cardiovascular diseases: molecular and genetic predictors. Acta Cardiol. 2022;77:805–14.
68. Chaix M-A, Parmar N, Kinnear C, Lafreniere-Roula M, Akinrinade O, Yao R, et al. Machine learning identifies clinical and genetic factors associated with Anthracycline cardiotoxicity in pediatric cancer survivors. JACC CardioOncol. 2020;2:690–706.
69. Bhatia S. Genetics of Anthracycline cardiomyopathy in cancer survivors: JACC: CardioOncology state-of-the-art review. JACC CardioOncol. 2020;2:539–52.
70. Blanco JG, Sun C-L, Landier W, Chen L, Esparza-Duran D, Leisenring W, et al. Anthracycline-related cardiomyopathy after childhood cancer: role of polymorphisms in carbonyl reductase genes--a report from the Children's Oncology Group. J Clin Oncol Off J Am Soc Clin Oncol. 2012;30:1415–21.
71. Tripaydonis A, Conyers R, Elliott DA. Pediatric Anthracycline-induced cardiotoxicity: mechanisms, pharmacogenomics, and Pluripotent stem-cell modeling. Clin Pharmacol Ther. 2019;105:614–24.

Multilevel Drivers of Disparities in Cardio-Oncology

Megan McCrohan, Tyren Busby, Dionisia Quiroga, Mary Branch, Arnethea L. Sutton, Sherry-Ann Brown, Sarju Ganatra, Sanam Ghazi, Onaopepo Kola-Kehinde, Mussammat Ferdousi, Karen P. Williams, Eric D. Miller, and Daniel Addison

Abbreviations

ICI	Immune checkpoint inhibitors
ADMA	Asymmetric dimethylarginine
CVD	Cardiovascular disease
HER2	Human epidermal growth factor 2
IL	Interleukin
LGBTQIA	Lesbian/gay/bisexual/transgender/queer/intersex/asexual
MRI	Magnetic resonance imaging
SEER	Surveillance epidemiology and end results
TNF	Tissue necrosis factor
US	United States

M. McCrohan
Ohio State University Wexner Medical Center, Department of Internal Medicine, Columbus, OH, USA

Cardio-Oncology Program, Division of Cardiology, Ohio State University Medical Center, Columbus, OH, USA
e-mail: megan.mccrohan@osumc.edu

T. Busby · O. Kola-Kehinde · M. Ferdousi
Cardio-Oncology Program, Division of Cardiology, Ohio State University Medical Center, Columbus, OH, USA
e-mail: tb884115@ohio.edu; onaopepokolakehinde@gmail.com; Mussammat.Ferdousi@utsouthwestern.edu

D. Quiroga
Division of Medical Oncology, Ohio State University Medical Center, Columbus, OH, USA
e-mail: dionisia.quiroga@osumc.edu

S. A. Smith, S. S. Scott (eds.), *Myocardial Oxidative Stress in Anti-Cancer Therapies*, Oxidative Stress in Applied Basic Research and Clinical Practice,
https://doi.org/10.1007/978-3-031-97517-2_2

Introduction

Successful development and implementation of novel anticancer therapies has led to rapid increase in cancer survivorship [1–4]. In the United States alone, nearly 20 million individuals are classified as cancer survivors—a number expected to

M. Branch
Division of Cardiology, Cone Health, Greensboro, NC, USA
e-mail: mary.branch@conehealth.com

A. L. Sutton
Department of Kinesiology and Health Sciences, Virginia Commonwealth University, Richmond, VA, USA
e-mail: abrahamal@vcu.edu

S.-A. Brown
Cardio-Oncology Program, Division of Cardiology, Medical College of Wisconsin, Milwaukee, WI, USA

S. Ganatra
Cardio-Oncology Program, Division of Cardiology, Lehigh Clinic Medical Center, Boston, MA, USA

S. Ghazi
Cardio-Oncology Program, Division of Cardiology, Ohio State University Medical Center, Columbus, OH, USA

Cardio-Oncology Program, University of Texas Southwestern Medical Center, Texas, Dallas, United States
e-mail: Sanam.Ghazi@UTSouthwestern.edu

K. P. Williams
Center for Women, Children and Youth, Ohio State University Medical Center, Columbus, OH, USA
e-mail: williams.5963@osu.edu

E. D. Miller
Department of Radiation Oncology, Ohio State University Medical Center, Columbus, OH, USA
e-mail: eric.miller@osumc.edu

D. Addison (✉)
Cardio-Oncology Program, Division of Cardiology, Ohio State University Medical Center, Columbus, OH, USA

Division of Cancer Prevention and Control, Department of Internal Medicine, College of Medicine, The Ohio State University, Columbus, OH, USA

Division of Cardiovascular Medicine, Davis Heart & Lung Research Institute, Columbus, OH, USA

Cardio-Oncology Program, University of Texas Southwestern Medical Center, Texas, Dallas, United States

Simmons Comprehensive Cancer Center, University of Texas Southwestern Medical Center, Texas, Dallas, United States
e-mail: daniel.addison@osumc.edu; daniel.addison@utsouthwestern.edu

dramatically rise over the next two decades [5]. In parallel, increasing evidence has established that a significant proportion of these contemporary therapies may carry frequently under-recognized adverse cardiotoxic risks [6, 7]. This is particularly true for populations wherein rigorous studies have been largely unavailable, including among women and underrepresented minorities [8–10]. For example, emerging data suggest that despite accounting for socioeconomic and insurance factors, Native American and Black patients see a nearly twofold increase in cardiotoxic risk after standard chemotherapy treatment for breast cancers [8, 9]. Although these disparities are in part accounted for by traditional measures of inequity, such as lack of consistent healthcare access and geographic distance, many of these observations are not well accounted for by traditional factors alone. Given the rapid increase in the global applications of many contemporary cancer treatment strategies to populations where initial consideration was not well represented, the gaps in potential cardiotoxicity risks and outcomes may only represent a portion of the magnitude of this serious issue.

Recent sentinel statements have attempted to address the factors and consequences of disparities in cardiotoxicity for women and underrepresented populations, with suggestions for improvement in management and resultant cardiovascular outcomes [9–12]. These prior efforts specifically call for improved equity in access to cardioprotective drugs and in patient disease surveillance, as well as elucidation of population-specific risks [10, 11]. Although impactful, these investigations have also shown the need for deeper consideration of the underlying drivers of cardio-oncologic toxicity disparities.

In this review, we aim to highlight the major contributors to risk differences among cancer patients developing cardiovascular disease (CVD). We specifically focus on social and biologic pathways that may pivotally contribute to disparate CVD risk and survival after a cancer diagnosis.

Established Data on Disparities

Sex Disparities

Chest radiotherapy is a known risk for development of cardiotoxicity—this is particularly well studied in long-term survivors of lymphoma [13, 14]. Studies have also focused on cardiotoxicity after radiotherapy in specific cancer types: non-small-cell lung cancer and risk for major adverse cardiac event [15], breast cancer and risk for heart failure [16], and esophageal cancer and risk for atrial fibrillation [17]. Given these associations, attention is being given to modalities to detect cardiotoxicity earlier in treatment cycles. For example, utilizing cardiac magnetic resonance imaging (MRI) to identify regional myocardial injury prior to irreversibility [18]. A recent meta-analysis identified a significantly higher risk for mediastinal radiotherapy-related cardiotoxicity, as well as all-cause mortality, for women undergoing treatment for lymphoma when compared to men [19].

Similar to radiation-based therapies, recent data suggests there exist sex disparities in regard to pharmacologic therapies, as well. Existing cancer drug data highlight sex disparities in development of cardiotoxicities, particularly with immune checkpoint inhibitors (ICI). A large study documented a 34% (66% for ICIs) increased risk for women compared to men in developing cardiotoxicity after anticancer drug therapy [20]. With non-ICI-based therapies, available data have been mixed on the potential for sex-specific differences in cardiotoxic risk. Additional prospective and mechanism-based studies are needed, as the mechanisms for these sex-specific differences in cardio-oncologic risk are not well elucidated.

For breast cancer, patients are often exposed to a variety of antineoplastic therapies with potential cardiovascular risks, including radiotherapy and systemic therapies (e.g., anthracycline chemotherapy, human epidermal growth factor 2 (HER2)-targeted therapies, and endocrine therapy) [5, 21, 22]. The spectrum of these risks may go unrecognized, but still contribute to higher morbidity and mortality for women treated for breast cancer through increased rates of heart failure, hypertension, and overall CVD risk [23, 24]. Management and surveillance for early and late cardiotoxicity dependent upon sex differences remains to be elucidated.

Disparities of Ethnicity, Race, and Underrepresented Groups

Prior studies have established disparities in risk for cardiotoxicities after anticancer therapy in Black/African American patients, particularly for anthracyclines and anti-HER2 agents (Table 1) [9, 25, 26]. These risk inequalities exist even after accounting for cancer stage at diagnosis, socioeconomic status, and patient-specific traditional cardiac factors [27]. In two Surveillance, Epidemiology, and End Results (SEER)-based studies including patients with multiple cancer types (e.g., breast, colon, lymphoma, prostate), Black patients saw up to a nearly twofold increase in CVD incidence and related mortality, despite adjustment for socioeconomic factors [27, 28]. Similarly, in a SEER-based evaluation of lymphoma survivors, Black patients saw a twofold increase in the risk of cardiovascular mortality (adjusted hazard ratio 1.97, $P < 0.01$) [29]. A recent large cohort study further elucidated this disparity, identifying the risk for cancer-related death and CVD-related death was significantly higher in the Black population compared to the White population [30]. Likewise, Hispanic/Latine, Asian and Pacific Islander, and Lesbian/Gay/Bisexual/Transgender/Queer/Intersex/Asexual (LGBTQIA) individuals are known to be at higher risk for cardiovascular disease compared to their White and non-LGBTQIA counterparts, respectively, with subsequent data suggesting elevated cardiotoxicity rates with anticancer drugs, as well [8, 12, 30, 31].

Pediatric Populations

Table 1 Focused summary of existing evidence of disparities in cardio-oncology

Disparities in cardiotoxicity	Evidence and contributors	References
Sex disparities		
Increased risk for adverse cardiotoxicities in women	Women show a 34% increased risk for cardiotoxicities, particularly with immune checkpoint inhibitors	Unger et al. 2022 [20]
Radiotherapy-related disparities		
Sex disparities in radiotherapy cardiotoxicity	Women undergoing mediastinal radiotherapy for lymphoma have higher risk of cardiotoxicity and mortality	Khalid et al. 2020 [19]
Breast cancer management disparities		
Cardiovascular risks in breast cancer treatment	Anthracyclines, trastuzumab, and endocrine therapies pose cardiovascular risks. Management guidelines need to address sex-specific cardiotoxicity	Mehta et al. 2018, 2019 [21, 24]
Sex and racial disparities in cardiotoxicities	Black/African American patients have higher risk for cardiotoxicities with certain agents, even after accounting for traditional cardiac factors	Husain et al. 2019, Prasad et al. 2020, Al-Sadawi et al. 2021 [25, 9, 26]
Pediatric patients' disparities		
Elevated cardiac-related mortality risk in pediatric patients	Pediatric cancer survivors face significantly higher risk of cardiac-related mortality and various cardiovascular diseases	Armstrong et al. 2009, Khanna et al. 2019, Suh et al. 2020, Addison et al. 2023 [32–34, 10]

Pediatric patients have been identified as another disparate group in relation to cardiotoxicity after treatment for primary cancer [11]. Long-term survivors see up to a ninefold increase in cardiac-related mortality risk compared to the general population [32, 33]. Risk for heart failure, valvular disease, coronary artery disease, hypertension, and dyslipidemia is also increased in survivors [34]. Similar to sex differences and underrepresented groups, pediatric patients' risk for cardiotoxicity after cancer therapy needs to be better understood to improve group-specific morbidity and mortality.

Nonbiologic Contributors

Social determinants of health, clinical trial representation, and structural racism serve as essential nonbiologic contributors to sex-based and underrepresented populations' disparities in regard to anticancer pharmacologic cardiotoxicities.

Social Determinants of Health

First, social determinants of health are key players in an individual's risk for development of high mortality diseases, particularly cardiovascular disease and cancer [35]. Recently, several indices—the social vulnerability index and area deprivation-based index—have explicated a graded relationship between social determinants (i.e., socioeconomic status, social discrimination, social support, living environment) and adverse health outcomes in underrepresented populations [12]. These data allow objective interpretation of potential differences even beyond the United States.

Diversity in Clinical Trials

It is well established that there is a tremendous lack of diversity in US clinical trial samples, particularly for those investigating oncology therapies [36]. To start, many published cancer trials have not reported race or ethnicity in their demographics [37]. Concurrently, translational research data on cardiotoxic risks have historically utilized biospecimens originating from non-Hispanic White patients [37]. This lack of representation of traditionally underrepresented groups (who often may have higher baseline risk factors for development of CVD) presents enhanced considerations for conquering these disparities.

Other Drivers

Structural racism contributes to disparities in various capacities. It significantly impacts one's environmental exposures and psychosocial stressors, which can have deleterious downstream effects on health outcomes [38]. Further, structural racism is heavily implicated in the lack of diversity amongst CVD and cancer professionals [24, 39]. Together, these nonbiologic contributors to risk of CVD, cancer development, and cardiotoxicity after anticancer drug use must be addressed to close the disparity gap for women and underrepresented populations.

Biologic Contributors

Several biologic contributors may also play a role in oncologic cardiotoxicity disparities (Table 2). The way in which genes are expressed and interact with environmental pressures or stressors (e.g., cancer, sex, or racial bias) has been postulated to be a key determinant in cardiotoxic disease development [11]. Utilized in the

Table 2 Potential contributors to disparities in cardio-oncology

Potential driver	Contribution in cardio-oncology	References
Nonbiologic contributors		
Social determinants of health impact disparities	Socioeconomic status, discrimination, and living environment contribute to adverse health outcomes in underrepresented populations	Harper et al. 2011 [35]
Lack of clinical trial representation	Trials often lack diversity, making it challenging to address disparities effectively	Gunaldi et al. 2016 [37]
Structural racism	Barriers due to structural racism limit access to healthcare and contribute to disparities	Ahmad et al. 2022 [38]
Biologic contributors		
Epigenetic stress	Epigenetic makeup affects disease propensity, potentially predisposing minority groups to cardiovascular risk	Mathews et al. 2011 [40]
Genomic markers and pathways	Clonal hematopoiesis and genetic factors play roles in cardiotoxic susceptibility	Jaiswal et al. 2017 [42]
Arginine-nitric oxide pathway	Alterations in ADMA pathway associated with greater risk in breast cancer patients	Finkelman et al. 2017 [45]
Inflammation	Inflammation plays a role in cardiovascular disease and cancer development, potentially contributing to disparities	Saban et al. 2014 [41]

Abbreviations: *ADMA* asymmetric dimethylarginine

context of these interactions, outside of actual genetic information, these factors driving expression differences have generally been coined as epigenetic stress [40]. One's response to environmental stress and propensity to disease (particularly cardiovascular disease) has increasingly been identified as a key driver of oncologic and CVD phenotype development (e.g., clonal hematopoiesis). This raises concern for how epigenetic stressors may predispose female and minority groups to disparate CVD risk in the setting of anticancer therapies [41]. Several genomic variants also likely play a role in cardiotoxic susceptibility. For example, clonal hematopoiesis of indeterminate potential in peripheral cells has been associated with a nearly 50% increase in risk for CVD [42]. Available hematologic malignancy-focused data have suggested no or some increased risk for the presence of CVD in Black and other minority populations [43]; however, additional studies are needed.

In addition, alterations in nitrate metabolism pathways such as arginine-nitric oxide metabolites (e.g., asymmetric dimethylarginine (ADMA)) are known drivers of increased CVD risk and have been identified as significantly correlating with cardiotoxic risk among breast cancer patients receiving doxorubicin therapy [44]. Notably, Black race is associated with a greater precancer treatment level of these metabolites [46], and concurrent noncancer data have increasingly suggested higher levels of ADMA and similar metabolites even in Black population progressing through pregnancy [47–49]. In general, inflammation (as measured by acute phase reactants/cytokines (e.g., IL-6, TNF) and activated immune cells) is known to play a role in development of both CVD and cancer [49] and has been reported to increase among non-White breast cancer patients. Given this relationship and rapidly growing appreciation for the role of immune response in CVD risk and development, the question arises of how inflammation from both biologic and nonbiologic stressors may contribute to disparities in cardio-oncology [41]. Together, these biologic domains create a foundation for increased risk of development of and poor CVD outcomes in the setting of anticancer drug therapies.

Conclusion

Cardio-oncology is increasingly a survivorship issue for patients requiring treatment with anticancer therapies. Disparities in cardiotoxic risk represent a significant limitation to effective long-term anticancer outcomes. The drivers of these disparities are increasingly multilevel, including social determinants, lack of clinical trial and workforce representation, structural racism, epigenetic, and inflammation. Given the multifaceted nature of differential risk, a several pronged and multilevel approach is needed. Clinical trials and healthcare must (1) be diversified to accurately represent the populations at risk, (2) be created with aims to appropriately incorporate social determinants of health into clinical care and diminish existing barriers to that care, (3) reduce personal and structural biases within the healthcare and research workforce, and (4) ensure equitable access of emerging technologies to effectively help reduce long-term CVD risks. As we progress further into the twenty-first century, intentional efforts to address these serious challenges are needed.

Acknowledgments The manuscript's content is solely the responsibility of the authors and does not necessarily represent the official views of the National Institutes of Health.

Disclosures This work was supported in part by an NIH P50-CA140158 grant. Dr. Addison is supported by NIH grant number K23-HL155890, R01HL170038, R01CA301579 and R01HL168045 and an American Heart Association-Robert Wood Johnson Foundation (Harold Amos) grant. Dr. Quiroga is supported by a Robert A. Winn Career Development Award. All other authors have reported that they have no relationships relevant to the contents of this paper to disclose.

References

1. Shanafelt TD, Wang XV, Kay NE, Hanson CA, O'Brien S, Barrientos J, et al. Ibrutinib-Rituximab or Chemoimmunotherapy for chronic lymphocytic leukemia. N Engl J Med. 2019;381(5):432–43.
2. San-Miguel J, Dhakal B, Yong K, Spencer A, Anguille S, Mateos MV, et al. Cilta-cel or standard care in Lenalidomide-refractory multiple myeloma. N Engl J Med. 2023;389(4):335–47.
3. Wakelee H, Liberman M, Kato T, Tsuboi M, Lee SH, Gao S, et al. Perioperative Pembrolizumab for early-stage non-small-cell lung cancer. N Engl J Med. 2023;389(6):491–503.

4. Westin JR, Oluwole OO, Kersten MJ, Miklos DB, Perales MA, Ghobadi A, et al. Survival with Axicabtagene Ciloleucel in large B-cell lymphoma. N Engl J Med. 2023;389(2):148–57.
5. Siegel RL, Miller KD, Wagle NS, Jemal A. Cancer statistics, 2023. CA Cancer J Clin. 2023;73(1):17–48.
6. Guha A, Gong Y, DeRemer D, Owusu-Guha J, Dent SF, Cheng RK, et al. Cardiometabolic consequences of targeted anticancer therapies. J Cardiovasc Pharmacol. 2022;80(4):515–21.
7. Herrmann J, Lenihan D, Armenian S, Barac A, Blaes A, Cardinale D, et al. Defining cardiovascular toxicities of cancer therapies: an International Cardio-Oncology Society (IC-OS) consensus statement. Eur Heart J. 2022;43(4):280–99.
8. Fazal M, Malisa J, Rhee JW, Witteles RM, Rodriguez F. Racial and ethnic disparities in cardio-oncology: a call to action. JACC Cardio Oncol. 2021;3(2):201–4.
9. Prasad P, Branch M, Asemota D, Elsayed R, Addison D, Brown S-A. Cardio-oncology preventive care: racial and ethnic disparities. Curr Cardiovasc Risk Rep. 2020;14(10):18.
10. Sadler D, Okwuosa T, Teske AJ, Guha A, Collier P, Moudgil R, et al. Cardio oncology: digital innovations, precision medicine and health equity. Front Cardiovasc Med. 2022;9:951551.
11. Addison D, Branch M, Baik AH, Fradley MG, Okwuosa T, Reding KW, et al. Equity in cardio-oncology care and research: a scientific statement from the American Heart Association. Circulation. 2023;148(3):297–308.
12. Ganatra S, Dani SS, Kumar A, Khan SU, Wadhera R, Neilan TG, et al. Impact of social vulnerability on comorbid cancer and cardiovascular disease mortality in the United States. JACC Cardio Oncol. 2022;4(3):326–37.
13. van Nimwegen FA, Schaapveld M, Janus CP, Krol AD, Petersen EJ, Raemaekers JM, et al. Cardiovascular disease after Hodgkin lymphoma treatment: 40-year disease risk. JAMA Intern Med. 2015;175(6):1007–17.
14. Wethal T, Nedregaard B, Andersen R, Fosså A, Lund MB, Günther A, et al. Atherosclerotic lesions in lymphoma survivors treated with radiotherapy. Radiother Oncol. 2014;110(3):448–54.
15. Atkins KM, Rawal B, Chaunzwa TL, Lamba N, Bitterman DS, Williams CL, et al. Cardiac radiation dose, cardiac disease, and mortality in patients with lung cancer. J Am Coll Cardiol. 2019;73(23):2976–87.
16. Saiki H, Petersen IA, Scott CG, Bailey KR, Dunlay SM, Finley RR, et al. Risk of heart failure with preserved ejection fraction in older women after contemporary radiotherapy for breast cancer. Circulation. 2017;135(15):1388–96.
17. Miller ED, Wu T, McKinley G, Slivnick J, Guha A, Mo X, et al. Incident atrial fibrillation and survival outcomes in esophageal cancer following radiotherapy. Int J Radiat Oncol Biol Phys. 2023;118:124.
18. Omidi A, Weiss E, Trankle CR, Rosu-Bubulac M, Wilson JS. Quantitative assessment of radiotherapy-induced myocardial damage using MRI: a systematic review. Cardio-Oncology. 2023;9(1):24.
19. Khalid Y, Fradley M, Dasu N, Dasu K, Shah A, Levine A. Gender disparity in cardiovascular mortality following radiation therapy for Hodgkin's lymphoma: a systematic review. Cardiooncology. 2020;6:12.
20. Unger JM, Vaidya R, Albain KS, LeBlanc M, Minasian LM, Gotay CC, et al. Sex differences in risk of severe adverse events in patients receiving immunotherapy, targeted therapy, or chemotherapy in cancer clinical trials. J Clin Oncol. 2022;40(13):1474–86.
21. Mehta LS, Watson KE, Barac A, Beckie TM, Bittner V, Cruz-Flores S, et al. Cardiovascular disease and breast cancer: where these entities intersect: a scientific statement from the American Heart Association. Circulation. 2018;137(8):e30–66.
22. Greenlee H, Iribarren C, Rana JS, Cheng R, Nguyen-Huynh M, Rillamas-Sun E, et al. Risk of cardiovascular disease in women with and without breast cancer: the pathways heart study. J Clin Oncol. 2022;40(15):1647–58.
23. Reding KW, Cheng RK, Vasbinder A, Ray RM, Barac A, Eaton CB, et al. Lifestyle and cardiovascular risk factors associated with heart failure subtypes in postmenopausal breast cancer survivors. JACC Cardio Oncol. 2022;4(1):53–65.

24. Mehta LS, Fisher K, Rzeszut AK, Lipner R, Mitchell S, Dill M, et al. Current demographic status of cardiologists in the United States. JAMA Cardiol. 2019;4(10):1029–33.
25. Husain M, Nolan TS, Foy K, Reinbolt R, Grenade C, Lustberg M. An overview of the unique challenges facing African-American breast cancer survivors. Support Care Cancer. 2019;27(3):729–43.
26. Al-Sadawi M, Hussain Y, Copeland-Halperin RS, Tobin JN, Moskowitz CS, Dang CT, et al. Racial and socioeconomic disparities in cardiotoxicity among women with HER2-positive breast cancer. Am J Cardiol. 2021;147:116–21.
27. Zhu C, Shi T, Jiang C, Liu B, Baldassarre LA, Zarich S. Racial and ethnic disparities in all-cause and cardiovascular mortality among cancer patients in the U.S. JACC Cardio Oncol. 2023;5(1):55–66.
28. Shi T, Jiang C, Zhu C, Wu F, Fotjhadi I, Zarich S. Insurance disparity in cardiovascular mortality among non-elderly cancer survivors. Cardiooncology. 2021;7(1):11.
29. Al-Kindi SG, Abu-Zeinah GF, Kim CH, Hejjaji V, William BM, Caimi PF, et al. Trends and disparities in cardiovascular mortality among survivors of Hodgkin lymphoma. Clin Lymphoma Myeloma Leuk. 2015;15(12):748–52.
30. Sung H, Nisotel L, Sedeta E, Islami F, Jemal A. Racial and ethnic disparities in survival among people with second primary cancer in the US. JAMA Netw Open. 2023;6(8):e2327429.
31. Caceres BA, Streed CG, Corliss HL, Lloyd-Jones DM, Matthews PA, Mukherjee M, et al. Assessing and addressing cardiovascular health in LGBTQ adults: a scientific statement from the American Heart Association. Circulation. 2020;142(19):e321–e32.
32. Armstrong GT, Liu Q, Yasui Y, Neglia JP, Leisenring W, Robison LL, et al. Late mortality among 5-year survivors of childhood cancer: a summary from the Childhood Cancer Survivor Study. J Clin Oncol. 2009;27(14):2328–38.
33. Khanna A, Pequeno P, Gupta S, Thavendiranathan P, Lee DS, Abdel-Qadir H, et al. Increased risk of all cardiovascular disease subtypes among childhood cancer survivors: population-based matched Cohort Study. Circulation. 2019;140(12):1041–3.
34. Suh E, Stratton KL, Leisenring WM, Nathan PC, Ford JS, Freyer DR, et al. Late mortality and chronic health conditions in long-term survivors of early-adolescent and young adult cancers: a retrospective cohort analysis from the Childhood Cancer Survivor Study. Lancet Oncol. 2020;21(3):421–35.
35. Harper S, Lynch J, Smith GD. Social determinants and the decline of cardiovascular diseases: understanding the links. Annu Rev Public Health. 2011;32:39–69.
36. Loree JM, Anand S, Dasari A, Unger JM, Gothwal A, Ellis LM, et al. Disparity of race reporting and representation in clinical trials leading to cancer drug approvals from 2008 to 2018. JAMA Oncol. 2019;5(10):e191870.
37. Gunaldi M, Duman BB, Afsar CU, Paydas S, Erkisi M, Kara IO, et al. Risk factors for developing cardiotoxicity of trastuzumab in breast cancer patients: an observational single-centre study. J Oncol Pharm Pract. 2016;22(2):242–7.
38. Ahmad J, Muthyala A, Kumar A, Dani SS, Ganatra S. Disparities in cardio-oncology: effects on outcomes and opportunities for improvement. Curr Cardiol Rep. 2022;24(9):1117–27.
39. Damp JB, Cullen MW, Soukoulis V, Tam MC, Keating FK, Smith SA, et al. Program directors survey on diversity in cardiovascular training programs. J Am Coll Cardiol. 2020;76(10):1215–22.
40. Mathews HL, Janusek LW. Epigenetics and psychoneuroimmunology: mechanisms and models. Brain Behav Immun. 2011;25(1):25–39.
41. Saban KL, Mathews HL, DeVon HA, Janusek LW. Epigenetics and social context: implications for disparity in cardiovascular disease. Aging Dis. 2014;5(5):346–55.
42. Jaiswal S, Natarajan P, Silver AJ, Gibson CJ, Bick AG, Shvartz E, et al. Clonal hematopoiesis and risk of atherosclerotic cardiovascular disease. N Engl J Med. 2017;377(2):111–21.
43. Peres LC, Colin-Leitzinger CM, Teng M, Dutil J, Alugubelli RR, DeAvila G, et al. Racial and ethnic differences in clonal hematopoiesis, tumor markers, and outcomes of patients with multiple myeloma. Blood Adv. 2022;6(12):3767–78.

44. Liu X, Hou L, Xu D, Chen A, Yang L, Zhuang Y, et al. Effect of asymmetric dimethylarginine (ADMA) on heart failure development. Nitric Oxide. 2016;54:73–81.
45. Finkelman BS, Putt M, Wang T, Wang L, Narayan H, Domchek S, et al. Arginine-nitric oxide metabolites and cardiac dysfunction in patients with breast cancer. J Am Coll Cardiol. 2017;70(2):152–62.
46. Sandrim VC, Palei AC, Metzger IF, Cavalli RC, Duarte G, Tanus-Santos JE. Interethnic differences in ADMA concentrations and negative association with nitric oxide formation in preeclampsia. Clin Chim Acta. 2010;411(19–20):1457–60.
47. Lima FV, Parikh PB, Zhu J, Yang J, Stergiopoulos K. Association of cardiomyopathy with adverse cardiac events in pregnant women at the time of delivery. JACC Heart Fail. 2015;3(3):257–66.
48. Ng AT, Duan L, Win T, Spencer HT, Lee MS. Maternal and fetal outcomes in pregnant women with heart failure. Heart. 2018;104(23):1949–54.
49. Prousi GS, Joshi AM, Atti V, Addison D, Brown SA, Guha A, et al. Vascular inflammation, cancer, and cardiovascular diseases. Curr Oncol Rep. 2023;25:955.

Cardiotoxicity, Cancer Therapies, and Cardiac Cell Signaling

Virang Desai and Ragavendra R. Baliga

Abbreviations

ROS	Reactive oxygen species
TOP-28	Topoisomerase Il β
CUT&Tag	Cleavage Under Targets and Tagmentation
CH	Clonal hematopoiesis
mtDNA	Mitochondrial DNA
Alas1	Aminolaevulinic acid synthase 1
5-ALA	5-Aminolaevulinic acid
ZBP1	Z-DNA-binding protein 1
IFN-1	Type 1 interferon
LVEF	Left ventricular ejection fraction
DRP1	Dynamin-related protein 1
PKM2	Pyruvate kinase muscle isozyme 2
TLR	Toll-like receptor
TLR4–/–	Toll-like receptor 4 knockout
TLR2–/–	Toll-like receptor 2 knockout
ASC	Apoptosis-associated Speck-like protein containing a caspase recruitment domain
NLRP3	NOD-, LRR-, and pyrin domain-containing protein 3
CYP1	Cytochrome P450 family 1
LRP-1/CD91	Low-density lipoprotein receptor–related protein 1/cluster of differentiation 91
BNP	Brain natriuretic peptide
DEGs	Differentially expressed genes
RRAA	Retinoid X receptor alpha a

V. Desai · R. R. Baliga (✉)
Department of Cardiovascular Medicine, The Ohio State University Wexner Medical Center, Columbus, OH, USA
e-mail: ragavendra.baliga@osumc.edu

S. A. Smith, S. S. Scott (eds.), *Myocardial Oxidative Stress in Anti-Cancer Therapies*, Oxidative Stress in Applied Basic Research and Clinical Practice,
https://doi.org/10.1007/978-3-031-97517-2_3

Nrf2	Nuclear factor erythroid 2-related factor 2
UCP2	Uncoupling protein 2
NRG-1	Neuregulin-1
EGFR	Epidermal growth factor receptor
TKIs	Tyrosine kinase inhibitors
HF	Heart failure
DNA	Deoxyribonucleic acid
hiPSC-ECs	Human-induced pluripotent stem cell-derived endothelial cells
hiPSC-CFs	Human-induced pluripotent stem cell-derived cardiac fibroblasts
VEGFR2	Vascular endothelial growth factor receptor 2
PDGFR	Platelet-derived growth factor receptor
IGF	Insulin-like growth factor
CTLA-4	Cytotoxic T lymphocyte antigen 4
PD-L;1	Programmed death ligand 1
irAE	Immune-related adverse event
MAPK	Mitogen-activated protein kinase
TNBC	Triple-negative breast cancer
CAR	Constitutive androstane receptor

Introduction

Cancer therapies, including chemotherapy, targeted therapy, immunotherapy, and radiation, have significantly improved cancer patient survival rates. However, they often come with unintended consequences, including cardiotoxicity [1, 2]. Cardiotoxicity refers to the damage or dysfunction of the heart and cardiovascular system caused by cancer treatments.

Broadly, there are two types of cardiotoxicity: (1) acute cardiotoxicity, which occurs shortly after treatment and is usually reversible, and (2) chronic cardiotoxicity, which develops gradually over time and may be irreversible, leading to long-term heart issues. Several mechanisms contribute to cardiotoxicity, including the following: (1) *Oxidative stress*: Overproduction of reactive oxygen species, damaging cardiac cells and leading to inflammation and apoptosis. (2) *Mitochondrial dysfunction*: Cardiotoxic substances that impact mitochondria encompass a range of commonly utilized anticancer medications such as cisplatin, anthracyclines (doxorubicin/Adriamycin), trastuzumab (Herceptin), sorafenib (Nexavar), imatinib (Gleevec), mitoxantrone (Novantrone), bevacizumab (Avastin), sunitinib (Sutent), and imatinib (Gleevec). Mitochondrial toxicity stems from an array of mechanisms, encompassing the disruption of the mitochondrial respiratory chain (such as uncoupling) or the suppression of crucial mitochondrial enzymes (namely, oxidative phosphorylation, Szent–Györgyi–Krebs cycle, mitochondrial DNA replication, and ADP/ATP translocator). (3) *Inflammatory responses*: Activation of inflammatory pathways that cause damage to cardiac tissue. (4) *Apoptosis*: Programmed cell death of cardiac cells due to various stressors induced by cancer therapies. (5) *Autophagy dysregulation*: Abnormal autophagy affects the removal of damaged cellular components and leads to cardiac dysfunction.

A group of researchers aimed to understand the autophagic variations in anthracycline-induced cardiotoxicity [3] and identify a mechanism-based therapy using genetic and pharmacological approaches. In an adult model denoting anthracycline-induced cardiotoxicity, the scholars discerned a dual-phase reaction in autophagy: initial mobilization and subsequent inhibition, characterized by a deterioration in cardiac performance. They conducted conditional genetic studies involving *atg7*, a gene encoding a critical autophagy core protein. Their findings demonstrated that overexpression of *atg7* exhibited therapeutic effects in the late phase but adverse effects in the early phase of adult anthracycline-induced cardiotoxicity. Subsequently, a zebra fish model of embryonic anthracycline-induced cardiotoxicity was employed to assess the therapeutic potential of 37 FDA-approved autophagy activators. The most promising drugs, spironolactone, pravastatin, and minoxidil, were identified. To ascertain their therapeutic efficacy, these drugs underwent testing in an adult anthracycline-induced cardiotoxicity model where they demonstrated effectiveness in the later phases of cardiotoxicity. Moreover, the researchers employed a mouse model to elucidate the time-dependent therapeutic effects of anthracycline-induced cardiotoxicity, unveiling the activation of autophagy through spironolactone and rapamycin in an atg7-dependent manner. These significant findings indicate that the activation of autophagy through atg7 presents a feasible therapeutic approach for mitigating the decline in cardiac function brought about by anthracycline-induced cardiotoxicity. Hence, it is essential to contemplate the temporal aspect of autophagy-focused treatment.

Cancer Therapies and Cardiotoxicity

I. *Anthracyclines*: A class of chemotherapy drugs, including doxorubicin, are known for their potent anticancer effects but are associated with dose-dependent cardiotoxicity.

   The mechanisms of anthracycline cardiotoxicity are complex and multifactorial, but they are thought to involve several factors, including the following:

   - The primary pathophysiological mechanisms responsible for anthracycline cardiotoxicity include the following:

   1. Heightened production of mitochondrial reactive oxygen species (ROS); ROS can damage DNA, proteins, and lipids, and can lead to cell death.
   2. The buildup of cardiotoxic metabolites, for example, anthracyclines, can increase the accumulation of iron in the heart, which can also contribute to oxidative stress.
   3. Nuclear DNA and chromatin damage:

      (a) The binding of anthracyclines with topoisomerase-2β (TOP-2β) is responsible for causing DNA damage, as this enzyme plays a crucial role in both DNA replication and repair. This can lead to DNA damage—double-stranded breaks (DSBs) and cell death. It has been suggested that the primary mediator of anthracycline cardiotoxicity is topoisomerase IIβ (encoded by TOP2B). Building on this mechanism, dexrazoxane, an iron chelator, was effectively formulated and stands as the anthracycline cardiotoxicity sole preventive compound approved by the U.S. Food and Drug Administration (FDA). A recent study reported that the benefits of dexrazoxane are long-lasting [4].

(b) While the conventional theory suggests that the drugs hinder the enzyme responsible for DNA untangling and repair during the replication of rapidly dividing tumor cells, it has been uncovered that doxorubicin also counteracts cancer cells by impairing chromatin through disrupting histones: the spherical proteins that serve as a foundation for DNA coiling. Such impairment of chromatin seems to hinder the transcription of genes into proteins as well as other crucial cellular processes.

Aclarubicin, an approved antitumor medication, and diMe-Doxo, a modified form of doxorubicin, were investigated by a team from Leiden University [6] for their potential to extract histones without inducing DNA damage. These compounds demonstrated comparable or even superior efficacy in killing cultured cancer cells and slowing tumor growth in mice compared to the original drug. Remarkably, mice that were prone to developing tumors and received aclarubicin treatment showed no indications of heart damage, indicating the possibility of avoiding these effects in human patients. Additionally, these mice were less prone to developing tumors later on, as reported by the Leiden team. Both aclarubicin and diMe-Doxo warrant thorough investigation—DiMe-Doxo demonstrates greater efficacy in targeting solid tumors, while aclarubicin shows promise in blood cancers.

One study has outlined that anthracycline-induced cardiotoxicity necessitates a dual interplay of cellular activities, specifically DNA damage and chromatin damage [5]. The use of agents like topoisomerase II poisons, when solely targeting one activity, does not induce any harm to the heart in mice and human cardiac microtissues, as evidenced by aclarubicin. Additionally, the study indicates the possibility of detoxifying anthracycline by chemically separating these dual activities. Anthracycline variants that can cause chromatin damage, but not double-strand breaks, have been found to be effective against cancer in cell lines, mice, and human acute myeloid leukemia patients. This suggests that chromatin damage is an important way that anthracyclines kill cancer cells. By using these anthracycline variants, which do not cause heart problems or therapy-related tumors, researchers were able to separate the negative side effects from the positive cancer-fighting effects. These results suggest that anthracycline variants that primarily cause chromatin damage could be used to treat cancer patients for a longer period of time, improving the quality of life for cancer survivors.

Anthracyclines disrupt the structure of chromatin by integrating into DNA and amplifying the turnover of nucleosomes. In order to understand the specific effects of anthracycline-induced changes on chromatin, Wooten and colleagues utilized the technique of Cleavage Under Targets and Tagmentation (CUT&Tag) to study the activity of RNA polymerase II in Drosophila cells during anthracycline treatment. According to their investigation, exposure to anthracycline aclarubicin was found to result in increased levels of RNA polymerase II and changes in chromatin accessibility. Notably, the proximity and orientation of promoters played a significant role in influencing chromatin modifications during aclarubicin treatment. Specifically, divergent promoter pairs in close proximity displayed more prominent chromatin alterations, in contrast to tandem promoters with the same orientation. The study identified that aclarubicin treatment led to changes in the structure of noncanonical DNA G-quadruplexes at promoters and G-rich pericentromeric repeats. These findings suggest that aclarubicin's effectiveness in targeting cancer cells originates from its capacity to disrupt nucleosomes and RNA polymerase II.

More recently, clonal hematopoiesis has been implicated in anthracycline cardiotoxicity [6]. Clonal hematopoiesis (CH) is the presence of somatic mutations in hematopoietic stem or progenitor cells in individuals without hematologic malignancy. This condition is linked to an increased risk of adverse cardiovascular events and overall worse outcomes. In a retrospective cohort study of 100 cancer patients undergoing doxorubicin treatment, it was observed that 25 of them developed arrhythmia, heart failure (HF), or decreased left ventricular ejection fraction. However, when accounting for variables such as age at chest radiation, diabetes, doxorubicin initiation, and dyslipidemia, the researchers discovered a significant correlation between a high cumulative dose of doxorubicin (>240 mg/m^2) and anthracycline cardiotoxicity (odds ratio [OR], 7.00; 95% CI, 1.77–27.74; $P = 0.0056$). Additionally, clonal hematopoiesis (OR, 8.58; 95% CI, 2.05–35.99; $P = 0.0033$) was also identified as a significant risk factor for anthracycline cardiotoxicity.

4. Mitochondrial DNA (mtDNA) damage:

Ferroptosis, a process within mitochondria driven by an excess of iron and extensive lipid peroxidation, is vital in the progression of cardiomyopathy induced by doxorubicin. Doxorubicin accumulates in mitochondria by inserting into mitochondrial DNA (mtDNA), instigating ferroptosis in a manner dependent on the content of mtDNA [7]. Furthermore, doxorubicin disrupts heme synthesis by reducing the levels of 5′-aminolevulinate synthase 1 (Alas1), a key enzyme in this process. This disruption impairs iron utilization, resulting in an accumulation of iron and subsequent ferroptosis within the mitochondria of cultured cardiomyocytes. Overexpressing Alas1 effectively prevents this adverse outcome. When 5-aminolevulinic acid

(5-ALA), a product of Alas1, is administered to cultured cardiomyocytes and mice, it alleviates iron overload and lipid peroxidation, effectively preventing ferroptosis induced by doxorubicin and consequent anthracycline cardiomyopathy. These findings underscore the cooperative role of doxorubicin and iron accumulation in mitochondria in inducing ferroptosis in cardiomyocytes and propose 5-ALA as a promising therapeutic option for addressing anthracycline cardiomyopathy.

The specific immunostimulatory traits of mitochondrial DNA and the timing of its recognition by intracellular nucleic acid sensors are not sufficiently defined, despite being a powerful activator of the innate immune system. Recent research findings have identified that a malfunction in the mitochondrial genome fosters the gradual buildup of Z-form DNA. Z-DNA-binding protein 1 (ZBP1) assumes a critical function as it stabilizes Z-form mtDNA and establishes a consequential cytosolic complex with RIPK3, cGAS, and RIPK1 within the cytosol. Following exposure to doxorubicin, cardiomyocytes exhibited heightened expression of ZBP1, increased levels of Z-form mtDNA, and elevated signaling of type I interferon (IFN-I), thereby sustaining the phosphorylation of STAT1 [8]. Intriguingly, mice lacking ZBP1 or IFN-I signaling demonstrated protection against anthracycline cardiotoxicity. These findings bring forth the critical involvement of ZBP1 as a collaborative associate alongside cGAS in upholding the immune responses initiated by IFN-I during episodes of mitochondrial genome instability. Furthermore, they emphasize the potential of ZBP1 as a possible target for therapeutic interventions aimed at managing HF and other ailments in which mtDNA stress contributes to the manifestation of interferon-related pathology.

5. Impairment of sarcomere structure and function: In a recent study, it was highlighted that there is a 5% reduction in left ventricular mass within 6 months of initiating anthracycline-based chemotherapy compared to non-anthracycline-based chemotherapies and healthy individuals. These reductions in LV mass align with increased symptoms of HF, resembling mild established HF. Notably, this decline in left ventricular mass and the associated HF symptoms occur independently of any decrease in left ventricular ejection fraction, which is the conventional imaging biomarker for cardiotoxicity. These findings underscore the possibility of an additional mechanism contributing to HF development, separate from a decline in LVEF [9].
6. Mitochondrial fission: Abnormal mitochondrial fission, directly influenced by the large GTPase dynamin-related protein 1 (DRP1), plays a significant and direct role in cardiovascular pathophysiology. Extensive research has characterized the restraint of pathological cardiac and vascular remodeling in mice through the genetic elimination of DRP1. Proof of concept has been further validated using pharmacological inhibitors. However, in the development of treatment strategies, careful consideration of the physiological roles of mitochondrial fission is essential. Mitochondrial fission is essential for processes such as mitochondrial division, which is vital for segregating

mitochondrial DNA and promoting cellular proliferation. Therefore, it may not be advisable to limit fission during certain developmental stages or for patients with heritable mitochondrial DNA mutations. Moreover, mitochondrial fission plays a crucial role in facilitating mitophagy. Hence, when considering pharmacological approaches to inhibit fission, careful consideration of dosage and duration is necessary to ensure proper mitochondrial quality control through mitophagy. Nevertheless, there are observations indicating that the inhibition of uncontrolled mitophagy during anthracycline cardiotoxicity also restrains excessive mitochondrial fission, thus highlighting the potential synergistic promotion of cardiovascular pathophysiology by coupling these two processes in specific scenarios. The evaluation of drug efficacy and safety becomes even more challenging when considering the requirement for cell-type specificity, specifically when targeting mitochondrial fission. The importance of careful drug design is emphasized by the role of macrophages and endothelial cells in promoting vascular inflammation through excessive fission activities. Given DRP1's vital role in immune cell maturation and phagocytosis, preserving physiological immune function when designing inhibitor regimens is paramount before transitioning animal study findings to clinical applications. By targeting particular DRP1 adaptor proteins, it becomes possible to regulate specific forms of mitochondrial fission, which contribute to the development of disease while maintaining the normal division of mitochondria. However, further investigation is necessary to confirm the viability of this approach. While there are currently no clinical trials investigating inhibitors for mitochondrial fission, a substantial amount of evidence indicates a specific pathological threshold for this process and the potential to repurpose several FDA-approved compounds, highlighting the need for increased practical implementation. The significance of mitochondrial fission in anthracycline cardiotoxicity is now clear, and continued research will likely open up new avenues for therapeutic approaches in patients.

7. *Other mechanisms*: Other mechanisms that may contribute to anthracycline cardiotoxicity include apoptosis and autophagy.
8. *Apoptosis*: Anthracycline cardiotoxicity arises due to the activation of p53, a protein known for inducing apoptosis in tumors and various organs, including the heart, leading to nonspecific tissue injury. One group of investigators [10] discovered a direct interaction between pyruvate kinase muscle 2 (PKM2) and p53. The tetrameric form of PKM2 undergoes preferential oxidation in the heart, acting to suppress apoptosis. Conversely, this same tetramer boosts p53 activity in environments with low oxidation, such as tumors. The authors identified a compound that stabilizes the tetrameric form of PKM2, thus safeguarding cardiomyocytes while aiding in the elimination of cancer cells through anthracyclines.
9. *Inflammation*: Our understanding of the role and mechanisms of sterile inflammation in anthracycline-induced cardiotoxicity remains limited. Recent studies have shed light on the beneficial effects of Toll-like receptor

(TLR) deletion, pivotal components of innate immune responses. For example, Riad et al. [14] demonstrated inflammatory responses, enhanced cardiac performance, and decreased oxidative stress in mice lacking TLR4 (TLR4−/−) following the administration of doxorubicin. Similarly, Nozaki et al. [15] noted reduced cardiac impairment and apoptotic cell demise in TLR2-deficient mice (TLR2−/−) treated with anthracycline. Evidence supports the notion that inflammasomes, which are intricate multiprotein complexes found within cells, play a role in sterile inflammation by regulating the secretion of the proinflammatory cytokine interleukin (IL)-1β.

NLRP3 inflammasomes, which belong to the nucleotide-binding oligomerization domain-like receptor family with pyrin domain, have been associated with sterile inflammation. The initiation of the processing of pro-IL-1β into its mature form occurs when caspase-1 is activated by NLRP3 inflammasomes, which consist of NLRP3, apoptosis-associated speck-like protein containing a caspase recruitment domain (ASC), and caspase-1. As a result, IL-1β is secreted, leading to inflammatory responses and tissue damage. NLRP3 inflammasomes detect nonmicrobial danger signals and cause the release of IL-1β, leading to sterile inflammation in cardiovascular disease.

Sterile inflammation plays a role in doxorubicin-induced cardiotoxicity, so scientists have examined the impact of NLRP3 inflammasomes in this context. NLRP3-deficient mice exhibited cardiac dysfunction and injury when exposed to low-dose anthracycline, indicating that NLRP3 deficiency increases susceptibility to Dox-induced cardiotoxicity regardless of IL-1β. Although there was no significant difference in inflammatory cell infiltration between the hearts of wild-type mice and NLRP3-deficient mice, NLRP3-deficient mice treated with anthracycline experienced a decrease in cardiac IL-10 production, with macrophages being the primary inflammatory cells in the hearts. Transplanting bone marrow revealed that bone marrow-derived cells contribute to the exacerbation of anthracycline-induced cardiotoxicity in NLRP3-deficient mice. In vitro experiments demonstrated a decrease in IL-10 production in macrophages lacking NLRP3. Additionally, delivery of IL-10 through adeno-associated virus reversed the aggravation of cardiotoxicity in NLRP3-deficient mice. These findings emphasize the role of NLRP3 in regulating macrophage IL-10 production and its involvement in the pathophysiology of anthracycline-induced cardiotoxicity, independent of IL-1β. This discovery uncovers a new aspect of NLRP3 and provides fresh insights into the underlying mechanisms of anthracycline cardiotoxicity.

In another study, the researchers aimed to enhance the potency of visnagin analogs, intending to utilize them as tools for elucidating the mechanisms of visnagin-induced cardioprotection [11]. They conducted structure–activity relationship investigations that were performed utilizing a zebrafish model of doxorubicin cardiomyopathy. By relocating the 5-carbonyl to the 7 position and introducing short ester side chains, visnagin analogs exhibited an extraordinary potency increase of 1000-fold in zebrafish and 250-fold in mice. Employing proteomics, this study unveiled a notable impact of doxorubicin on the activation of Cytochrome P450 family 1 (CYP1), an effect that was mitigated by the inclusion of visnagin and its potent analog. The investigation demonstrated that administering a range of CYP1 inhibitors, combined with the suppression of CYP1A, effectively averted the development of doxorubicin-induced cardiomyopathy in zebrafish. These discoveries strongly endorse the pivotal role of CYP1 in doxorubicin-related cardiotoxicity, underscoring the potential benefits of manipulating this pathway in clinical scenarios.

10. Hemopexin is a glycoprotein found in the bloodstream, weighing 57 kDa and synthesized in the liver. It possesses a strong binding affinity with heme, an iron-containing molecule known for its pro-oxidant and inflammatory properties. The interaction between heme and hemopexin initiates receptor-mediated endocytosis in circulating macrophages. This process is facilitated by a protein called low-density lipoprotein receptor–related protein 1 (LRP-1/CD91). As a result, heme recycling occurs within the reticuloendothelial system in the spleen. Recent studies have emphasized the potential benefits of administering external hemopexin in murine models of sickle cell disease. This strategy could potentially alleviate heme-induced endothelial toxicity and cardiomyopathy, offering therapeutic value in this context. While intracellular iron has conventionally been recognized as a mediator of anthracycline-induced cardiac toxicity, the involvement of heme metabolism in this context has not been explored. A study sheds light on the induction of circulating hemopexin as a cardioprotective mechanism applicable to both patients and mice undergoing anthracycline treatment. The study provides evidence supporting the induction of hemopexin as a protective mechanism activated during cases of chronic cardiomyopathy induced by doxorubicin. Despite the link between increased plasma hemopexin levels and heightened cardiac toxicity in patients, our preclinical data indicate that higher levels of hemopexin in anthracycline-treated patients may offer physiological benefits and cardioprotection. Circulating hemopexin has the potential to be used as a blood marker for diagnosis and prognosis, similar to brain natriuretic peptide (BNP) in HF cases. BNP is released when the heart is overwhelmed and helps reduce congestion by promoting urine production and widening blood vessels. Similarly, it is hypothesized that circulating hemopexin levels increase in response to anthracycline-induced cardiac dysfunction in both patients and mice. However, this study did not find any elevation in heme levels in the blood or establish a link between hemopexin and the acute-phase inflammatory response. Further research should investigate the factors that trigger the induction of hemopexin in patients experiencing anthracycline-related cardiac toxicity.
11. Epigenetic modifications of DNA: One potential explanation for the delayed myocardial damage seen after short-term anthracycline therapy is the epigenetic modification of DNA.

    A team of researchers [19] utilized RNA-sequencing on left ventricular biopsies obtained from human endomyocardial samples along with mass spectrometry on genomic DNA to examine the progression of epigenetic regulators during the onset and later phases of anthracycline-induced cardiotoxicity. The identified differentially regulated genes were further validated through RT-qPCR. Furthermore, an in vitro study was performed to delve into specific aspects of epigenetic memory in the development of cardiotoxicity induced by anthracycline. The connection between the expression of genes in the early and late stages of cardiotoxicity demonstrated an exceptionally strong correlation with an R^2 value of 0.98, highlighting the presence of 369 differentially expressed genes (DEGs, FDR < 0.05). Out of these, a significant majority of 72% ($n = 266$) were found to be upregulated, while the remaining 28% ($n = 103$) were downregulated in cases of later-onset cardiotoxicity compared to earlier-onset cases. An examination of gene ontology revealed a noteworthy enrichment of genes linked to methyl-CpG DNA binding, chromatin remodeling, transcription regulation, and apoptosis promotion. The differential mRNA expression of genes connected to DNA methylation metabolism was then confirmed in endomyocardial biopsies using RT-qPCR. In a larger set of biopsies, it was observed that Tet2 had significantly higher expression in samples from individuals with cardiotoxicity compared to control biopsies and patients with nonischemic cardiomyopathy. Furthermore, an in vitro investigation was conducted using H9c2 cells, which were exposed to short-term doxorubicin treatment and then cultured and passed once they reached a confluency of 70%–80%. In comparison, cells treated solely with a vehicle showed substantial upregulation of Nppa, Nppb, Tet1/2, and other genes involved in active DNA demethylation 3 weeks after the short-term doxorubicin treatment. These changes were consistent with a decrease in DNA methylation and an increase in hydroxymethylation, providing a reflection of the epigenetic modifications observed in the endomyocardial biopsies.

12. *Nonmyocardial cells*: Traditionally viewed as the primary cellular target, cardiomyocytes have historically been considered the main site of impact. However, emerging evidence now suggests that abnormalities in noncardiomyocytes [12], such as endothelial cells [13], vascular smooth muscle cells [14], and cardiac fibroblasts [15, 16], significantly contribute to the pathogenesis of anthracycline cardiotoxicity. Despite this, therapeutic approaches focused on these noncardiomyocytes have not been thoroughly investigated or explored. Researchers recently developed a method using zebrafish to explore the genetic basis of anthracycline cardiotoxicity. In their study, they investigate a mutant called GBT0419, which affects a protein called retinoid x receptor alpha a (rxraa) and its molecular and cellular properties. They discovered that activating RXRA in endothelial cells, rather than in heart cells, can protect against anthracycline-induced cardiotoxicity. They also found that two FDA-approved drugs called isotretinoin and bexarotene, both RXRA agonists, have cardioprotective effects, but only if given during the early stages of cardiotoxicity or as a preventative measure, not during later stages. The positive effects of RXRA activation are thought to be due to the

repair of damaged endothelial cell barriers through the regulation of a protein called zonula occludens-1. Overall, this study provides strong evidence that RXRA could be a potential target for treating anthracycline cardiotoxicity and identifies a previously overlooked mechanism that should be considered in future research.

There is no cure for anthracycline cardiotoxicity, but there are several strategies that can be used to prevent or reduce its severity. These include the following:

- *Limiting the cumulative dose of anthracyclines:* The higher the cumulative dose of anthracyclines, the greater the risk of cardiotoxicity.
- *Using cardioprotective agents:* There are several drugs that can be used to protect the heart from anthracycline toxicity. These include antioxidants, such as dexrazoxane, and beta-blockers, such as carvedilol.
- *Monitoring for signs of cardiotoxicity:* Patients receiving anthracycline therapy should be closely monitored for signs of cardiotoxicity, such as decreased left ventricular ejection fraction (LVEF) and arrhythmias.

Researchers are actively investigating new ways to prevent and treat anthracycline cardiotoxicity. Some of the most promising targets include the following:

- *Mitochondrial dynamics:* Anthracyclines can disrupt mitochondrial dynamics, which can lead to mitochondrial dysfunction and cell death. Targeting mitochondrial dynamics is a promising approach for preventing and treating anthracycline cardiotoxicity.
- *Nuclear factor erythroid-2-related factor 2 (Nrf2):* Nrf2 is responsible for controlling the activation of antioxidant genes through transcriptional regulation. Targeting Nrf2 activation is a potential way to increase the heart's antioxidant defenses and protect against anthracycline toxicity.
- *Uncoupling protein 2 (UCP2):* UCP2 is a mitochondrial protein that helps to regulate mitochondrial ROS production. Targeting UCP2 is a potential way to reduce mitochondrial ROS production and protect the heart from anthracycline cardiotoxicity.

II. *Trastuzumab*: A targeted therapy used for breast cancer, known to cause cardiac dysfunction. Gaining insight into the intricate molecular mechanisms driving the pathological alterations in cardiomyocytes subsequent to trastuzumab treatment holds significant importance.

Trastuzumab binds to the extracellular domain of the HER2 protein, which is expressed in both tumor cells and cardiomyocytes. Inhibition of downstream signaling disrupts crucial protective pathways in the heart that are activated in response to heightened cardiovascular stress, as a result of preventing HER2 from heterodimerizing with HER3/4 receptors in response to neuregulin (NRG-1) signals.

One of the most important protective pathways activated by NRG-1 signaling via HER2 is the RAS/RAF/ERK1/2 pathway. This pathway upregulates ERK1/2, which stabilizes myofibril structure and inhibits apoptosis. Another important protective pathway is the PI3K/AKT pathway. This pathway heightens the eNOS activity, mitochondrial respiration, and production of reactive oxygen species (ROS). All of these processes are essential for maintaining cardiac health.

Conditional deletion of HER2 in mice led to dilated cardiomyopathy (DCM), highlighting the significance of HER2-mediated NRG-1 signaling in cardiac protection. Similarly, studies in human fetal cardiomyocytes and rat cardiomyoblasts have revealed trastuzumab's disruption of downstream ERK1/2 and AKT activation. Consequently, the perturbation of these protective pathways initiates diverse pathological alterations, encompassing both energetic and structural domains, potentially culminating in HF.

HER2, a receptor with tyrosine kinase activity situated in the cell membrane, is part of the EGFR (epidermal growth factor receptor) family. When dimerization occurs, whether it is homogenous or heterogenous, it triggers signaling pathways that promote cell growth, proliferation, survival, and angiogenesis. These effects primarily involve amplified PI3K/Akt and MEK/ERK signaling.

Trastuzumab and pertuzumab exert their effects by binding to the extracellular region of the HER2 receptor, interrupting signal transmission, and hindering various processes such as dimerization, antibody-dependent cytotoxicity, and downregulation of HER2. The endothelium releases NRG-1, a ligand that activates prosurvival and growth signaling in cardiomyocytes. Considerable investigation has been conducted in animal models and iPSC-CM to evaluate the protective role of NRG-1/HER2 signaling in the heart.

As decompensated cardiomyopathy develops during pressure overload in live animals, the initial levels of HER2 and HER4 remain steady but later decline by around 40%–50%. The knockdown of HER2 specifically in cardiomyocytes led to the development of dilated cardiomyopathy in mice, and isolated cardiomyocytes exhibited increased susceptibility to anthracyclines.

Recent research employing human iPSC-CMs suggests that altered metabolism may contribute to the cardiac toxicity induced by trastuzumab. iPSC-CMs treated with trastuzumab or lapatinib, a small molecule inhibitor of HER1/HER2, displayed reduced expression of genes involved in small molecule metabolism and decreased glucose uptake. These cells also demonstrated impaired contractile and calcium handling abilities, as well as diminished autophagy, glucose uptake, and mitochondrial function. Analysis of RNA-seq data revealed distinct gene expression patterns enriched in mitochondrial oxidative phosphorylation and pathways associated with cardiac dysfunction and enlargement, including critical metabolism genes. Enhancing AMPK activity through the use of drugs like metformin improved contraction velocity and metabolism, while treatment with the mTOR inhibitor rapamycin did not yield the same outcome. Further investigations have demonstrated decreased autophagy and dysfunction in mitochondria in trastuzumab-induced cardiac toxicity. These findings offer valuable insights into potential mechanisms for targeted prevention and treatment of trastuzumab-induced cardiac toxicity. Here are some specific examples of the pathological changes that may occur in cardiomyocytes as a result of trastuzumab-mediated cardiomyopathy.

- *Energetic changes:* Decreased mitochondrial respiration, decreased ATP production, and increased ROS production.
- *Structural changes:* Myofibril disarray, decreased contractility, and chamber dilation.
- *Cell death:* Apoptosis and necrosis.

These pathological changes can lead to several clinical manifestations of HF, including shortness of breath, fatigue, and HF hospitalization.

It is important to note that trastuzumab-mediated cardiomyopathy is a complex phenomenon with multiple contributing factors. The molecular mechanisms described above are just one part of the puzzle. Further research is needed to fully understand the underlying mechanisms of this adverse effect and to develop strategies to mitigate its risk.

III. *Tyrosine Kinase Inhibitors (TKIs)*:

Tyrosine kinase inhibitors (TKIs) hold the responsibility for the phosphorylation of substances containing tyrosine residues, achieved by transferring a phosphate group from adenosine triphosphate (ATP). These enzymes play an indispensable role in vital biological processes like cell proliferation, differentiation, and metabolism. Small molecule TKIs are carefully crafted for anticancer therapy, aiming to impede the utilization of ATP by tyrosine kinases, thereby impeding the proliferation of cancer cells. However, it is essential to recognize that TKIs not only obstruct tyrosine kinases in cancerous cells but also impede protein kinases in noncancerous cells, including cardiomyocytes. This unintended consequence disrupts crucial molecular pathways, potentially leading to cardiac dysfunction due to the exhaustion of ATP, which serves as the primary energy source for cardiomyocytes. These targeted therapies have the potential to induce hypertension, HF, or arrhythmias.

A team of researchers examined the gene expression patterns of primary cardiomyocyte-like cells derived from the human heart. These cells were treated with 26 different FDA-approved TKIs. They categorized the impacts of these TKIs on specific subcellular pathways and processes. In assessing cardiotoxicity, the researchers employed patient reports sourced from the FDA Adverse Event Reporting System pertaining to these TKIs, calculating relative risk scores. These scores were subsequently merged with transcriptomic datasets from cell lines using elastic net regression analysis. This comprehensive approach facilitated the identification of a gene signature capable of predicting the risk of cardiotoxicity. Furthermore, the researchers established associations between the risk of cardiotoxicity and the structural/binding profiles of individual TKIs. These findings suggest that the immediate changes in gene expression observed in cell-based experiments, in combination with drug substructures, hold considerable predictive value in assessing the likelihood of cardiotoxicity caused by TKI treatment.

A separate research team [17] conducted an investigation utilizing cardiomyocytes derived from human induced pluripotent stem cells (hiPSC-CMs). These cells were obtained from 11 healthy individuals and 2 cancer patients undergoing treatment. The primary objective was to assess the cardiotoxic effects of tyrosine kinase inhibitors (TKIs) that have been approved by the U.S. Food and Drug Administration (FDA). The researchers embarked on a comprehensive analysis, evaluating alterations in various aspects of cardiomyocyte function, including viability, contractility, electrophysiology, calcium handling, and signaling. Based on the data obtained, a "cardiac safety index" was developed as a reliable indicator of the cardiotoxicity associated with different TKIs. The findings revealed that TKIs with low cardiac safety indices were correlated with cardiotoxicity in patients. Additionally, the researchers investigated the specific cardiotoxicities exhibited by different cell types by generating endothelial cells (hiPSC-ECs) and cardiac fibroblasts (hiPSC-CFs). Through the use of high-throughput screening, it was determined that TKIs inhibiting vascular endothelial growth factor receptor 2 (VEGFR2) and platelet-derived growth factor receptor (PDGFR) induced cardiotoxicity in hiPSC-CMs, hiPSC-ECs, and hiPSC-CFs. Further examination of phosphoproteins unveiled a compensatory increase in cardioprotective insulin and insulin-like growth factor (IGF) signaling in hiPSC-CMs treated with these specific TKIs. Augmenting cardioprotective signaling through the addition of exogenous insulin or IGF1 enhanced the viability of hiPSC-CMs when cotreated with cardiotoxic VEGFR2/PDGFR-inhibiting TKIs. Consequently, hiPSC-CMs demonstrated efficacy in identifying cardiovascular toxicities associated with anticancer TKIs, closely paralleling the clinical manifestations. This groundbreaking approach yielded unforeseen insights, particularly regarding the potential mitigation of toxicity through the activation of cardioprotective insulin/IGF signaling. Although hypertension is a recognized side effect of TKIs, it remains uncertain whether it directly causes TKI-induced cardiotoxicity or is a distinct complication. Numerous scientific inquiries have delved into the correlation between TKI-induced hypertension and cardiotoxicity. Notably, a study unearthed a connection between dasatinib-induced hypertension and heightened oxidative stress and impaired cardiac function in mice, which could potentially contribute to its adverse effect on the heart [16]. On the contrary, other research posits that TKIs have the ability to directly affect the heart, resulting in cardiotoxicity, irrespective of their influence on blood pressure. For instance, one study exhibited how the TKI sunitinib discouraged cardiac contractility in isolated rat hearts, totally unrelated to its effect on blood pressure. The authors put forth the idea that sunitinib-induced cardiotoxicity might originate from direct interference with cardiac ion channels and contractile proteins. Consequently, the task of determining whether hypertension directly causes TKI-induced cardiotoxicity or represents a distinct complication can be quite daunting. Evidence indicates that TKIs have the capacity to directly impact the heart, contributing to cardiotoxicity independently of hypertension. Furthermore, recent investigations propose that TKI-induced cardiotoxicity may be mediated through the activation of intracellular signaling pathways in cardiomyocytes. For instance, researchers discovered that the TKI ponatinib activated the ERK1/2 signaling pathway in human cardiomyocytes, a process that was linked to diminished contractility and an increase in cell death. The authors speculated that the activation of the ERK1/2 pathway could potentially serve as a mechanism for TKI-induced cardiotoxicity.

IV. *Immune checkpoint inhibitors*: These monoclonal antibodies serve the purpose of regulating immune responses through the inhibition of specific molecules present on T cells (CTLA-4 and PD1) as well as antigen-presenting cells (PD-L1). While they have proven effective in treating various types of cancer, it is important to note that they can lead to detrimental immune reactions (known as irAEs) in different tissues, including potentially life-threatening cardiac irAEs like myocarditis and pericarditis.

   Monoclonal antibodies known as ICIs function by bolstering existing immune responses via the blockade of inhibitory molecules on T cells (CTLA-4 and PD-1) and antigen-presenting cells (PD-L1). While they elicit a robust anti-tumor reaction, it has been observed that they can also trigger immune-related adverse events (known as irAEs) in a wide array of tissues, including cardiac irAEs which have the potential to be particularly fatal.

V. *Radiation therapy*: This can cause damage to the heart and surrounding vasculature, leading to long-term cardiovascular issues.

Cardiac Cell Signaling in Cardiotoxicity

Understanding the signaling pathways involved in cardiac cells is crucial for addressing cardiotoxicity:

1. *MAPK (mitogen-activated protein kinase) pathway*: Modulates cellular differentiation, proliferation, and apoptosis, the disturbance of which can result in detrimental cardiac consequences.
2. *PI3K/Akt/mTOR pathway*: This plays a key role in cardiac cell survival and growth; its alteration can lead to cardiotoxic effects.
3. *Nrf2 (nuclear factor erythroid 2-related factor 2) pathway*: Protects cells from oxidative stress; its activation can mitigate oxidative damage in cardiotoxicity.

   Doxorubicin and cyclophosphamide play essential roles in the administration of chemotherapy for triple-negative breast cancer (TNBC). However, the effectiveness of these drugs is often hindered by the development of drug resistance or the occurrence of intolerable side effects. Utilizing a synergistic blend of high-throughput screening and chemical modification methodologies, scientists have effectively crafted CN06, an exceptional compound having adeptness to activate two vital receptors, namely, the constitutive androstane receptor (CAR) and the nuclear factor erythroid 2-related factor 2 (Nrf2) [18]. CN06 facilitates the CAR-mediated conversion of cyclophosphamide, a prodrug, into its active form (4-hydroxy-cyclophosphamide) by stimulating the production of CYP2B6 in the liver. Additionally, CN06 protects cardiomyocytes from the toxic effects of doxorubicin by activating the Nrf2-antioxidant signaling pathway

in vitro. Through the use of a multicellular coculture model comprising human primary hepatocytes, TNBC cells, and cardiomyocytes, the study demonstrates that CN06 enhances the death of TNBC cells induced by cyclophosphamide/doxorubicin through increased CAR-dependent CYP2B6 induction. This simultaneous action allows for the activation of cyclophosphamide while shielding cardiomyocytes from doxorubicin-induced cardiotoxicity. Furthermore, CN06 maintains the viability and function of human cardiomyocytes derived from induced pluripotent stem cells (iPSCs) by regulating antioxidant defenses, reducing apoptosis, and improving contraction and relaxation kinetics. In conclusion, these findings highlight the potential of CAR and Nrf2 as promising therapeutic targets and emphasize CN06's ability to improve the balance between the effectiveness and toxicity of cyclophosphamide/doxorubicin-based chemotherapy.
4. *Calcium signaling*: Dysregulation of calcium signaling can lead to cardiac arrhythmias and dysfunction.

Mitigation and Management

1. *Early detection*: It is essential to continuously monitor the heart and promptly identify any cardiac dysfunction during and after cancer treatment.
2. *Cardioprotective agents*: Administering cardioprotective drugs alongside cancer treatments to mitigate cardiac damage.
3. *Lifestyle modifications*: Encouraging healthy lifestyles, including regular exercise and a heart-healthy diet, to reduce cardiac risk factors.
4. *Research and innovation*: Continued research to develop targeted cancer therapies with reduced cardiotoxicity and a better understanding of cardiac signaling pathways.

Future

Transcription factors crucial for the survival of cardiomyocytes are suppressed during HF. In a particular investigation, Lorenzana-Carrillo and colleagues [17] delved into the function of pyruvate kinase M2 (PKM2), a glycolytic enzyme, in modulating the transcription factors GATA4/6 and P53. Their research revealed that mice lacking PKM2 in their cardiomyocytes exhibited reduced GATA4/6 levels but an increase in P53 levels, ultimately leading to the development of dilated cardiomyopathy. Within adult cardiomyocytes, direct interaction between nuclear PKM2 and these transcription factors was noted. PKM2 played a role in preserving GATA4/6 integrity by impeding cleavage and degradation while promoting the degradation of P53. The E3 ubiquitin ligase TRIM35 was identified as responsible for breaking down nuclear PKM2. When overexpressed in cardiomyocytes in mice, TRIM35 resulted in cardiac dysfunction. Moreover, TRIM35 was upregulated in human HF tissue samples. The results of this study provide insight into the importance of PKM2 in promoting the survival of cardiomyocytes during HF.

The way forward is to target epigenetic changes in cardiac myocytes, pericytes, fibroblasts, and novel transcription factors like *sertad4* (SERTA domain-containing protein 4) [18, 19]. A combinatorial approach of in silico modeling, cell culture, and animal models, high-throughput genetic screens, and omics technologies will be required to conduct discovery and hypothesis-guided research.

Conclusion

The advent of cancer therapies has indisputably precipitated a paradigm shift in the treatment of cancer; however, due to their potential cardiotoxic consequences, it is imperative to delve further into the intricate mechanisms underlying these effects and devise effective strategies to mitigate them. By comprehensively examining the cellular signaling pathways and mechanisms involved in cardiotoxicity, we can work toward minimizing cardiac damage and improving the overall well-being of cancer patients undergoing treatment.

Disclosures The authors have no relevant affiliations or financial involvement to disclose.

References

1. Baliga RR, Addison D, Lenihan DJ, Narula J, Bossone E. When helping hurts: restricting cardiotoxicity of cancer therapeutics with accurate cardiovascular risk assessment works! Eur J Prev Cardiol. 2023; https://doi.org/10.1093/eurjpc/zwad273. Epub 20230925. PubMed PMID: 37747453
2. Baliga RR, Mensah GA, Bossone E. Translational cardio-oncology research to promote better outcomes for one and all. Heart Fail Clin. 2022;18(3):xv–xvii. https://doi.org/10.1016/j.hfc.2022.04.001. PubMed PMID: 35718424
3. Wang Y, Lu X, Wang X, Qiu Q, Zhu P, Ma L, et al. Based autophagy activation reverses Doxorubicin-induced cardiotoxicity. Circ Res. 2021;129(8):e166–82. https://doi.org/10.1161/CIRCRESAHA.121.319104. Epub 20210813. PubMed PMID: 34384247; PubMed Central PMCID: PMC8484060
4. Chow EJ, Aggarwal S, Doody DR, Aplenc R, Armenian SH, Baker KS, et al. Dexrazoxane and long-term heart function in survivors of childhood cancer. J Clin Oncol. 2023;41(12):2248–57. https://doi.org/10.1200/JCO.22.02423. Epub 20230120. PubMed PMID: 36669148; PubMed Central PMCID: PMC10448941
5. Qiao X, van der Zanden SY, Wander DPA, Borràs DM, Song JY, Li X, et al. Uncoupling DNA damage from chromatin damage to detoxify doxorubicin. Proc Natl Acad Sci USA. 2020;117(26):15182–92. https://doi.org/10.1073/pnas.1922072117. Epub 20200617. PubMed PMID: 32554494; PubMed Central PMCID: PMC7334570
6. Mammadova J, Colin-Leitzinger C, Nguyen D, Mhaskar R, Ganesan S, Tang YH, et al. Clonal Hematopoiesis as a molecular risk factor for Doxorubicin-induced cardiotoxicity: a proof-

of-concept study. JCO Precis Oncol. 2023;7:e2300208. https://doi.org/10.1200/PO.23.00208. PubMed PMID: 37738545
7. Abe K, Ikeda M, Ide T, Tadokoro T, Miyamoto HD, Furusawa S, et al. Doxorubicin causes ferroptosis and cardiotoxicity by intercalating into mitochondrial DNA and disrupting Alas1-dependent heme synthesis. Sci Signal. 2022;15(758):eabn8017. https://doi.org/10.1126/scisignal.abn8017. Epub 20221101. PubMed PMID: 36318618
8. Lei Y, VanPortfliet JJ, Chen YF, Bryant JD, Li Y, Fails D, et al. Cooperative sensing of mitochondrial DNA by ZBP1 and cGAS promotes cardiotoxicity. Cell. 2023;186(14):3013–3032.e22. https://doi.org/10.1016/j.cell.2023.05.039. Epub 20230622. PubMed PMID: 37352855; PubMed Central PMCID: PMC10330843
9. Jordan JH, Castellino SM, Meléndez GC, Klepin HD, Ellis LR, Lamar Z, et al. Left ventricular mass change after Anthracycline chemotherapy. Circ Heart Fail. 2018;11(7):e004560. https://doi.org/10.1161/CIRCHEARTFAILURE.117.004560. PubMed PMID: 29991488; PubMed Central PMCID: PMC6729136
10. Saleme B, Gurtu V, Zhang Y, Kinnaird A, Boukouris AE, Gopal K, et al. Tissue-specific regulation of p53 by PKM2 is redox dependent and provides a therapeutic target for anthracycline-induced cardiotoxicity. Sci Transl Med. 2019;11(478):eaau8866. https://doi.org/10.1126/scitranslmed.aau8866. PubMed PMID: 30728290
11. Asnani A, Zheng B, Liu Y, Wang Y, Chen HH, Vohra A, et al. Highly potent visnagin derivatives inhibit Cyp1 and prevent doxorubicin cardiotoxicity. JCI Insight. 2018;3(1):e96753. https://doi.org/10.1172/jci.insight.96753. Epub 20180111. PubMed PMID: 29321375; PubMed Central PMCID: PMC5821184
12. Spetz JKE, Florido MHC, Fraser CS, Qin X, Choiniere J, Yu SJ, et al. Heightened apoptotic priming of vascular cells across tissues and life span predisposes them to cancer therapy-induced toxicities. Sci Adv. 2022;8(45):eabn6579. https://doi.org/10.1126/sciadv.abn6579. Epub 20221109. PubMed PMID: 36351019; PubMed Central PMCID: PMC9645721
13. Wilkinson EL, Sidaway JE, Cross MJ. Cardiotoxic drugs Herceptin and doxorubicin inhibit cardiac microvascular endothelial cell barrier formation resulting in increased drug permeability. Biol Open. 2016;5(10):1362–70. https://doi.org/10.1242/bio.020362. Epub 20161015. PubMed PMID: 27543060; PubMed Central PMCID: PMC5087671
14. Bielak-Zmijewska A, Wnuk M, Przybylska D, Grabowska W, Lewinska A, Alster O, et al. A comparison of replicative senescence and doxorubicin-induced premature senescence of vascular smooth muscle cells isolated from human aorta. Biogerontology. 2014;15(1):47–64. https://doi.org/10.1007/s10522-013-9477-9. Epub 20131116. PubMed PMID: 24243065; PubMed Central PMCID: PMC3905196
15. Cappetta D, Esposito G, Piegari E, Russo R, Ciuffreda LP, Rivellino A, et al. SIRT1 activation attenuates diastolic dysfunction by reducing cardiac fibrosis in a model of anthracycline cardiomyopathy. Int J Cardiol. 2016;205:99–110. https://doi.org/10.1016/j.ijcard.2015.12.008. Epub 20151215. PubMed PMID: 26730840
16. Zhan H, Aizawa K, Sun J, Tomida S, Otsu K, Conway SJ, et al. Ataxia telangiectasia mutated in cardiac fibroblasts regulates doxorubicin-induced cardiotoxicity. Cardiovasc Res. 2016;110(1):85–95. https://doi.org/10.1093/cvr/cvw032. Epub 20160209. PubMed PMID: 26862121; PubMed Central PMCID: PMC4798048
17. Lorenzana-Carrillo MA, Gopal K, Byrne NJ, Tejay S, Saleme B, Das SK, et al. TRIM35-mediated degradation of nuclear PKM2 destabilizes GATA4/6 and induces P53 in cardiomyocytes to promote heart failure. Sci Transl Med. 2022;14(669):eabm3565. https://doi.org/10.1126/scitranslmed.abm3565. Epub 20221102. PubMed PMID: 36322626
18. Stratton MS, McKinsey TA. Epigenetic regulation of cardiac fibrosis. J Mol Cell Cardiol. 2016;92:206–13. https://doi.org/10.1016/j.yjmcc.2016.02.011. Epub 20160212. PubMed PMID: 26876451; PubMed Central PMCID: PMC4987078
19. Stratton MS, Bagchi RA, Felisbino MB, Hirsch RA, Smith HE, Riching AS, et al. Dynamic chromatin targeting of BRD4 stimulates cardiac fibroblast activation. Circ Res. 2019;125(7):662–77. https://doi.org/10.1161/CIRCRESAHA.119.315125. Epub 20190814. PubMed PMID: 31409188; PubMed Central PMCID: PMC7310347

Phytochemicals in Cardioprotection and Cardiotoxicity

Shane S. Scott, Cora Albers, Maryam Iqbal, Israel Ailemen, Sakima A. Smith, and Narasimham L. Parinandi

Abbreviations

Akt	Protein kinase B
AMPK	AMP-activated protein kinase
ANP	Atrial natriuretic peptide
BNP	Brain natriuretic peptide
CAD	Coronary artery disease
CAMKII	Calcium/calmodulin-dependent protein kinase II
CAT	Catalase
CGTF	Connective tissue growth factor
CHF	Congested heart failure
CK	Creatinine kinase
CK-MB	Creatine kinase-MB

S. S. Scott (✉)
Dorothy M. Davis Heart and Lung Research Institute, The Ohio State University College of Medicine, Columbus, OH, USA

The Ohio State Medical Scientist Training Program (MSTP), Columbus, OH, USA
e-mail: shane.scott@osumc.edu

C. Albers · I. Ailemen · S. A. Smith
Dorothy M. Davis Heart and Lung Research Institute, The Ohio State University College of Medicine, Columbus, OH, USA
e-mail: cora.albers@osumc.edu; israel.ailemen@osumc.edu; Sakima.smith@osumc.edu

M. Iqbal
Virginia Tech Carilion School of Medicine, Roanoke, VA, USA
e-mail: miqbal@vt.edu

N. L. Parinandi
Division of Pulmonary, Critical Care and Sleep Medicine Davis Heart and Lung Research Institute, The Ohio State University College of Medicine, Columbus, OH, USA
e-mail: narasimham.parinandi@osumc.edu

S. A. Smith, S. S. Scott (eds.), *Myocardial Oxidative Stress in Anti-Cancer Therapies*, Oxidative Stress in Applied Basic Research and Clinical Practice,
https://doi.org/10.1007/978-3-031-97517-2_4

Col	Collagen
CoQ	Coenzyme Q, ubiquinone
COX-2	Cyclooxygenase 2
CRP	C-reactive protein
CTS	Cryptotanshinone
CVD	Cardiovascular disease
DL	D-Limonene
DOXO	Doxorubicin
EGCG	Epigallocatechin gallate
EGFR	Epidermal growth factor receptor
ET-1R	Endothelin receptor 1
GSH	Glutathione
HDL	High-density lipoprotein
HF	Heart failure
IL	Interleukins
*i*NOS	Inducible nitric oxide synthase
JNK	Jun N-terminal kinase
LDH	Lactate dehydrogenase
LDL	Low-density lipoprotein
LV	Left ventricle
MAPK	Mitogen-activated protein kinase
MCP-1	Monocyte chemoattractant protein-1
MDA	Malondialdehyde
MHC	Myosin heavy chain
MI	Myocardial infarction
mTOR	Mammalian target of rapamycin
NF-κB	Nuclear factor kappa B
NLRP3	NOD-like receptor protein 3
NO	Nitric oxide
NOX	NADPH oxidase
Nrf2	Nuclear factor erythroid 2-related factor 2
OH	Hydroxyl radicals
PI3K	Phosphoinositide 3-kinase
PQQ	Pyrroloquinoline quinone
RCT	Randomized controlled trial
ROS	Reactive oxygen species
RyR2	Ryanodine receptor 2
SIRT1	Silent information regulator 1
SMA	Smooth muscle actin
SOD	Superoxide dismutase
STAT3	Signal transducer and activator of transcription 3
TNF-α	Tumor necrosis factor alpha

Introduction

Plants are reservoirs of a myriad of bioactive phytochemicals, which have garnered significant attention for their potential applications in the treatment of cardiovascular diseases (CVDs) and cardiotoxicity. Oxidative stress, which is characterized by overproduction of reactive oxygen species (ROS) and nitric oxide free radicals, is a central mediator of myocardial injury and the development and progression of CVDs. ROS, including superoxide anions ($O_2^{\cdot -}$), hydrogen peroxide (H_2O_2), and hydroxyl radicals (•OH), trigger cardiomyocyte oxidative stress and apoptosis, which impair cardiac function and promote arrhythmogenesis [1]. Phytochemical compounds universally target oxidative stress-associated cardiac remodeling and subsequent development of heart failure (HF). Polyphenols, flavonoids, alkaloids, terpenes, terpenoids, and quinones are subclasses of phytochemicals with unique structures conferring both distinct and overlapping biological actions. Significant in vitro and in vivo studies support that these compounds reduce oxidative stress primarily through the donation of electrons, chelation of metal ions, and modulation of redox-sensitive pathways [2]. As such, most phytochemicals are renowned for their potent antioxidant actions through neutralization of ROS and upregulation of antioxidant enzymes, including superoxide dismutase (SOD) and catalase (CAT) [3].

Many phytochemicals also exhibit anti-inflammatory, anti-apoptotic, anti-arrhythmic, and lipid-regulating properties, which target critical pathways involved in myocardial toxicity and CVD progression [4, 5]. They exhibit cardiovascular protection through vasodilation, modulation of cardiac ion signaling, regulation of endothelial function, and prevention of low-density lipoprotein (LDL) oxidation, a key step in atherosclerosis development. Polyphenols can also modulate transcription factors such as nuclear factor erythroid 2-related factor 2 (Nrf2) and nuclear factor kappa B (NF-κB) to combat oxidative and inflammatory damage, respectively [3, 5]. In particular, Nrf2 upregulates antioxidant response elements (AREs), which increase the expression of antioxidant and cytoprotective detoxifying enzymes that suppress oxidative stress [1]. Polyphenols, including quercetin and anthocyanins, protect against endothelial dysfunction, another precursor to atherosclerosis, by scavenging ROS and preventing the formation of oxidized-LDL (ox-LDL) and macrophage foam cells [6, 7]. Terpenes and terpenoids also have anti-inflammatory and lipid-lowering properties [8, 9].

While the role of alkaloids in CVD is less studied compared to polyphenols and flavonoids, some alkaloids have been implicated in modulating blood pressure and protecting against myocardial injury through antioxidant mechanisms [10]. Furthermore, the berberine alkaloid demonstrates anti-atherogenic and anti-arrhythmic actions [11, 12]. Quinones, including ubiquinone (i.e., coenzyme Q10 (CoQ), play a vital role in mitochondrial bioenergetics and stabilization of mitochondrial membranes, further underscoring their impact in attenuating CVD [3]. Interestingly, phytochemicals from all classes have demonstrated efficacy in reducing cardiotoxicity induced by chemotherapeutic/anticancer drugs by inhibiting mitochondrial dysfunction and apoptosis, preventing HF, and mitigating myocardial remodeling in models of ischemic-reperfusion injury and myocardial infarction (MI) [13, 14].

In this chapter, the chemical properties, biological actions, and cellular targets of various phytochemicals are described. Compounds with well-studied roles in preventing and treating CVDs and cardiotoxicity are also highlighted. In addition to their antioxidant properties, certain phytochemicals have exhibited their ability to target multifaceted mechanisms of inflammation, lipid metabolism, and endothelial function, demonstrating that they are well poised to impact CVD management. Future research should focus on elucidating their precise molecular pathways and optimizing their therapeutic applications through randomized control clinical trials.

Oxidative Stress

Imbalances between the production and neutralization of ROS, including $O_2^{\cdot-}$ and H_2O_2, are hallmarks of oxidative stress. ROS are predominantly generated during mitochondrial cellular respiration via complexes I and III of the electron transport chain and the tricarboxylic acid cycle [15, 16]. During cellular respiration, the reduction of molecular oxygen gives rise to superoxide anions. SOD then rapidly converts the superoxide species into H_2O_2, which is then converted to water by cellular glutathione (GSH). Cellular glutathione (GSH) further converts H_2O_2 to water. Accumulation of cellular ROS can be deleterious. Excess H_2O_2 molecules may act as signaling molecules and mediators of oxidative stress, triggering further ROS generation via Fenton reaction in the presence of iron (Fe^{2+}), subsequently producing reactive hydroxyl radicals ($^{\cdot}OH$). Free hydroxyl radicals damage cellular membranes via lipid peroxidation and generate toxic by-products such as malondialdehyde (MDA) that further augment cellular dysfunction [17–19]. Disruption in the function of the antioxidant protective enzymes and excessive ROS overwhelm the cellular antioxidant defenses and subsequently lead to oxidative damage to lipids, proteins, and DNA [20, 21].

ROS can also impair enzymatic activity through oxidation of amino acids, formation of protein carbonyls, and disruption of iron-sulfur clusters. These oxidative modifications of key metabolic enzymes or structural proteins lead to the loss of cellular function. In the TCA cycle, aconitase and citrate synthase enzymes are particularly susceptible to ROS-induced inactivation, disrupting mitochondrial energy metabolism. Mitochondrial dysfunction further exacerbates ROS accumulation, disrupts the mitochondrial membrane potential, and impairs energy production. ROS also triggers cellular apoptosis and necrosis [16]. Moreover, damage to antioxidant enzymes, including SOD, catalase (CAT), and GSH, also contributes to an imbalance of ROS production and neutralization [22]. This creates a vicious cycle of oxidative damage [20, 21, 23]. In addition, changes in regulation of cellular signaling pathways can also exacerbate ROS. The most well-studied mechanisms that can lead to excess ROS include activation of the phosphoinositide 3-kinase (PI3K)/protein kinase B (Akt), mitogen-activated protein kinase (MAPK), and epithelial growth factor receptor (EGFR) signaling pathways [24, 25]. ROS also regulates transcription factors such as Nrf2 and hypoxia-inducible factor 1α (HIF-1α), which can influence cell survival and apoptosis.

Cellular protein alterations can also upregulate pro-inflammatory signaling pathways such as NF-κB and activator protein-1 (AP-1), leading to increased production of cytokines like tumor necrosis factor-alpha (TNF-α) and interleukin-6 (IL-6) that perpetuate inflammation [26–28]. There is a bidirectional relationship between oxidative stress and inflammation. ROS generated by inflammatory cells can exacerbate tissue damage through the oxidization of cellular enzymes. Simultaneously, inflammatory cytokines stimulate cellular ROS production, which promotes chronic inflammation and exacerbates cellular injury. This interdependence contributes to the pathogenesis of cancer, CVDs, and cardiotoxicity [26–29]. As such, pharmacotherapies targeting these pathways represent an important therapeutic avenue for curtailing atherosclerosis, HF, and anticancer therapy-induced cardiotoxicity.

Cardiovascular Disease and Cardiotoxicity

CVDs affect the heart and blood vessels and are the leading causes of global morbidity and mortality [30, 31]. One of the primary types of CVDs is atherosclerosis, which promotes coronary artery disease (CAD). The consequences of CAD include angina, MI, and HF [32]. Due to the aging population and lifestyle factors such as smoking, poor diet, physical inactivity, and obesity, the incidence of CVDs has risen. Mortality remains high despite advances in prevention and treatment; for instance, sudden cardiac death is common in patients with advanced HF or arrhythmias [33, 34].

While the pathogenesis of CVD is multifactorial, atherosclerosis is the central mechanism underlying most CVDs. Hypertension, hyperlipidemia, and smoking trigger endothelial dysfunction, promoting arterial lipid accumulation and plaque formation. In the presence of chronic inflammation and oxidative stress, plaque growth and instability are exacerbated [34, 35]. Subsequent plaque rupture and thrombosis result in MI. In addition, hyperglycemia-induced oxidative stress, endothelial damage, and systemic inflammation seen in diabetes also accelerate atherosclerosis [36]. In hypertension-induced CVDs, untreated high systemic blood pressure triggers vascular remodeling and left ventricular hypertrophy, increasing the risk of HF and arrhythmias.

Cardiotoxicity or "cancer therapy-related cardiac dysfunction," is mediated by several classes of anticancer drugs, including anthracyclines, immune checkpoint inhibitors, and tyrosine kinase inhibitors [37–40]. These anticancer treatments are associated with a wide range of adverse cardiovascular side effects, such as hypertension, QT prolongation, arrhythmia, cardiomyopathy, and HF [41, 42]. Studies evaluating the mechanisms underlying cardiotoxicity have implicated ROS overproduction, oxidative stress, inflammation, and cardiomyocyte apoptosis as central mediators [43, 44]. Despite progress in understanding these mechanisms and developing therapeutic strategies, continued research into prevention, protection, and treatment approaches is imperative. Phytochemicals offer a promising avenue for cardioprotection by targeting ROS production and modulation of antioxidant defenses via modulation of several key cellular signaling pathways.

Types of Phytochemicals

Phytochemicals (natural products) are bioactive molecules/compounds derived from plants that exhibit antioxidant and cell signaling modulatory properties, making them critical in mitigating myocardial oxidative stress in a variety of pathophysiological states. ROS are central to myocardial damage during ischemia-reperfusion injury, and phytochemicals act as ROS scavengers to neutralize oxidative stress, leading to the protection of mitochondrial function and prevention of cardiomyocyte apoptosis.

Polyphenols

Polyphenols are abundant secondary plant-derived metabolites that are found in vegetables, grains, and fruits [45–47]. Structurally, polyphenols are aromatic compounds that contain more than one phenolic group and are generated from several cellular metabolism pathways, including pentose phosphate, phenylpropanoid, and shikimate pathways [48, 49]. As a family, phenolic compounds have gained considerable attention due to their antioxidant, anti-inflammatory, anticancer, neuroprotective, and cardioprotective actions [5]. The polyphenols also encompass two major groups: flavonoids and non-flavonoids (Fig. 1) [48].

Flavonoids are the most abundant polyphenols present in the human diet and are structurally characterized by their heterocyclic C6–C3–C6 backbone (Fig. 1) [47, 48]. Flavonoids are subclassified according to their major constituent components, such as flavanones, flavones, flavanols, and anthocyanins. While they can be found in almost all vegetables, they are found with the highest yield in apples, onions, red wine, grapes, citrus fruits, tea, berries, and olive oil. Quercetin is the most common flavonoid in plants, and within the flavonol subclass of flavonoids, myricetin, kaempferol, and quercetin are among the most well studied. Among the flavanols, catechins are the most abundant and primarily found in tea leaves. Flavanones are mainly represented by taxifolin, naringenin, and hesperitin and are ubiquitous to citrus fruits [2, 47]. While the molecular activities of flavonoids are vast, significant literature supports inhibition of transcription factors such as NF-κB and AP-1, as well as activation of Nrf2 [2, 50].

Among the non-flavonoid polyphenols, caffeic acid, gallic acid, epigallocatechin gallate (EGCG), resveratrol, and vanillic acid have garnered significant interest and

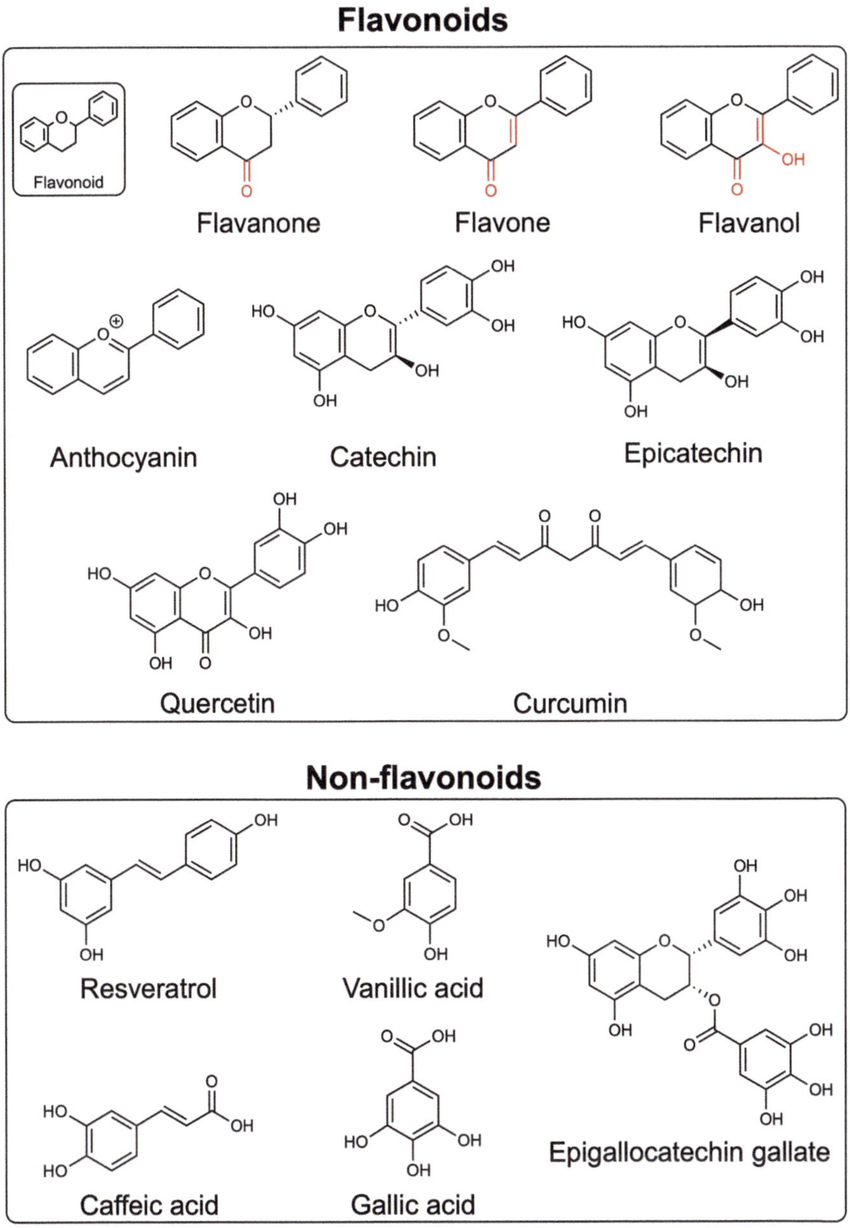

Fig. 1 Major groups of polyphenols. The polyphenols encompass two major groups including flavonoids and non-flavonoids. Flavonoids are polyphenolic secondary metabolites of plants. Flavonoids possess the chemical structure of a 15-carbon skeleton of two phenyl rings and a heterocyclic ring (see Inset). Non-flavonoid polyphenols are a broad class of polyphenolic compounds found in plants; however, they do not possess the 15-carbon skeleton

are found in plants and tea. Gallic acid, regarded as the purest natural polyphenolic compound, is extracted primarily from *Rheum palmatum* and *Eucalyptus robusta* plants [51]. EGCG is the most abundant and potent antioxidant found in green tea for cancer chemoprevention [52–54]. Preclinical and clinical studies have demonstrated that polyphenols, including resveratrol and quercetin, enhance endogenous antioxidant enzymes like SOD, CAT, and GSH while simultaneously reducing ROS and MDA production [55, 56]. Polyphenols have also been shown to inhibit the oxidation of LDL, reducing the cytotoxic effects caused by ox-LDL on endothelial cells [57]. Furthermore, anti-inflammatory and antimicrobial effects of polyphenols have also been documented [58].

Alkaloids

Alkaloids represent another group of organic compounds that consist of a nitrogen-containing cyclic ring structure (Fig. 2) [59, 60]. Alkaloids are synthesized from amino acids and are found as naturally occurring plant- and animal-based secondary metabolites. Despite the wide distribution of alkaloids in both plants and animals, plant sources have been the main sources of screening, identification, and discovery of pharmacologically active alkaloids for disease therapy and management [60, 61]. Relevant plants as sources of alkaloids include the members of families such as *Solanaceae*, *Ranunculaceae*, *Loganiaceae*, *Menispermaceae*, *Amaryllidaceae,* and *Papaveraceae* [62]. The vinca alkaloids, such as vincristine and vinblastine, are among the first identified plant-derived alkaloids and are clinically approved for the treatment of several cancers [60, 63]. Other plant alkaloids with anti-neoplastic properties including, chaetocochin J, neopapillarine, coclaurine, reflexin A, 3,10-dibromofascaplysin, and neferine, have also been identified [60]. Collectively, these drugs disrupt microtubule dynamics, causing mitotic arrest and cell death [63].

Nicotine represents another naturally occurring alkaloid found in dried leaves of *Nicotiana tabacum* belonging to the family *Solanaceae* [62]. The biological perturbations caused by nicotine are diverse and are mediated through specific pathways downstream of nicotinic receptors [64]. Nicotine is introduced to the body via tobacco smoking, which has been linked to increased CAD, diabetes, and hypertension mortality [65]. Nicotine promotes endothelial injury, insulin resistance, inflammation, and myocardial oxygen demand mismatch, all of which contribute to the development of CVD and HF [65–67]. On the contrary, recently discovered alkaloids such as amino alcohol-diterpenoid alkaloid, berberine, rhynchophylline, rutaecarpine, and hippophae rhamnoides have also been evaluated for their cardioprotective effects [68–73].

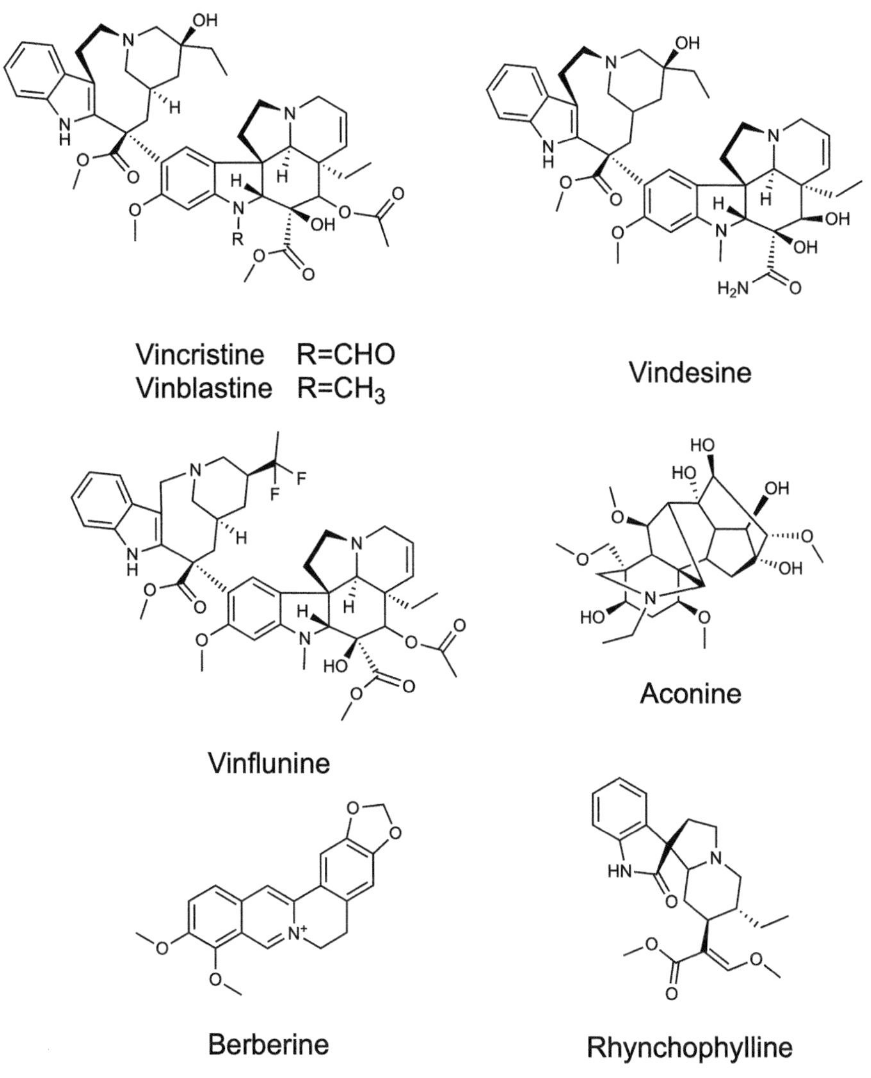

Fig. 2 Alkaloids. Alkaloids are primarily plant metabolites containing organic nitrogen-containing bases and are among the most studied phytochemicals for their range of therapeutic and toxic effects

Terpenes and Terpenoids

Terpenes are the largest family of plant-derived secondary metabolites found primarily in the leaves, stems, roots of conifers, citrus, and eucalyptus [74]. Structurally, terpenes consist of isoprene units (C5H8), a 5-carbon compound that forms terpene backbones and are classified into mono-, sesqui-, di-, tri-, and tetra-terpenes based on carbon count (Fig. 3) [75]. Terpenoids, a subclass of terpenes, are generated through oxygenation, hydrogenation, or dehydrogenation of terpenes [74, 75]. Cellularly, both terpenes and terpenoids are synthesized via two independent pathways, the mevalonate or the methylerythritol phosphate pathways [74–76]. Carotenoids such as lycopene have also been studied, which exert their actions primarily through quenching oxygen radicals and reducing lipid peroxidation [15, 56].

Collectively, terpenes have been shown to reduce oxidative stress and ROS. Selected mechanisms of their antioxidant behavior include suppression of silent information regulator 1 (SIRT1) and Nrf2 expression as well as increased SOD and glutathione peroxidase (GSH-Px) production [8, 9, 77]. In addition,

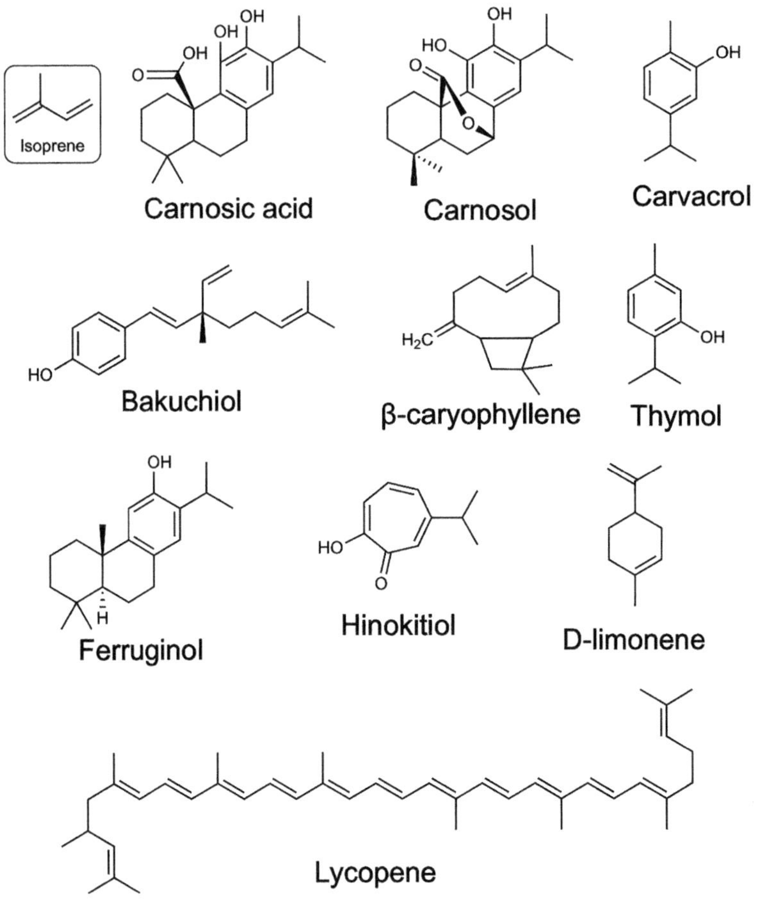

Fig. 3 Terpenes and terpenoids. Terpenes are volatile and aromatic unsaturated hydrocarbons occurring in found in plants, mostly in essential oils. Structurally, terpenes consist of isoprene units (C5H8, See Inset), a 5-carbon compound that forms the terpene backbones. Terpenoids are terpene-derived compounds with different structural modifications giving rise to hemiterpenoids, monoterpenoids, diterpenoids, and sesquiterpenoids. Terpenoids exist as linear, acyclic, monocyclic, bicyclic, tricyclic, tetracyclic, pentacyclic, or macrocyclic natural products of plants

terpenes are able to suppress the release of pro-inflammatory cytokines, such as TNF-α, IL-6, monocyte chemoattractant protein-1 (MCP-1), and NF-κB [8, 77, 78].

Quinones

Quinones are a class of aromatic biologically active compounds primarily generated through simple oxidation of hydroquinones, catechols, and polyphenols catalyzed by oxidative enzymes including cytochrome P450s, cyclooxygenase-2 (COX-2), peroxidase, tyrosinase, xanthine oxidase, and monoamine oxidase [79–82]. Quinones can be found in plants, fungi, bacteria, algae, and variably in animals [79, 83, 84]. Types of quinones include xenobiotics (e.g., anthracyclines and mitomycin C), endogenous eukaryotic molecules (e.g., ubiquinone, CoQ), and reactive intermediates (benzoquinone) [79, 81, 85]. Quinones containing the cyclohexadienedione structure (Fig. 4), in either the *p*–or *o*–form, determine the stability and reactivity of the compound.

Quinones are not created equal in terms of their biological actions and uses in health and disease. On the one hand, quinones have exhibited cytoprotection via detoxification enzymes and anti-inflammatory actions. However, quinones can also promote oxidative stress and cytotoxicity through increased ROS production [79, 81, 86]. For example, these compounds can contribute to the cardiotoxicity of anthracyclines such as doxorubicin through the presence of a quinone moiety that generates free radicals. These free radicals react with oxygen to form ROS, subsequently damaging cardiac mitochondria and leading to cardiomyopathy [86–90]. In adult rat heart cells, the anthracycline and mitomycin C quinones stimulate cardiac production of hydrogen peroxide, contributing to cardiotoxicity [86]. On the other hand, coenzyme Q10 has emerged as a cardioprotective agent that is able to mitigate mitochondrial damage and oxidative stress without compromising its anticancer efficacy [85].

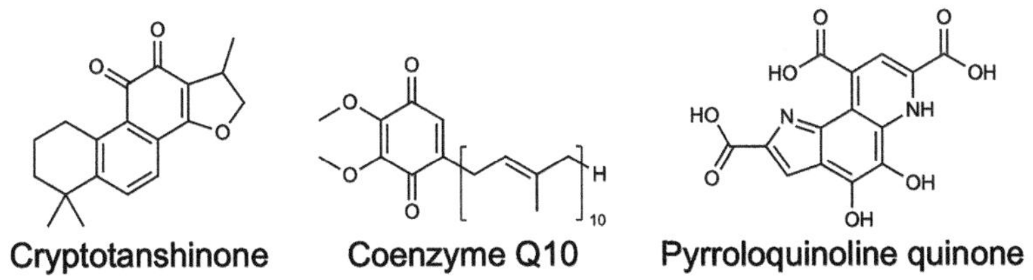

Fig. 4 Quinones. Quinones are natural products belonging to a class of cyclic organic compounds possessing two carbonyl groups positioned next to each other or separated by a vinylene group in a six-membered unsaturated ring

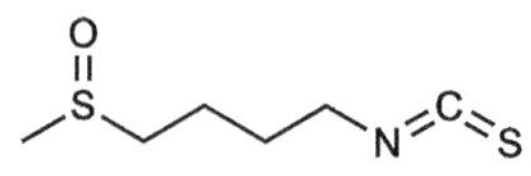

Sulforaphane

Fig. 5 Sulforaphane. Sulforaphane is found naturally in cruciferous vegetables, particularly broccoli sprouts, and belongs to the isothiocyanate group of phytochemicals. Sulforaphane exerts its potent antioxidant and anti-inflammatory effects through activation of the Nrf2 pathway, which stimulates the production of detoxifying and antioxidant enzymes

Miscellaneous Phytochemicals

Several other phytochemicals have been studied for their role in health and treatment of disease. For example, sulforaphane (SFN, Fig. 5) is an isothiocyanate plant compound found in cruciferous vegetables such as broccoli, brussels sprouts, cauliflower, and kale. Mechanistically, SFN activates Nrf2 through upregulation of several kinases, including extracellular signal-related kinase 1/2 (ERK1/2), p38-mitogen-activated protein kinase (p38-MAPK), c-Jun N-terminal kinases (JNK), PI3K, and Akt [91].

Phytochemicals Implicated in Cardioprotection

Polyphenols including resveratrol, quercetin, catechins, and curcumin, are among the most well-studied phytochemicals in CVD (Table 1) [55, 56].

Resveratrol

Inflammation and endothelial dysfunction in the setting of high serum cholesterol promote plaque development and formation of the fibrous cap, both hallmarks of atherosclerosis. Therefore, antagonism of this process is a necessary step in atherosclerosis management. Brito and colleagues have demonstrated that resveratrol blocks the proliferation of smooth muscle cells required for the fibrous cap through inhibition of the mammalian target of rapamycin (mTOR) mitogenic signaling pathway triggered by ox-LDL. Specifically, resveratrol blocks the ox-LDL-induced phosphorylation and activation of the PI3K/Akt/mTOR/p70S6K pathway and strongly inhibits DNA synthesis [57]. In mice with pressure overload-induced HF, short-term exposure to resveratrol treatment improves diastolic function and mitochondrial energy and metabolism [14].

Ahmet and colleagues have explored the long-term actions of low-dose dietary resveratrol in rats with chronic heart failure (CHF). The authors have observed a reduction in structural and functional changes with improved left ventricular

Table 1 Cardioprotective effects of phytochemicals

Phytochemical	Cardioprotective role	References
Polyphenols		
Resveratrol	Blocked the proatherogenic phosphorylation and activation of the PI3K/Akt/mTOR/p70S6K pathway, which reduces proliferation of smooth muscle cells and inhibits DNA synthesis.	Brito et al. (2009) [57]
	Improved survival, ventricular and vascular function, and striated muscle energy metabolism in a hypertensive rat model of HF. Increased mitochondrial mass and biogenesis through preservation of mitochondrial fatty acid oxidation and PPARα expression.	Rimbaud et al. (2011) [93]
	Reduced structural and functional cardiac deterioration in CHF rats, as evidenced by improved LV systolic function and AV coupling.	Ahmet et al. (2017) [92]
	Increased cardiac mitochondrial oxidative function, CAMP-activated protein kinase activation, and myocardial insulin sensitivity. Increased survival and HF phenotype in pressure-overload mice by improving diastolic function, attenuating cardiac fibrosis, and reducing hypertrophic and extracellular matrix markers.	Sung et al. (2015) [14]
Quercetin	Blocked progression of experimental autoimmune myocarditis by suppression of oxidative and endoplasmic reticulum stress via endothelin-1/MAPK signaling in a rat model of experimental autoimmune myocarditis.	Arumugam et al. (2012) [95]
	Reduced doxorubicin-induced apoptosis, mitochondrial dysfunction, ROS generation, and DNA damage through decreased expression of Bid, p53, p47, and NOX, as well as increased expression of Bcl-2 and Bmi-1 in H9c2 cells. Protected mice from doxorubicin-induced cardiac dysfunction in vivo and was associated with reduced ROS levels and lipid peroxidation but increased expression of Bmi-1 and SOD.	Dong et al. (2014) [13]
	Restored serum CK, LDH, and lipid profile to normal levels when in the presence of candesartan in the setting of cardiomyopathy. Reduced levels of lipid peroxides and serum troponin-T and increased CAT, GSH, and SOD proteins.	Majhi et al. (2022) [100]

(continued)

Table 1 (continued)

Phytochemical	Cardioprotective role	References
	Preserved the LV hemodynamic parameters and prevented remote myocardium damage and fibrosis via suppression of ROS, increased Nrf2, and enhanced SOD and GSH levels in the LVs of the control and MI model rats. Reduced angiotensin II, NF-κβ activity, and protein expression of TGF-β1, α-SMA, and total/phospho-SMAD3 in the LV of both groups. Increased LV SMAD7 and BMP7 expression blocking MI-induced LV remodeling.	Albadrani et al. (2021) [97]
	Prevented cardiac function decline and serum elevation in NT pro-BNP, cardiac troponin I, and CK-MB following chronic doxorubicin treatment. Prevented doxorubicin-induced SOD1 depletion and DNA damage.	Dulf et al. (2024) [99]
Flavonoids		
Catechin	Improved vascular function and reduces oxidative stress markers	Ashraf et al. (2024) [15]
	Protected against doxorubicin-induced cardiotoxicity by improving antioxidant enzyme activity, reducing oxidative stress markers (NO, H_2O_2, MDA), and reversing histopathological damage such as myofibril loss and mitochondrial degeneration.	Saleh Ahmed (2022) [105]
	Prevented doxorubicin-induced cardiotoxicity in rats by attenuating electrographic changes and myocardial dysfunction. For example, catechin reduced the incidence of QT prolongation and significantly reduced cytoplasmic vacuolization, myofibril loss, and mitochondrial swelling compared to DOXO-treated controls. In addition, isolated atria from mice treated with catechin and DOXO had more spontaneous contractility compared to DOXO-treated controls.	Kozluca et al. (1996) [106]
Epicatechin	Improved cardiac electrophysiology parameters, including reduced resting membrane potential, action potential amplitude, force of contraction, and action potential duration, via an opioid receptor-independent mechanism.	MacRae et al. (2019) [104]
Epigallocatechin gallate	Blocked doxorubicin-induced cardiotoxicity by modulating pro-survival pathways, increasing ErbB2 expression, and reducing apoptosis.	Saeed et al. (2015) [107]
Curcumin	Protected against acute adriamycin (ADR)-induced myocardial toxicity in rats, which was accompanied by reduced serum CK and LDH levels, as well as reduced lipid peroxidation. Improved antioxidant levels and inhibited early signs of cardiotoxicity, suggesting curcumin could be a potential adjunct to ADR chemotherapy for mitigating free radical-induced damage.	Venkatesan et al. (1998) [114]

(continued)

Table 1 (continued)

Phytochemical	Cardioprotective role	References
	Prevented doxorubicin-induced cardiotoxicity by modulating TP53, TNF-α, and Akt enzymes toward attenuating apoptosis, inflammation, and cell metabolism, respectively.	Hu (2024) [115]
	Significantly reduced cardiac enzyme levels, LDH, oxidative stress biomarkers, and histological damage to myocardial fibers, preventing chronic chlorpyrifos-mediated cardiac damage.	Abdollahi-Karizno et al. (2024) [116]
Alkaloids		
Aconine	Ameliorated heart failure through reduction of hypertrophic markers (e.g., ANP, NT-proBNP, β-MHC, MMP-2, and MMP-9), fibrosis markers (e.g., α-SMA, Col I, and Col III), and inflammatory cytokines (e.g., TNF-α and IL-6). Proteomic analysis revealed that aconine also regulate the calcium signaling pathway by reducing expression of RyR2 and SERCA2a.	Xu et al. (2021) [68]
Berberine	Inhibited the formation of lipid-laden macrophages (i.e., foam cells) by reducing accumulation of ox-LDL, through activation of the AMPK-SIRT1-PPARγ pathway in macrophages.	Chi et al. (2014) [141]
Rhynchophylline	Improved endothelium-dependent relaxation in renal arteries from spontaneously hypertensive rats, in a time- and concentration-dependent manner without affecting endothelium-independent relaxation. This effect was mediated by the activation of Src-PI3K/Akt-eNOS pathways, promoting increased in NO production and relaxation.	Hao et al. (2017) [122]
	Protected against myocardial ischemia-reperfusion injury by enhancing cell viability, reducing apoptosis, and modulating mitochondrial function in cardiomyocytes. Decreased OS and mitochondrial permeability transition pore levels.	Qin et al. (2019) [123]
N-*p*-coumaroyl-4-aminobutan-1-ol, hippophamide	Inhibition of doxorubicin-induced apoptosis in H9C2 cardiac muscle cells by reducing apoptosis and improving mitochondrial function, with hippophamide offering stronger protection	Zhou et al. (2021) [73]
Lycopene	Inhibited cardiac hypertrophy in phenylephrine-induced and pressure overload-induced cardiac hypertrophy in vitro and in vivo, respectively. Attenuated antioxidant response element (ARE) activity and activated ARE-driven expression of antioxidant genes.	Zeng et al. (2019) [155]
	Suppressed doxorubicin-induced myocyte damage and prevented interstitial collagen accumulation but did not prevent doxorubicin-induced left ventricular systolic dysfunction in male Wistar rats.	Anjos Ferreira et al. (2007) [156]

(continued)

Table 1 (continued)

Phytochemical	Cardioprotective role	References
	Protected against adriamycin-induced cardiotoxicity and nephrotoxicity by regulating oxidant–antioxidant status.	Yilmaz et al. (2006) [157]
	Ameliorated serum increases in CK-MB, LDH, and cardiac troponin-T, and tissue levels increases of NO and MDA in tulathromycin- or DFS- (individually or in combination) treated Swiss albino mice. Reduced pathological expression of caspase-3-, Bax-, and CK-positive cells.	Abdel-Daim et al. (2018) [158]
Terpenes and terpenoids		
β-Caryophyllene	Prevented of in vivo myocardial damage in an isoproterenol (ISO)-induced myocardial infarction rat model by reducing inflammation through inhibiting HSP-60/TLR/MyD88/NFκB signaling pathways.	Younis and Mohamed (2019) [172]
	Prevented diabetic cardiomyopathy in mice via reduced oxidative stress, cardiac lipid accumulation, inflammatory cytokines, and inflammatory enzyme mediators through activation of the TLR4/NF-κB/MAPK signaling pathway.	Mamdouh Hashiesh et al. (2023) [173]
	Blocked and/or reversed cardiac hypertrophy in isoprenaline-induced H9c2 cells and mice via the P13K-Akt signaling pathway when used in combination with 1,8-cineole.	Bai et al. (2022) [174]
Bakuchiol	Attenuated cardiomyocyte apoptosis and oxidative stress in diabetic in vivo and in vitro models through activation of SIRT1/Nrf2 signaling pathway. Inhibited fibrotic-protein expression in the myocardium of diabetic mice through reducing expression of α-SMA and Smad3 proteins.	Ma et al. (2020) [146]
	Ameliorated hypertrophic gene expression (ANP, BNP, and β-MHC) in vivo and in vitro. Reduced fibrosis and inflammatory responses via the NF-κB pathway, resulting in reduced Col I & III, CGTF, TNF-α, IL-6, and MCP-1 expression.	Wang et al. (2018) [78]
Carvacrol	Reduced lipid peroxidation and attenuated apoptosis in a rat model of myocardial ischemic perfusion injury through activation of MAPK/ERK and Akt-eNOS signaling pathways.	Chen et al. (2017) [147]
	Blocked apoptosis in pressure overload rats by disrupting the balance of pro-apoptotic and apoptotic markers.	Sadeghzadeh et al. (2018) [148]
Carnosol	Inhibited apoptosis by downregulating caspase-3 activity. Decreased ROS production by upregulating expression of GSH, GST, CAT, and Nrf2.	Ou et al. (2017) [175]

(continued)

Table 1 (continued)

Phytochemical	Cardioprotective role	References
Carnosic Acid	Prevented pressure overload-induced cardiac remodeling by reducing the expression of hypertrophic genes (e.g., ANP, BNP, and β-MHC), reducing cardiac fibrotic markers (Col 1, Col 3, CTGF, and α-SMA), suppressing oxidative damage, and apoptosis via the Akt/GSK3β/NOX4 signaling pathway.	Wei et al. (2020) [176]
	Attenuated doxorubicin-induced cardiotoxicity when used alone and in combination with carvedilol by suppressing oxidative stress, reducing inflammatory cytokines (COX2, TNF-α, IL-6, IL-1β, and IL-18) via inactivation of NF-κB signaling, autophagy, and apoptosis.	Zhang et al. (2019) [177]
D-limonene	Abolished oxidative stress and apoptosis through restoration of superoxide dismutase activity, and suppressed pro-apoptotic enzymes, respectively, in an isoproterenol-induced murine model of myocardial infarction.	Durço et al. (2019) [178]
	Promoted persistent bradycardia with associated hypotension and antiarrhythmic activity using in vivo approaches.	Nascimento et al. (2019) [149]
	Decreased the occurrence of delayed afterdepolarizations and triggered activities, and reduced cardiac arrhythmias in vivo, probably via inhibition of CaMKII hyperactivation (i.e., decreased *p*-CAMKII and *ox*-CAMKII)	Durço et al. (2023) [151]
Ferruginol	Decreased cardiac oxidative damage and injury myocardial infarction via modulation of inflammatory response (e.g., reduced TNF-α, IL-6, and NF-κB), upregulation of antioxidant enzymes (e.g., SOD, GPx, CAT, and GSH), and reduced cTnT, cTnI, CK, and CK-MB.	Zhang et al. (2021) [179]
	Ameliorated myocardial injury, ROS, and apoptosis in doxorubicin-induced cardiotoxicity by preserving the mitochondria function by restoring the SIRT1-PGC-1α pathway, which mitochondrial biogenesis and fatty acid oxidation.	Li et al. (2021) [180]
Hinokitiol	Mitigated cardiomyocyte oxidative injury by inhibiting GSK-3β-mediated autophagy and subsequent apoptosis in an ischemic-reperfusion injury cellular model.	Xiao et al. (2022) [181]
Thymol	Reduced oxidative stress, inflammation, and apoptosis in a rat model of adrenaline-induced myocardial infarction by upregulating antioxidant enzymes, reducing expression of NF-κB and IL-1β, and reducing expression of caspase-3.	El-Marasy et al. (2020) [182]

(continued)

Table 1 (continued)

Phytochemical	Cardioprotective role	References
Quinones		
Cryptotanshinone	Modulated NF-κB and MAPK pathways in cardiovascular diseases, PI3K/Akt/mTOR and JAK/STAT in cancer, and Nrf2/HO-1 in neurological conditions.	Zheng et al. (2024) [168]
	Reduced angiotensin II-induced cardiac fibrosis markers and oxidative stress in vitro and in vivo. Inhibited the upregulation of fibronectin, connective tissue growth factor, ERK1/2, NOX-2/4, and ROS in cardiac fibroblasts and attenuated damage in a rat model of acute MI.	Ma et al. (2014) [169]
	Protected against doxorubicin-induced damage in H9c2 cells by increasing cell viability, reducing ROS levels, inhibiting apoptosis, and maintaining mitochondrial membrane integrity. Activated the Akt-GSK-3β-pathway, which promotes GSK-3β-ANT interaction and reduces the formation of the ANT-CypD complex, preventing mPTP opening.	Wang et al. (2021) [166]
Coenzyme Q10	Attenuated CHF, improving symptoms, reducing BNP levels, and decreasing major adverse cardiovascular events, including hospitalizations and mortality.	Mortenson et al. (2014) [171]
Pyrroloquinoline quinone	Prevented mitochondrial dysfunction and cardiac damage by increasing NCLX expression to reduce calcium overload and ROS production. Enhanced PGC-1α and TFAM expression promoting mitochondrial biogenesis and mitigating pressure-overload-induced heart failure.	Xu et al. (2020) [162]
Others		
Sulforaphane	Improved cardiac function and myocardial tissue remodeling by reducing ROS (e.g., MDA, SOD), hypertrophy markers (e.g., BNP, ANP), fibrotic marker expression levels (Col I, Col III), inflammation (TNF-α, IL-6), and apoptotic index.	Ma et al. (2018) [183]
	Increased the cardiac contractility and relaxation indexes following ischemic/reperfusion injury by modulating expression of calcium-handling proteins, including reduced RyR2 and *p*-PLB/PLB ratio, without changes in SERCA expression. Decreased RyR2 and ROS expression in rat MI model.	Poletto Bonetto et al. (2022) [184]
	Preserved cardiac function, decreased infarct size, oxidative stress, and inflammation through activation of common protective pathways including AhR and Nrf2 in vivo.	Silva-Palacios et al. (2019) [91]

(continued)

Table 1 (continued)

Phytochemical	Cardioprotective role	References
	Enhanced mitochondrial respiration and reduced cardiac oxidative stress due to lipid peroxidation, the activation of Nrf2, and associated antioxidant enzymes in doxorubicin-cardiotoxicity in a rat model. In rat neonatal cardiomyocytes and multiple rat and human breast cancer cell lines treated with doxorubicin cardiotoxicity SFN protected against cardiac stress and cytotoxicity but did not potentiate doxorubicin-mediated cancer death.	Bose et al. (2018) [185]
	Reduced free radicals (O2-, OH-), lipid peroxidation MDA, cholesterol, and cardiac troponin and increased antioxidant enzymes (SOD, CAT, and GSH), through activation of the PI3K/Akt/Nrf2 pathway in vitro and in vivo.	Thangapandiyan et al. (2024) [186]

systolic function [92]. In a rat model of hypertension and subsequent HF, resveratrol improves survival, hemodynamics, and energetics [93]. Together, these data provide support for further evaluation of clinical trials. Currently, there are several ongoing clinical trials evaluating the role of resveratrol in cardiac disease. For example, a unique collaboration among cardiologists, cardiac surgeons, and basic scientists is spearheading the SIRT-CVS (Short Interval Resveratrol Trial in Cardiovascular Surgery, NCT03762096) trial, which is studying the effect of resveratrol on the microvascular function of diabetic patients undergoing surgical revascularization with coronary artery bypass grafting.

Quercetin

Several preclinical studies are examining the therapeutic mechanisms and potential clinical translation of quercetin in CVDs (Table 1) [94]. In an experimental autoimmune myocarditis (EAM) model, quercetin protects against the progression of EAM to dilated cardiomyopathy by reducing myocardial fibrosis and endoplasmic reticulum stress, modulating MAPK signaling, and suppressing myocardial endothelin-1 [95]. The mechanisms underlying the protective effects of quercetin following trauma-induced secondary cardiac injury have been evaluated. Pretreatment with quercetin significantly blocks trauma-induced apoptosis and cardiac dysfunction in vivo and inhibits increases in TNF-α, ROS overproduction, and cardiomyocyte intracellular Ca^{2+} overload in H9c2 cells [96]. In the setting of MI, quercetin reduces cardiac fibrosis and oxidative stress by promoting ROS scavenging, suppressing NF-κB, and stimulating Nrf2, SMAD7, and BMP7, resulting in reduced myocardial TGF-β1 and ECM deposition [97]. Additionally, the treatment of Sprague-Dawley rats with quercetin following myocardial ischemia/reperfusion stimulates the PI3K/Akt signaling pathway, promoting cellular survival and reducing cardiomyocyte apoptosis [98].

Dong et al. have evaluated underlying quercetin-mediated protection against doxorubicin-induced cardiotoxicity using H9c2 cardiomyocytes and C57BL/6 mice [13]. In vitro, quercetin disrupted doxorubicin-induced apoptosis, mitochondrial dysfunction, ROS generation, and DNA damage through upregulation of Bmi-1 expression [13]. In vivo, quercetin prevents doxorubicin-induced ventricular dysfunction, and reduces ROS levels and lipid peroxidation but enhances the expression of Bmi-1 and SOD [13]. In a separate study, quercetin prevents cardiac dysfunction and elevation of NT pro-BNP, troponin I, and CK-MB in rats chronically treated with doxorubicin. Furthermore, quercetin blocks doxorubicin-induced depletion of SOD and DNA damage through Nrf2 activation [99]. Majhl and colleagues have explored the synergistic effect of quercetin combined with candesartan in protection against doxorubicin-induced cardiomyopathy [100]. Pre-cotreatment of candesartan and quercetin has resulted in decreased levels of CK, LDH, and cardiac troponins, and altered lipid profiles [100]. Despite these compelling findings, continued clinical study is required to appraise the value of quercetin in the management of CVD.

Catechins

Among the flavonoids, catechins, known antioxidant and iron-chelating agents, have also been studied in cardioprotection [15, 56, 101–103]. MacRae et al. evaluated cardioprotective actions of epicatechin on isolated rat conduit arteries, resistant vessels, and cardiac electrophysiology [104]. The authors have shown that epicatechin induces vasorelaxation through opioid receptors and increased nitric oxide (NO) production. In addition, epicatechin reduces cardiomyocyte resting membrane potential, action potential amplitude, and force of contraction through activation of potassium and calcium channels, improving cardiac electrophysiology [104].

In adult albino rats, catechin reverses the doxorubicin-induced loss of myofibrils and intercalated disks, as well as mitochondrial degeneration [105]. In another study, catechin has been shown to protect against doxorubicin-induced cardiotoxicity by improving myocardial contractility and reducing oxidative stress [106]. Clinical and experimental studies have also implicated catechins in the regulation of lipid metabolism, protection of vascular endothelium, and lowering of blood pressure. These actions of catechin are promising in the prevention of CVDs, contributing to antihypertensive benefits, restoring vascular health, and atherosclerosis prevention [101, 102].

Saeed et al. have studied the role of epigallocatechin-3-gallate (EGCG), the main green tea catechin, in protecting against doxorubicin-induced cardiotoxicity [107]. In the study, EGCG significantly ameliorates doxorubicin-evoked oxidative stress injury by blocking glutathione depletion and lipid peroxidation as well as increasing antioxidant enzyme activities. EGCG also blocks doxorubicin-mediated inflammation and apoptosis, as evidenced by reduced expression of NF-κB, tumor suppressor protein p53, calpain 2, caspase-3, and caspase-12 [107]. In contrast, in diabetic mice, EGCG paradoxically elevates both serum CK-MB and troponin-I levels, which are accompanied by microscopic myocardial damage [108]. In addition, EGCG increases myocardial inflammatory markers (e.g., NF-κB and TNF-α) and promotes overexpression of inducible nitric oxide synthase (iNOS) and activated caspase-3 [108].

Curcumin

The protective mechanisms of curcumin in CVD are well studied and documented by Yang et al. [109]. Preclinical in vitro and in vivo studies implicate that curcumin prevents cardiomyocyte ischemic-reperfusion injury, inhibits myocardial fibrosis, attenuates atherosclerosis progression, and improves ventricular remodeling while reducing vascular dysfunction [109–111]. In a rat model of myocardial ischemia, curcumin has been shown to protect against myocardial necrosis by scavenging ROS and increasing expression of detoxifying enzymes such as glutathione-S-transferase, which reduces oxidative stress and cardiac damage [112].

Several clinical trials have evaluated the use of curcumin in metabolic, inflammatory, and oxidative stress indices in patients with metabolic syndrome [113]. In a recent meta-analysis with 785 participants from 13 RCTs, Qiu et al. have demonstrated that patients with metabolic syndrome treated with curcumin have statistically significant reductions in fasting blood sugar, diastolic blood pressure, TNF-α, C-reactive protein (CRP), and MDA, as well as increased high-density lipoprotein cholesterol (HDL-C) [113]. While these data support the promising potential of curcumin in attenuating inflammatory and oxidative stress markers in metabolic syndrome, they may be confounded by the high heterogeneity in the compared studies. Therefore, additional studies are required to confirm these findings.

In terms of cardiotoxicity, curcumin has been first shown to attenuate adriamycin-induced cardiotoxicity by preventing lipid peroxidation products, reducing MDA, and preventing serum increases in CK and LDH [114]. Recently, in a network pharmacology and molecular docking study, enrichment analysis has revealed that curcumin could reduce doxorubicin-induced oxidative stress, inflammation, and apoptosis by targeting pathways such as NF-κB, NOD-like receptor protein 3 (NLRP3) inflammasome, signal transducer and activator of transcription 3 (STAT3), and Akt signaling [115]. The role of curcumin in protecting against chlorpyrifos-induced cardiotoxicity has also been studied. In the study, Wistar rats treated with chlorpyrifos, the most commonly used organophosphate pesticide, showed that curcumin significantly reduced cardiac damage, blocked myocardial oxidative stress, and normalized cardiac antioxidant enzyme levels. The authors have also shown that curcumin protects against myocardial histopathological changes and prevents chlorpyrifos-induced disorganization and degeneration in myocardial fibers and cytoplasmic vacuolization of cardiac muscle fibers [116].

Alkaloids

Plant alkaloids have demonstrated effects on cardiovascular and metabolic diseases (Table 1).

Water-soluble alkaloids from *Aconiti Radix lateralis,* including aconine and hypaconine, have been shown to improve cardiac function in CHF. In rats subjected to suprarenal abdominal aortic coarctation-induced CHF, aconine prevents myocardial fibrosis, cardiomyocyte hypertrophy, and apoptosis, resulting in improved LV function as compared to controls [68]. Mechanistically, the authors further elucidated that aconine alters the expression of HF markers such as ANP, NT-proBNP, and β-MHC; reduces the expression of α-SMA, Col I, and Col III fibrotic markers; and reduces expression of inflammatory proteins such as TNF-α and IL-6. In addition, proteomic characterization reveals that aconine regulates calcium signaling by the altered expression of RyR2 and SERCA2a [68].

Rhynchophylline

Rhynchophylline (Rhy), a tetracyclic oxindole alkaloid, has been shown to protect the cardiovascular system through several signaling pathways [71]. As previously mentioned, inflammation, endothelial dysfunction, and macrophage-mediated ox-LDL accumulation contribute to the development and progression of atherosclerosis. Owing to this, Guo et al. have investigated the role of a racemic mixture of Rhy (G2) in ameliorating diabetic-induced endothelial dysfunction in rat vasculature [117]. The authors have elucidated that G2-upregulated endothelial nitric oxide synthase (eNOS) expression has increased NO that is required for vascular relaxation [117]. In a separate study, the Rhy analog, Y396, has been shown to prevent OS-mediated endothelial dysfunction through inhibition of the EGFR and to promote vascular relaxation by downregulating NADPH oxidase 2 and 4 (NOX2 and NOX4) [118]. Since the EGFR is activated in both diabetes and atherosclerosis, these findings suggest that the Rhy analog could protect against oxidative stress in the progression of CVD.

The pathogenesis of hypertension involves vascular damage and dysfunction, which leads to target organ injury. Several preclinical studies have reported antihypertensive activities of Rhy and its analogs [71, 119]. Its vasodilatory effect is mediated through the modulation of calcium signaling and inhibition of the endothelin-1 receptor in vascular smooth muscle cells [120, 121]. In renal arteries of spontaneous hypertensive rats, Rhy has promoted vascular relaxation and ameliorated endothelial dysfunction in SHRs through the upregulation and activation of the Src-PI3K/Akt-eNOS signaling cascade [122].

Rhy may also be useful in abating myocardial dysfunction in the setting of ischemia-reperfusion injury and overt inflammation. In vitro experiments by Qin et al. have demonstrated that Rhy prevents cell death, maintains mitochondrial membrane potential, improves calcium handling, and reduces ROS formation following rat cardiomyocyte ischemia-reperfusion injury [123].

Berberine

Several alkaloids from the *Rhizoma Coptidis* (*Caoptis chinensis*) plant, including berberine, have been shown to reduce cardiovascular risks through anti-arrhythmic, anti-inflammatory, anti-atherosclerotic, and other antioxidant mechanisms, which are well described by Feng et al. [12]. Berberine protects against cardiac ischemic-reperfusion injury via inhibition of myocardial apoptosis. In neonatal rat cardiomyocytes, berberine attenuated apoptosis by increasing the Bcl-2/Bax ratio, decreasing caspase-3 expression, and activating PI3K/Akt and AMPK signaling following hypoxic stimulation [124, 125]. Similarly, in a rat model of ischemic-reperfusion injury, berberine treatment increases AMPK in the non-ischemic areas but decreases AMPK and Bcl-2-like protein 1 in the infarct risk areas as compared to controls [126].

Berberine also demonstrates antihypertensive and vasodilatory actions [12, 127, 128]. Given that the integrity of coronary arteries is paramount in the prevention of myocardial ischemic-reperfusion injury, agents that promote vasodilation and vascular flow through these arteries are essential. Berberine has been shown to inhibit α1-adrenoreceptors on vascular smooth muscle cells, promoting vasodilation [129, 130]. Berberine increases coronary artery blood flow in anesthetized canines and isolated guinea pig hearts in ventricular fibrillation while protecting rabbit hearts from hormonal-induced ischemic electrographic changes [131]. Berberine was also shown to promote hypotension via increased endothelium-dependent release of NO [132, 133].

Huang et al. were among the first to document berberine's anti-arrhythmic properties using an ischemic arrhythmia canine model [11]. Berberine suppresses ventricular premature beats and ventricular tachycardias following ligation of the left anterior descending coronary artery [11]. In a diabetic MI model, fewer arrhythmia scores have been recorded in berberine-treated rats [134]. In another study, berberine has been shown to inhibit acetylcholine-induced atrial fibrillation (AF) in anesthetized rabbits by prolonging the action potential duration and effective refractory period of atrial myocytes [135]. Recently, the impact of berberine on postoperative AF has been assessed in a randomized, double-blind, placebo-controlled trial with patients who have undergone isolated coronary artery bypass grafting [136]. The investigators have shown that berberine treatment decreases serum levels of inflammatory markers including lipopolysaccharide, CRP, and IL-6 [136].

Mechanistically, the anti-arrhythmic effects of berberine are mediated through several cardiac ion channel targets, including cardiac slow (*I*Ks) and rapid (*I*Kr) delayed rectifier K+ channels, ATP-sensitive K+ channel (K_{ATP}), inwardly rectifying K+ channel (IK_1), and L-type Ca^{2+} (*I*Ca) [134, 137]. Berberine reduces the expression of *I*Kr channel in rat ventricular tissues and inhibits the human ether-a-go-go-related gene (hERG) channel in an HEK293 overexpression model [134, 138]. Berberine also reduces sinoatrial nodal expression of human hyperpolarization-activated cyclic nucleotide-gated 4 channels in Xenopus oocytes [139]. Despite these promising data, Martini et al. have recently shown that berberine affects the development of zebrafish cardiac morphogenesis and functionality [140]. The authors also demonstrated that these actions of berberine are accompanied by increased vascular endothelial growth factor expression, bradycardia, and reduced cardiac output, atrial contractility, and stroke volume [140].

Researchers have shown that the anti-atherogenic actions of berberine are mediated through upregulation of AMPK signaling, preventing lipid accumulation and foam cell formation in the subendothelium [12, 141, 142]. In human hepatocytes, berberine treatment significantly inhibits LDL-cholesterol (LDL-C) synthesis by increasing AMPK phosphorylation and activity [143]. In macrophages, the lectin-like oxidized low-density lipoprotein receptor-1 (LOX-1) mediated the uptake of ox-LDL, promoting foam cell formation. In vitro and in vivo work by Chi et al. demonstrated that berberine in combination with atorvastatin downregulates LOX-1 expression on macrophages through endothelin 1 receptor (ET-1R) [144]. In vitro, berberine diminishes the uptake of ox-LDL in macrophages, preventing foam cell formation through activation of the AMPK-SIRT1-PPAR-γ pathway [141]. Inflammation is also a necessary requirement for atherosclerotic progression. In obese mice, berberine has also been shown to suppress the expression of pro-inflammatory genes such as TNF-α, IL-1β, IL-6, MCP-1, *i*NOS, and COX-2 through AMPK activation in macrophages [142].

Rutaecarpine, Aminoalcohol-diterpenoid, N-p-coumaroyl-4-aminobutan-1-ol, Hippophamide

Rutaecarpine is an alkaloid from the *Evodia rutaecarpa* family and exhibits vasodilatory, anti-inflammatory, and lipid-lowering effects [72]. In a search for active compounds from *Aconitum carmichaelii* (Ranunculaceae), known as "Fuzi," 14 amino alcohol-diterpenoid alkaloids have been isolated. All these compounds have demonstrated cardioprotective effects against doxorubicin-induced toxicity in vitro [69]. Recently, Zhou et al. have found that two alkaloids, N-*p*-coumaroyl-4-aminobutan-1-ol, an organic amide alkaloid, and hippophamide, a rare pyridoindole alkaloid obtained from sea buckthorn seed residues, have potential cardioprotective effects [73]. In embryonic rat H9c2 cells, both compounds attenuate doxorubicin-induced ROS production, reduce caspase-3 expression, and activate JNK, leading to reduced apoptosis and improved mitochondrial function [73].

Terpenes and Terpenoids

Biologically, terpenes are most commonly known for their antimicrobial, anti-inflammatory, and antitumor activities [75]. In the scenario of CVD, terpenes help mitigate oxidative stress, inflammation, and apoptosis [8, 77, 145]. The most well-studied terpenes and terpenoids in CVD and cardiotoxicity include monoterpenes (e.g., bakuchiol, carvacrol, and D-Limonene), diterpenoids (e.g., ferruginol, carnosic acid, and carnosol), carotenoids (e.g., lycopene), and thymol (Table 1).

Bakuchiol

Bakuchiol has been studied in several models of CVD. In diabetic in vivo and in vitro models, bakuchiol attenuates cardiomyocyte apoptosis and oxidative stress through activation of the SIRT1/Nrf2 signaling pathway. In the same study, bakuchiol also inhibits fibrotic-protein expression in the myocardium of diabetic mice through reducing expression of α-SMA and Smad3 proteins [146]. In another study, bakuchiol protects against pathological cardiac hypertrophy as evidenced by reduced hypertrophic gene expression (ANP, BNP, and β-MHC) in pressure overload mice [78]. In the same study, in vitro studies using neonatal rat cardiomyocytes support a role for NF-κB signaling in the promotion of bakuchiol-mediated protective effects against cardiac hypertrophy [78]. Furthermore, a reduction in Col I, Col III, CGTF, TNF-α, IL-6, and MCP-1 promotes reduced fibrosis and inflammatory responses via the NF-κB pathway, in vivo and in vitro [78]. In summary, these studies show that bakuchiol has significant potential in the treatment of several cardiac pathologies, suggesting that bakuchiol is a promising therapeutic molecule for CVD management.

Carvacrol

Carvacrol is a phenolic monoterpene present in the essential oil of oregano. The in vivo and in vitro cardiovascular effects of carvacrol have been well studied. In a rat model of myocardial ischemic perfusion injury, carvacrol reduces lipid peroxidation and attenuates apoptosis through activation of MAPK/ERK and Akt-eNOS signaling pathways [147]. In another study, Sadeghzadeh et al. have shown that carvacrol blocks apoptosis in a pressure overload rat model by disrupting the balance of pro-apoptotic and apoptotic markers [148].

D-Limonene

The monoterpene D-Limonene (DL) is also a promising cardioprotective agent due to its bradycardic and antiarrhythmic properties [149, 150]. In a study by Nascimento et al., persistent bradycardia, hypotension, and antiarrhythmic activity were observed in rat hearts following treatment with DL [149]. Similarly, a complex of d-limonene plus hydroxypropyl-β-cyclodextrin (HβDL) has been shown to reduce cardiac arrhythmias. The investigators also showed a decrease in delayed afterdepolarizations, triggered activities, and inhibition of CaMKII hyperactivation (i.e., decreased pCAMKII and oxCAMKII) [151]. While terpinolene and α-pinene have not been specifically studied in the heart, they have been shown to reduce oxidative damage and inflammation, which are key factors in CVD prevention [8, 152].

Lycopene

Lycopene is a carotenoid compound that protects against oxidative stress in CVD (Table 1) [153]. In a meta-analysis of observational studies investigating the association between lycopene and CVD risk, Song et al. have demonstrated that higher lycopene exposure is associated with a lower risk of CVD and stroke [154]. However, this study is limited by heterogeneity in the studies used. It has been shown that oxidative stress may promote cardiomyocyte hypertrophy. Zeng et al. have demonstrated that lycopene prevents cardiac hypertrophy in vitro and in vivo [155]. These changes are accompanied by preservation of antioxidant response element (ARE) activity and activated ARE-driven expression of antioxidant genes and suppression of ROS-dependent mechanisms.

In male Wistar rats cotreated with doxorubicin and lycopene, the investigators have observed suppression of doxorubicin-induced myocyte damage and prevention of interstitial collagen accumulation. However, lycopene does not prevent doxorubicin-induced left ventricular systolic dysfunction [156]. Yilmaz et al. have also shown that lycopene protects rats from adriamycin-induced cardiotoxicity and nephrotoxicity through reduction of oxidative stress, normalizing cardiac enzymes, and preventing tissue damage [157]. With respect to the cardiotoxic effect of the macrolide antibiotic (tulathromycin) and diclofenac sodium (DFS) alone or in combination, Abdel-Daim and colleagues have demonstrated cardioprotective effects of lycopene. In the study, mice treated with tulathromycin or DFS (alone or in combination) showed significantly elevated serum CK-MB, LDH, and cardiac troponin-T; increased tissue NO and MDA levels; as well as decreased tissue GSH-Px, SOD, and CAT antioxidant enzyme activities. However, these changes have been significantly reduced in the presence of lycopene [158].

Quinones

Specific quinones with cardioprotective effects include pyrroloquinoline quinone (PQQ), cryptotanshinone (CTS), and ubiquinone (CoQ) [159, 160]. PQQ is a strong antioxidant, and its cellular depletion leads to reduced mitochondrial content and function [161]. Xu et al. have demonstrated that PQQ pretreatment of cells and pressure overload-induced HF in a Sprague-Dawley rat model result in reduced cell hypertrophy and cardiac remodeling, preventing the development of HF [162]. Moreover, PQQ protects against HF through several mechanisms: reducing mitochondrial ROS production, preventing mitochondrial calcium overload, and increasing mitochondrial biogenesis through upregulation of proliferator-activated receptor gamma coactivator 1-α and transcription factor A mitochondrial (TFAM) protein expression [162].

The protective actions of CTS are largely attributed to its ability to decrease ROS levels, inhibit apoptosis, and protect mitochondrial membrane integrity. Several signaling mechanisms have been described to mediate its cardioprotective effects [163–166]. In terms of anti-inflammatory actions, CTS has been shown to block Ca^{2+} signaling and reduce mitochondrial ROS, which are critical upstream activators of the NLRP3 inflammasome [163]. In combination with direct binding to NLRP3, CTS may reduce inflammation and oxidative stress in CVDs [163, 167]. CTS has also been shown to target the NF-κB and MAPK pathways [168].

In vitro, CTS reduced the angiotensin-II-induced upregulation of fibrosis markers such as fibronectin, connective tissue growth factor, and COX-2 [169]. In addition, Yunzi et al. have demonstrated that oral administration of CTS attenuates fibrotic pathological changes accompanying MI. This attenuation is facilitated through inhibition of ERK1/2, angiotensin-II-stimulated upregulation of NOX2 and NOX4, and reduced ROS production [169]. In a separate study, CTS has been shown to inhibit fibrosis and improve cardiac function by suppressing the STAT3 pathway, matrix metalloproteinase-9, and connective tissue growth factor [165].

Using the rat in vivo model, CTS has been shown to attenuate doxorubicin-induced cardiac toxicity and HF through modulation of the p38 pathway [164]. Recently, in vitro studies by Wang et al. have also demonstrated that CTS prevents doxorubicin-induced cardiotoxicity by targeting the Akt-GSK-3β-mPTP pathway [166].

CoQ is another well-studied cardioprotective quinone. As previously mentioned, CoQ acts as a potent antioxidant, stabilizing the mitochondria and reducing oxidative damage in cardiotoxicity [159, 167, 170]. In the randomized controlled multicenter Q-SYMBIO trial with HF patients, daily treatment with CoQ, in addition to standard therapy, reduces cardiovascular mortality, all-cause mortality, and hospital stays as compared to placebo [171].

Miscellaneous Phytochemicals

Sulforaphane

Organosulfur compounds have also demonstrated antioxidant defense effects, with potential for treating CVD [15]. Among them, sulforaphane (SFN), an isothiocyanate compound, has been shown to have a documented role in experimental models of CVD and cardiotoxicity [184, 185]. Sulforaphane prevents cellular ischemia and myocardial remodeling secondary to MI. In an ischemic-reperfusion rat model, SFN improves left ventricle function, along with contractility and relaxation indexes. These functional changes have been associated with decreased expression of RyR2 and increased *p*-phospholamban/phospholamban ratio, suggesting a role for SFN in calcium regulations. SFN has also been shown to reduce remodeling in a rat MI model through modulation of RyR2 expression and ROS levels [184]. In a separate study using a rat MI model, SFN has been shown to preserve cardiac function and decrease infarct size, oxidative stress, and inflammation, possibly through activation of Nrf2 and aryl hydrocarbon receptor [91].

SFN has also been studied in CHF models. Ma et al. have demonstrated that in rabbits with CHF, subcutaneous SFN treatment improves LV function and reduces cardiac remodeling by inhibiting oxidative stress and inflammation. Specifically, reduced plasma BNP and ANP, reduced myocardial tissue levels of Col I, Col III, TNF-α, IL-6, and MDA, and increased SOD activity have been observed in this study [183].

In terms of cardiotoxicity, SFN has been shown to reduce cytotoxic effects of doxorubicin and arsenic through activation of the Nrf2 pathway [185, 186]. When comparing SFN effects on rat neonatal cardiomyocytes and multiple rat and human breast cancer cell lines, SFN treatment protects cardiac cells but potentiates doxorubicin toxicity in cancer cells [185]. In the same study, SFN treatment of doxorubicin-treated rats results in reduced myocardial oxidative stress via inhibition of lipid peroxidation and activation of Nrf2 and associated antioxidant enzymes, enhancing mitochondrial integrity and function [185]. Recently, Thangapandiyan and colleagues have shown that SFN can ameliorate arsenic-induced cardiotoxicity. Mechanistically, SFN blocks in vitro and in vivo arsenic-induced cardiotoxicity through reduction of free radicals, lipid oxidation, lipid profiles, cardiac troponins I and increased phase II antioxidants via activation of the PI3K/Akt/Nrf2 pathway [186].

Conclusions

In summary, while phytochemicals have long been revered in traditional medicine, there is increasing evidence supporting their integration in Western health care. Phytochemicals offer a promising preventive and therapeutic strategy for cardioprotection through targeting ROS production and modulation of antioxidant defenses and several critical cellular signaling pathways. These natural compounds, including polyphenols, flavonoids, alkaloids, terpenes, and quinones, exhibit antioxidant, anti-inflammatory, and antiapoptotic actions that mitigate oxidative stress, enhance mitochondrial function, and attenuate cardiac remodeling. By influencing key signaling pathways including, Nrf2, NF-κB, PI3K/Akt, and MAPK, phytochemicals have been shown to exhibit therapeutic potential in managing myocardial oxidative stress and favorably impact cardiac biomarkers like BNP, ANP, and inflammatory cytokines. Moreover, their efficacy in reducing chemotherapeutic-induced cardiotoxicity while preserving myocardial function further cements their clinical relevance. Given their wide range of benefits, low toxicity, and natural origin, phytochemicals present a compelling case for their use as adjunctive therapies in CVD, warranting further research and clinical trials.

Disclosures The authors have no relevant affiliations or financial involvement to disclose.

Funding This work was supported by National Institute of Health grant T32HL134616 to SSS and NIH R01 HL160590 to SAS.

References

1. Zhou S, Sun W, Zhang Z, Zheng Y. The role of Nrf2-mediated pathway in cardiac remodeling and heart failure. Oxidative Med Cell Longev. 2014;2014:260429.
2. Serafini M, Peluso I, Raguzzini A. Flavonoids as anti-inflammatory agents. Proc Nutr Soc. 2010;69:273–8.
3. Upadhyay S, Dixit M. Role of polyphenols and other phytochemicals on molecular signaling. Oxidative Med Cell Longev. 2015;2015:504253.
4. Alotaibi BS, Ijaz M, Buabeid M, Kharaba ZJ, Yaseen HS, Murtaza G. Therapeutic effects and safe uses of plant-derived polyphenolic compounds in cardiovascular diseases: a review. Drug Des Devel Ther. 2021;15:4713–32.
5. Hedayati N, Yaghoobi A, Salami M, Gholinezhad Y, Aghadavood F, Eshraghi R, et al. Impact of polyphenols on heart failure and cardiac hypertrophy: clinical effects and molecular mechanisms. Front Cardiovasc Med. 2023;10:1174816.
6. Ciumărnean L, Milaciu MV, Runcan O, Vesa Ştefan C, Răchişan AL, Negrean V, et al. The effects of flavonoids in cardiovascular diseases. Mol Basel Switz. 2020;25:4320.
7. Roy A, Khan A, Ahmad I, Alghamdi S, Rajab BS, Babalghith AO, et al. Flavonoids a bioactive compound from medicinal plants and its therapeutic applications. Biomed Res Int. 2022;2022:5445291.
8. Del Prado-Audelo ML, Cortés H, Caballero-Florán IH, González-Torres M, Escutia-Guadarrama L, Bernal-Chávez SA, et al. Therapeutic applications of Terpenes on inflammatory diseases. Front Pharmacol. 2021;12:704197.
9. Kim T, Song B, Cho KS, Lee I-S. Therapeutic potential of volatile Terpenes and Terpenoids from forests for inflammatory diseases. Int J Mol Sci. 2020;21:2187.
10. Haines DD, Cowan FM, Tosaki A. Evolving strategies for use of phytochemicals in prevention and long-term management of cardiovascular diseases (CVD). Int J Mol Sci. 2024;25:6176.
11. Huang WM, Wu ZD, Gan YQ. Effects of berberine on ischemic ventricular arrhythmia. Zhonghua Xin Xue Guan Bing Za Zhi. 1989;17(300–1):319.
12. Feng X, Sureda A, Jafari S, Memariani Z, Tewari D, Annunziata G, et al. Berberine in cardiovascular and metabolic diseases: from mechanisms to therapeutics. Theranostics. 2019;9:1923–51.
13. Dong Q, Chen L, Lu Q, Sharma S, Li L, Morimoto S, et al. Quercetin attenuates doxorubicin cardiotoxicity by modulating Bmi-1 expression. Br J Pharmacol. 2014;171:4440–54.
14. Sung MM, Das SK, Levasseur J, Byrne NJ, Fung D, Kim TT, et al. Resveratrol treatment of mice with pressure-overload-induced heart failure improves diastolic function and cardiac energy metabolism. Circ Heart Fail. 2015;8:128–37.
15. Ashraf MV, Khan S, Misri S, Gaira KS, Rawat S, Rawat B, et al. High-altitude medicinal plants as promising source of phytochemical antioxidants to combat lifestyle-associated oxidative stress-induced disorders. Pharm Basel Switz. 2024;17:975.
16. Hancock JT, Desikan R, Neill SJ. Role of reactive oxygen species in cell signalling pathways. Biochem Soc Trans. 2001;29:345–50.
17. Jomova K, Raptova R, Alomar SY, Alwasel SH, Nepovimova E, Kuca K, et al. Reactive oxygen species, toxicity, oxidative stress, and antioxidants: chronic diseases and aging. Arch Toxicol. 2023;97:2499–574.
18. Ayala A, Muñoz MF, Argüelles S. Lipid peroxidation: production, metabolism, and signaling mechanisms of malondialdehyde and 4-hydroxy-2-nonenal. Oxidative Med Cell Longev. 2014;2014:360438.
19. Del Rio D, Stewart AJ, Pellegrini N. A review of recent studies on malondialdehyde as toxic molecule and biological marker of oxidative stress. Nutr Metab Cardiovasc Dis NMCD. 2005;15:316–28.
20. Hajam YA, Rani R, Ganie SY, Sheikh TA, Javaid D, Qadri SS, et al. Oxidative stress in human pathology and aging: molecular mechanisms and perspectives. Cells. 2022;11:552.

21. Schieber M, Chandel NS. ROS function in redox signaling and oxidative stress. Curr Biol CB. 2014;24:R453–62.
22. Valko M, Leibfritz D, Moncol J, Cronin MTD, Mazur M, Telser J. Free radicals and antioxidants in normal physiological functions and human disease. Int J Biochem Cell Biol. 2007;39:44–84.
23. Aranda-Rivera AK, Cruz-Gregorio A, Arancibia-Hernández YL, Hernández-Cruz EY, Pedraza-Chaverri J. RONS and oxidative stress: an overview of basic concepts. Oxygen. 2022;2:437–78.
24. Averill-Bates D. Reactive oxygen species and cell signaling. Review. Biochim Biophys Acta, Mol Cell Res. 2024;1871:119573.
25. Liu R, Bian Y, Liu L, Liu L, Liu X, Ma S. Molecular pathways associated with oxidative stress and their potential applications in radiotherapy (review). Int J Mol Med. 2022;49:65.
26. Biswas SK. Does the interdependence between oxidative stress and inflammation explain the antioxidant paradox? Oxidative Med Cell Longev. 2016;2016:5698931.
27. Federico A, Morgillo F, Tuccillo C, Ciardiello F, Loguercio C. Chronic inflammation and oxidative stress in human carcinogenesis. Int J Cancer. 2007;121:2381–6.
28. Papaharalambus CA, Griendling KK. Basic mechanisms of oxidative stress and reactive oxygen species in cardiovascular injury. Trends Cardiovasc Med. 2007;17:48–54.
29. Belch JJ, Bridges AB, Scott N, Chopra M. Oxygen free radicals and congestive heart failure. Br Heart J. 1991;65:245–8.
30. Farley A, McLafferty E, Hendry C. The cardiovascular system. Nurs Stand R Coll Nurs G B. 1987;2012(27):35–9.
31. Benjamin EJ, Virani SS, Callaway CW, Chamberlain AM, Chang AR, Cheng S, et al. Heart Disease and stroke statistics—2018 update: a report from the American Heart Association. Circulation [Internet]. 2018 [cited 2025 Apr 5];137. Available from: https://www.ahajournals.org/doi/10.1161/CIR.0000000000000558.
32. Olvera Lopez E, Ballard BD, Jan A. Cardiovascular disease. StatPearls [Internet]. Treasure Island (FL): StatPearls Publishing; 2025 [cited 2025 Apr 4]. Available from: http://www.ncbi.nlm.nih.gov/books/NBK535419/.
33. Zoccali C, Mallamaci F, Adamczak M, de Oliveira RB, Massy ZA, Sarafidis P, et al. Cardiovascular complications in chronic kidney disease: a review from the European Renal and Cardiovascular Medicine Working Group of the European Renal Association. Cardiovasc Res. 2023;119:2017–32.
34. Frąk W, Wojtasińska A, Lisińska W, Młynarska E, Franczyk B, Rysz J. Pathophysiology of cardiovascular diseases: new insights into molecular mechanisms of atherosclerosis, arterial hypertension, and coronary artery disease. Biomedicines. 2022;10:1938.
35. Scott J. Pathophysiology and biochemistry of cardiovascular disease. Curr Opin Genet Dev. 2004;14:271–9.
36. Shah A, Isath A, Aronow WS. Cardiovascular complications of diabetes. Expert Rev Endocrinol Metab. 2022;17:383–8.
37. Hutchins E, Yang EH, Stein-Merlob AF. Inflammation in chemotherapy-induced cardiotoxicity. Curr Cardiol Rep. 2024;26:1329–40.
38. Curigliano G, Cardinale D, Dent S, Criscitiello C, Aseyev O, Lenihan D, et al. Cardiotoxicity of anticancer treatments: epidemiology, detection, and management. CA Cancer J Clin. 2016;66:309–25.
39. Rawat PS, Jaiswal A, Khurana A, Bhatti JS, Navik U. Doxorubicin-induced cardiotoxicity: an update on the molecular mechanism and novel therapeutic strategies for effective management. Biomed Pharmacother. 2021;139:111708.
40. Gao F, Xu T, Zang F, Luo Y, Pan D. Cardiotoxicity of anticancer drugs: molecular mechanisms, clinical management and innovative treatment. Drug Des Devel Ther. 2024;18:4089–116.
41. Moslehi JJ. Cardiovascular toxic effects of targeted cancer therapies. N Engl J Med. 2016;375:1457–67.

42. Shah CP, Moreb JS. Cardiotoxicity due to targeted anticancer agents: a growing challenge. Ther Adv Cardiovasc Dis. 2019;13:1753944719843435.
43. Morelli MB, Bongiovanni C, Da Pra S, Miano C, Sacchi F, Lauriola M, et al. Cardiotoxicity of anticancer drugs: molecular mechanisms and strategies for Cardioprotection. Front Cardiovasc Med. 2022;9:847012.
44. Force T, Krause DS, Van Etten RA. Molecular mechanisms of cardiotoxicity of tyrosine kinase inhibition. Nat Rev Cancer. 2007;7:332–44.
45. Quideau S, Deffieux D, Douat-Casassus C, Pouységu L. Plant polyphenols: chemical properties, biological activities, and synthesis. Angew Chem Int Ed Engl. 2011;50:586–621.
46. Bolat E, Sarıtaş S, Duman H, Eker F, Akdaşçi E, Karav S, et al. Polyphenols: secondary metabolites with a biological impression. Nutrients. 2024;16:2550.
47. Manach C, Scalbert A, Morand C, Rémésy C, Jiménez L. Polyphenols: food sources and bioavailability. Am J Clin Nutr. 2004;79:727–47.
48. de Paulo FD, Neri-Numa IA, de Araújo FF, Pastore GM. A critical review of some fruit trees from the Myrtaceae family as promising sources for food applications with functional claims. Food Chem. 2020;306:125630.
49. de la Rosa LA, Moreno-Escamilla JO, Rodrigo-García J, Alvarez-Parrilla E. Chapter 12 - Phenolic Compounds. In: Yahia EM, editor. Postharvest Physiology and Biochemistry of Fruits and Vegetables [Internet]. Woodhead Publishing; 2019. p. 253–71. Available from: https://www.sciencedirect.com/science/article/pii/B9780128132784000129
50. Suraweera LT, Rupasinghe HPV, Dellaire G, Xu Z. Regulation of Nrf2/ARE pathway by dietary flavonoids: a friend or foe for cancer management? Antioxid Basel Switz. 2020;9:973.
51. Houle JJ, Hoffmann EM, Esser AF. Restriction of cell lysis by homologous complement: II. Protection of erythrocytes against lysis by newly activated complement. Blood. 1988;71:287–92.
52. Nakazato T, Ito K, Ikeda Y, Kizaki M. Green tea component, catechin, induces apoptosis of human malignant B cells via production of reactive oxygen species. Clin Cancer Res. 2005;11:6040–9.
53. Lambert JD, Elias RJ. The antioxidant and pro-oxidant activities of green tea polyphenols: a role in cancer prevention. Arch Biochem Biophys. 2010;501:65–72.
54. Du G-J, Zhang Z, Wen X-D, Yu C, Calway T, Yuan C-S, et al. Epigallocatechin Gallate (EGCG) is the most effective cancer chemopreventive polyphenol in green tea. Nutrients. 2012;4:1679–91.
55. Pagliaro B, Santolamazza C, Simonelli F, Rubattu S. Phytochemical compounds and protection from cardiovascular diseases: a state of the art. Biomed Res Int. 2015;2015:918069.
56. Muscolo A, Mariateresa O, Giulio T, Mariateresa R. Oxidative stress: the role of antioxidant phytochemicals in the prevention and treatment of diseases. Int J Mol Sci. 2024;25:3264.
57. Brito PM, Devillard R, Nègre-Salvayre A, Almeida LM, Dinis TCP, Salvayre R, et al. Resveratrol inhibits the mTOR mitogenic signaling evoked by oxidized LDL in smooth muscle cells. Atherosclerosis. 2009;205:126–34.
58. Zhang Y, Lu P, Qin H, Zhang Y, Sun X, Song X, et al. Traditional Chinese medicine combined with pulmonary drug delivery system and idiopathic pulmonary fibrosis: rationale and therapeutic potential. Biomed Pharmacother. 2021;133:111072.
59. Penners BM, Krämer M, Grüner O. Sudden cardiac death following psychoemotional stress. Z Rechtsmed. 1986;96:151–7.
60. Olofinsan K, Abrahamse H, George BP. Therapeutic role of alkaloids and alkaloid derivatives in cancer management. Mol Basel Switz. 2023;28:5578.
61. Heinrich M, Mah J, Amirkia V. Alkaloids used as medicines: structural phytochemistry meets biodiversity-an update and forward look. Mol Basel Switz. 2021;26:1836.
62. Debnath B, Singh WS, Das M, Goswami S, Singh MK, Maiti D, et al. Role of plant alkaloids on human health: a review of biological activities. Mater Today Chem. 2018;9:56–72.
63. Banyal A, Tiwari S, Sharma A, Chanana I, Patel SKS, Kulshrestha S, et al. Vinca alkaloids as a potential cancer therapeutics: recent update and future challenges. 3 Biotech. 2023;13:211.

64. Sansone L, Milani F, Fabrizi R, Belli M, Cristina M, Zagà V, et al. Nicotine: from discovery to biological effects. Int J Mol Sci. 2023;24:14570.
65. Lakier JB. Smoking and cardiovascular disease. Am J Med. 1992;93:8S–12S.
66. Benowitz NL, Burbank AD. Cardiovascular toxicity of nicotine: implications for electronic cigarette use. Trends Cardiovasc Med. 2016;26:515–23.
67. Prabhakaran D, Anand S, Gaziano TA, Mbanya J-C, Wu Y, Nugent R, editors. Disease control priorities, Third Edition (Volume 5): cardiovascular, respiratory, and related disorders [Internet]. The World Bank; 2017 [cited 2025 Apr 6]. Available from: http://elibrary.worldbank.org/doi/book/10.1596/978-1-4648-0518-9
68. Xu X, Xie X, Zhang H, Wang P, Li G, Chen J, et al. Water-soluble alkaloids extracted from Aconiti Radix lateralis praeparata protect against chronic heart failure in rats via a calcium signaling pathway. Biomed Pharmacother. 2021;135:111184.
69. Wang X-Y, Zhou Q-M, Guo L, Dai O, Meng C-W, Miao L-L, et al. Cardioprotective effects and concentration-response relationship of aminoalcohol-diterpenoid alkaloids from Aconitum carmichaelii. Fitoterapia. 2021;149:104822.
70. Tan H-L, Chan K-G, Pusparajah P, Duangjai A, Saokaew S, Mehmood Khan T, et al. Rhizoma Coptidis: A Potential Cardiovascular Protective Agent. Front Pharmacol. 2016;7:362.
71. Geetha RG, Ramachandran S. Recent advances in the anti-inflammatory activity of plant-derived alkaloid Rhynchophylline in neurological and cardiovascular diseases. Pharmaceutics. 2021;13:1170.
72. Tian K-M, Li J-J, Xu S-W. Rutaecarpine: A promising cardiovascular protective alkaloid from Evodia rutaecarpa (Wu Zhu Yu). Pharmacol Res. 2019;141:541–50.
73. Zhou W, Ouyang J, Hu N, Li G, Wang H. Protective effect of two alkaloids from Hippophae rhamnoides Linn. against Doxorubicin-induced toxicity in H9c2 Cardiomyoblasts. Mol Basel Switz. 2021;26:1946.
74. Ninkuu V, Zhang L, Yan J, Fu Z, Yang T, Zeng H. Biochemistry of terpenes and recent advances in plant protection. Int J Mol Sci. 2021;22:5710.
75. Masyita A, Mustika Sari R, Dwi Astuti A, Yasir B, Rahma Rumata N, Emran TB, et al. Terpenes and terpenoids as main bioactive compounds of essential oils, their roles in human health and potential application as natural food preservatives. Food Chem X. 2022;13:100217.
76. Oldfield E, Lin F-Y. Terpene biosynthesis: modularity rules. Angew Chem Int Ed Engl. 2012;51:1124–37.
77. Abdul Ghani MA, Ugusman A, Latip J, Zainalabidin S. Role of Terpenophenolics in modulating inflammation and apoptosis in cardiovascular diseases: a review. Int J Mol Sci. 2023;24:5339.
78. Wang Z, Gao L, Xiao L, Kong L, Shi H, Tian X, et al. Bakuchiol protects against pathological cardiac hypertrophy by blocking NF-κB signaling pathway. Biosci Rep. 2018;38:BSR20181043.
79. Ma W, Long Y-T. Quinone/hydroquinone-functionalized biointerfaces for biological applications from the macro- to nano-scale. Chem Soc Rev. 2014;43:30–41.
80. Devi SS, Mehendale HM. Quinone. Encycl Toxicol [Internet]. Elsevier; 2014 [cited 2025 Apr 6]. p. 26–8. Available from: https://linkinghub.elsevier.com/retrieve/pii/B978012386454300350X.
81. Bolton JL, Dunlap T. Formation and biological targets of Quinones: cytotoxic versus Cytoprotective effects. Chem Res Toxicol. 2017;30:13–37.
82. Thomson RH. Distribution of naturally occurring quinones. Pharm Weekbl Sci. 1991;13:70–3.
83. Gong H, He Z, Peng A, Zhang X, Cheng B, Sun Y, et al. Effects of several quinones on insulin aggregation. Sci Rep. 2014;4:5648.
84. Kristensen SB, van Mourik T, Pedersen TB, Sørensen JL, Muff J. Simulation of electrochemical properties of naturally occurring quinones. Sci Rep. 2020;10:13571.
85. Conklin KA. Coenzyme q10 for prevention of anthracycline-induced cardiotoxicity. Integr Cancer Ther. 2005;4:110–30.

86. Doroshow JH. Effect of anticancer Quinones on reactive oxygen production by adult rat heart myocytes. Oxidative Med Cell Longev. 2020;2020:8877100.
87. Horenstein MS, Vander Heide RS, L'Ecuyer TJ. Molecular basis of anthracycline-induced cardiotoxicity and its prevention. Mol Genet Metab. 2000;71:436–44.
88. Hahn VS, Lenihan DJ, Ky B. Cancer therapy–induced cardiotoxicity: basic mechanisms and potential cardioprotective therapies. J Am Heart Assoc Cardiovasc Cerebrovasc Dis. 2014;3:e000665.
89. Serrano J, Palmeira CM, Kuehl DW, Wallace KB. Cardioselective and cumulative oxidation of mitochondrial DNA following subchronic doxorubicin administration. Biochim Biophys Acta. 1999;1411:201–5.
90. Bhatia S. Genetics of Anthracycline cardiomyopathy in cancer survivors: JACC: CardioOncology state-of-the-art review. JACC Cardio Oncol. 2020;2:539–52.
91. Silva-Palacios A, Ostolga-Chavarría M, Sánchez-Garibay C, Rojas-Morales P, Galván-Arzate S, Buelna-Chontal M, et al. Sulforaphane protects from myocardial ischemia-reperfusion damage through the balanced activation of Nrf2/AhR. Free Radic Biol Med. 2019;143:331–40.
92. Ahmet I, Tae H-J, Lakatta EG, Talan M. Long-term low dose dietary resveratrol supplement reduces cardiovascular structural and functional deterioration in chronic heart failure in rats. Can J Physiol Pharmacol. 2017;95:268–74.
93. Rimbaud S, Ruiz M, Piquereau J, Mateo P, Fortin D, Veksler V, et al. Resveratrol improves survival, hemodynamics and energetics in a rat model of hypertension leading to heart failure. PLoS One. 2011;6:e26391.
94. Papakyriakopoulou P, Velidakis N, Khattab E, Valsami G, Korakianitis I, Kadoglou NP. Potential pharmaceutical applications of quercetin in cardiovascular diseases. Pharm Basel Switz. 2022;15:1019.
95. Arumugam S, Thandavarayan RA, Arozal W, Sari FR, Giridharan VV, Soetikno V, et al. Quercetin offers cardioprotection against progression of experimental autoimmune myocarditis by suppression of oxidative and endoplasmic reticulum stress via endothelin-1/MAPK signalling. Free Radic Res. 2012;46:154–63.
96. Jing Z, Wang Z, Li X, Li X, Cao T, Bi Y, et al. Protective effect of Quercetin on posttraumatic cardiac injury. Sci Rep. 2016;6:30812.
97. Albadrani GM, BinMowyna MN, Bin-Jumah MN, El-Akabawy G, Aldera H, Al-Farga AM. Quercetin prevents myocardial infarction adverse remodeling in rats by attenuating TGF-β1/Smad3 signaling: different mechanisms of action. Saudi. J Biol Sci. 2021;28:2772–82.
98. Wang Y, Zhang ZZ, Wu Y, Ke JJ, He XH, Wang YL. Quercetin postconditioning attenuates myocardial ischemia/reperfusion injury in rats through the PI3K/Akt pathway. Braz J Med Biol Res Rev Bras Pesqui Medicas E Biol. 2013;46:861–7.
99. Dulf PL, Coadă CA, Florea A, Moldovan R, Baldea I, Dulf DV, et al. Mitigating Doxorubicin-induced cardiotoxicity through Quercetin intervention: an experimental study in rats. Antioxid Basel Switz. 2024;13:1068.
100. Majhi S, Singh L, Yasir M. Evaluation of ameliorative effect of quercetin and Candesartan in Doxorubicin-induced cardiotoxicity. Vasc Health Risk Manag. 2022;18:857–66.
101. Chen X-Q, Hu T, Han Y, Huang W, Yuan H-B, Zhang Y-T, et al. Preventive effects of Catechins on cardiovascular disease. Mol Basel Switz. 2016;21:1759.
102. Babu PVA, Liu D. Green tea catechins and cardiovascular health: an update. Curr Med Chem. 2008;15:1840–50.
103. Mangels DR, Mohler ER. Catechins as potential mediators of cardiovascular health. Arterioscler Thromb Vasc Biol. 2017;37:757–63.
104. MacRae K, Connolly K, Vella R, Fenning A. Epicatechin's cardiovascular protective effects are mediated via opioid receptors and nitric oxide. Eur J Nutr. 2019;58:515–27.
105. Saleh Ahmed AS. Potential protective effect of catechin on doxorubicin-induced cardiotoxicity in adult male albino rats. Toxicol Mech Methods. 2022;32:97–105.

106. Kozluca O, Olcay E, Sürücü S, Güran Z, Kulaksiz T, Uskent N. Prevention of doxorubicin induced cardiotoxicity by catechin. Cancer Lett. 1996;99:1–6.
107. Saeed NM, El-Naga RN, El-Bakly WM, Abdel-Rahman HM, Salah ElDin RA, El-Demerdash E. Epigallocatechin-3-gallate pretreatment attenuates doxorubicin-induced cardiotoxicity in rats: a mechanistic study. Biochem Pharmacol. 2015;95:145–55.
108. Rasheed NOA, Ahmed LA, Abdallah DM, El-Sayeh BM. Paradoxical cardiotoxicity of intraperitoneally-injected epigallocatechin gallate preparation in diabetic mice. Sci Rep. 2018;8:7880.
109. Yang C, Zhu Q, Chen Y, Ji K, Li S, Wu Q, et al. Review of the protective mechanism of Curcumin on cardiovascular disease. Drug Des Devel Ther. 2024;18:165–92.
110. Wongcharoen W, Phrommintikul A. The protective role of curcumin in cardiovascular diseases. Int J Cardiol. 2009;133:145–51.
111. Singh L, Sharma S, Xu S, Tewari D, Fang J. Curcumin as a natural remedy for atherosclerosis: a pharmacological review. Mol Basel Switz. 2021;26:4036.
112. Miriyala S, Panchatcharam M, Rengarajulu P. Cardioprotective effects of curcumin. Adv Exp Med Biol. 2007;595:359–77.
113. Qiu L, Gao C, Wang H, Ren Y, Li J, Li M, et al. Effects of dietary polyphenol curcumin supplementation on metabolic, inflammatory, and oxidative stress indices in patients with metabolic syndrome: a systematic review and meta-analysis of randomized controlled trials. Front Endocrinol. 2023;14:1216708.
114. Venkatesan N. Curcumin attenuation of acute adriamycin myocardial toxicity in rats. Br J Pharmacol. 1998;124:425–7.
115. Hu Z. Exploring the mechanism of curcumin in the treatment of doxorubicin-induced cardiotoxicity based on network pharmacology and molecular docking technology. Medicine (Baltimore). 2024;103:e36593.
116. Abdollahi-Karizno M, Chahkandi M, Rajabi S, Roshanravan B, Jafari-Nozad AM, Aschner M, et al. The protective effect of Curcumin against cardiotoxic effects induced by chronic exposure to Chlorpyrifos. Curr Mol Med. 2024;24:676–82.
117. Guo W, Zhu H, Wang Z, Chen J-A, Wu J, Zhu Y, et al. Novel rhynchophylline analogues as microvascular relaxation agents for the treatment of microvascular dysfunction caused by diabetes. Eur J Med Chem. 2017;139:657–64.
118. Wang Z-J, Chang L-L, Wu J, Pan H-M, Zhang Q-Y, Wang M-J, et al. A novel Rhynchophylline analog, Y396, inhibits endothelial dysfunction induced by oxidative stress in diabetes through epidermal growth factor receptor. Antioxid Redox Signal. 2020;32:743–65.
119. Zhou J-Y, Zhou S-W. Isorhynchophylline: a plant alkaloid with therapeutic potential for cardiovascular and central nervous system diseases. Fitoterapia. 2012;83:617–26.
120. Li P-Y, Zeng X-R, Cheng J, Wen J, Inoue I, Yang Y. Rhynchophylline-induced vasodilation in human mesenteric artery is mainly due to blockage of L-type calcium channels in vascular smooth muscle cells. Naunyn Schmiedeberg's Arch Pharmacol. 2013;386:973–82.
121. Hao H-F, Liu L-M, Liu Y-Y, Liu J, Yan L, Pan C-S, et al. Inhibitory effect of rhynchophylline on contraction of cerebral arterioles to endothelin 1: role of rho kinase. J Ethnopharmacol. 2014;155:147–53.
122. Hao H-F, Liu L-M, Pan C-S, Wang C-S, Gao Y-S, Fan J-Y, et al. Rhynchophylline Ameliorates endothelial dysfunction via Src-PI3K/Akt-eNOS Cascade in the cultured Intrarenal arteries of spontaneous hypertensive rats. Front Physiol. 2017;8:928.
123. Qin Q-J, Cui L-Q, Li P, Wang Y-B, Zhang X-Z, Guo M-L. Rhynchophylline ameliorates myocardial ischemia/reperfusion injury through the modulation of mitochondrial mechanisms to mediate myocardial apoptosis. Mol Med Rep. 2019;19:2581–90.
124. Chen K, Li G, Geng F, Zhang Z, Li J, Yang M, et al. Berberine reduces ischemia/reperfusion-induced myocardial apoptosis via activating AMPK and PI3K-Akt signaling in diabetic rats. Apoptosis Int J Program Cell Death. 2014;19:946–57.

125. Wang Y, Liu J, Ma A, Chen Y. Cardioprotective effect of berberine against myocardial ischemia/reperfusion injury via attenuating mitochondrial dysfunction and apoptosis. Int J Clin Exp Med. 2015;8:14513–9.
126. Chang W, Zhang M, Li J, Meng Z, Xiao D, Wei S, et al. Berberine attenuates ischemia-reperfusion injury via regulation of adenosine-5′-monophosphate kinase activity in both non-ischemic and ischemic areas of the rat heart. Cardiovasc Drugs Ther. 2012;26:467–78.
127. Fatehi-Hassanabad Z, Jafarzadeh M, Tarhini A, Fatehi M. The antihypertensive and vasodilator effects of aqueous extract from Berberis vulgaris fruit on hypertensive rats. Phytother Res PTR. 2005;19:222–5.
128. Chun YT, Yip TT, Lau KL, Kong YC, Sankawa U. A biochemical study on the hypotensive effect of berberine in rats. Gen Pharmacol. 1979;10:177–82.
129. Olmez E, Ilhan M. Evaluation of the alpha-adrenoceptor antagonistic action of berberine in isolated organs. Arzneimittelforschung. 1992;42:1095–7.
130. Cheng B, Luo LY, Fang DC, Jiang MX. Cardiovascular aspects of pharmacology of berberine: I. Alpha-adrenoceptor blocking action of berberine in isolated rat anococcygeus muscle and rabbit aortic strip. J Tongji Med Univ Tong Ji Yi Ke Xue Xue Bao. 1987;7:239–41.
131. Huang W. The role and mechanism of berberine on coronary arteries. Zhonghua Xin Xue Guan Bing Za Zhi. 1990;18(231–4):254–5.
132. Wong KK. Mechanism of the aortic relaxation induced by low concentrations of berberine. Planta Med. 1998;64:756–7.
133. Ko WH, Yao XQ, Lau CW, Law WI, Chen ZY, Kwok W, et al. Vasorelaxant and antiproliferative effects of berberine. Eur J Pharmacol. 2000;399:187–96.
134. Wang L, Yu C, Fu Y, Li Q, Sun Y. Berberine elicits anti-arrhythmic effects via IK1/Kir2.1 in the rat type 2 diabetic myocardial infarction model. Phytother Res PTR. 2011;25:33–7.
135. Zhou Z-W, Zheng H-C, Zhao L-F, Li W, Hou J-W, Yu Y, et al. Effect of berberine on acetylcholine-induced atrial fibrillation in rabbit. Am J Transl Res. 2015;7:1450–7.
136. Zhang J, Wang Y, Jiang H, Tao D, Zhao K, Yin Z, et al. Preventive effect of Berberine on postoperative atrial fibrillation. Circ Arrhythm Electrophysiol. 2022;15:e011160.
137. Lau CW, Yao XQ, Chen ZY, Ko WH, Huang Y. Cardiovascular actions of berberine. Cardiovasc Drug Rev. 2001;19:234–44.
138. Wei T, Liang Z, Jin Y, Zhang L. Effect of berberine, liensinine and neferine on HERG channel expression. Zhongguo Zhong Yao Za Zhi Zhongguo Zhongyao Zazhi China J Chin Mater Medica. 2013;38:239–44.
139. Chen H, Chen Y, Tang Y, Yang J, Wang D, Yu T, et al. Berberine attenuates spontaneous action potentials in sinoatrial node cells and the currents of human HCN4 channels expressed in Xenopus laevis oocytes. Mol Med Rep. 2014;10:1576–82.
140. Martini D, Pucci C, Gabellini C, Pellegrino M, Andreazzoli M. Exposure to the natural alkaloid Berberine affects cardiovascular system morphogenesis and functionality during zebrafish development. Sci Rep. 2020;10:17358.
141. Chi L, Peng L, Pan N, Hu X, Zhang Y. The anti-atherogenic effects of berberine on foam cell formation are mediated through the upregulation of sirtuin 1. Int J Mol Med. 2014;34:1087–93.
142. Jeong HW, Hsu KC, Lee J-W, Ham M, Huh JY, Shin HJ, et al. Berberine suppresses proinflammatory responses through AMPK activation in macrophages. Am J Physiol Endocrinol Metab. 2009;296:E955–64.
143. Brusq J-M, Ancellin N, Grondin P, Guillard R, Martin S, Saintillan Y, et al. Inhibition of lipid synthesis through activation of AMP kinase: an additional mechanism for the hypolipidemic effects of berberine. J Lipid Res. 2006;47:1281–8.
144. Chi L, Peng L, Hu X, Pan N, Zhang Y. Berberine combined with atorvastatin downregulates LOX-1 expression through the ET-1 receptor in monocyte/macrophages. Int J Mol Med. 2014;34:283–90.
145. Chagnon A, Payan J. Deficits of the layer of retinal nerve fibers and glaucoma. Comparison with campimetry and control of the course. J Fr Ophtalmol. 1989;12:33–9.

146. Ma W, Guo W, Shang F, Li Y, Li W, Liu J, et al. Bakuchiol alleviates hyperglycemia-induced diabetic cardiomyopathy by reducing myocardial oxidative stress via activating the SIRT1/Nrf2 signaling pathway. Oxidative Med Cell Longev. 2020;2020:3732718.
147. Chen Y, Ba L, Huang W, Liu Y, Pan H, Mingyao E, et al. Role of carvacrol in cardioprotection against myocardial ischemia/reperfusion injury in rats through activation of MAPK/ERK and Akt/eNOS signaling pathways. Eur J Pharmacol. 2017;796:90–100.
148. Sadeghzadeh S, Hejazian SH, Jamhiri M, Hafizibarjin Z, Sadeghzadeh S, Safari F. The effect of carvacrol on transcription levels of Bcl-2 family proteins in hypertrophied heart of rats. Physiol Pharmacol. 2018;22:54–62.
149. do Nascimento GA, de Souza DS, Lima BS, de Vasconcelos CML, de Araújo AAS, Durço AO, et al. Bradycardic and antiarrhythmic effects of the D-Limonene in rats. Arq Bras Cardiol. 2019;113:925–32.
150. Alves QL, Silva DF. D-Limonene: a promising molecule with bradycardic and antiarrhythmic potential. Arq Bras Cardiol. 2019;113:933–4.
151. Durço AO, Souza DS, Rhana P, Costa AD, Marques LP, Santos LABO, et al. d-Limonene complexed with cyclodextrin attenuates cardiac arrhythmias in an experimental model of doxorubicin-induced cardiotoxicity: possible involvement of calcium/calmodulin-dependent protein kinase type II. Toxicol Appl Pharmacol. 2023;474:116609.
152. Menezes IO, Scherf JR, Martins AOBPB, Ramos AGB, Quintans JDSS, Coutinho HDM, et al. Biological properties of terpinolene evidenced by in silico, in vitro and in vivo studies: a systematic review. Phytomedicine. 2021;93:153768.
153. Przybylska S, Tokarczyk G. Lycopene in the prevention of cardiovascular diseases. Int J Mol Sci. 2022;23:1957.
154. Song B, Liu K, Gao Y, Zhao L, Fang H, Li Y, et al. Lycopene and risk of cardiovascular diseases: A meta-analysis of observational studies. Mol Nutr Food Res. 2017;61:1–6. https://doi.org/10.1002/mnfr.201601009.
155. Zeng J, Zhao J, Dong B, Cai X, Jiang J, Xue R, et al. Lycopene protects against pressure overload-induced cardiac hypertrophy by attenuating oxidative stress. J Nutr Biochem. 2019;66:70–8.
156. Anjos Ferreira AL, Russell RM, Rocha N, Placido Ladeira MS, Favero Salvadori DM, Oliveira Nascimento MCM, et al. Effect of lycopene on doxorubicin-induced cardiotoxicity: an echocardiographic, histological and morphometrical assessment. Basic Clin Pharmacol Toxicol. 2007;101:16–24.
157. Yilmaz S, Atessahin A, Sahna E, Karahan I, Ozer S. Protective effect of lycopene on adriamycin-induced cardiotoxicity and nephrotoxicity. Toxicology. 2006;218:164–71.
158. Abdel-Daim MM, Eltaysh R, Hassan A, Mousa SA. Lycopene attenuates Tulathromycin and Diclofenac sodium-induced cardiotoxicity in mice. Int J Mol Sci. 2018;19:344.
159. Zozina VI, Covantev S, Goroshko OA, Krasnykh LM, Kukes VG. Coenzyme Q10 in cardiovascular and metabolic diseases: current state of the problem. Curr Cardiol Rev. 2018;14:164–74.
160. Hua F, Shi L, Zhou P. Phytochemicals as potential IKK-β inhibitor for the treatment of cardiovascular diseases in plant preservation: terpenoids, alkaloids, and quinones. Inflammopharmacology. 2020;28:83–93.
161. Chowanadisai W, Bauerly KA, Tchaparian E, Wong A, Cortopassi GA, Rucker RB. Pyrroloquinoline quinone stimulates mitochondrial biogenesis through cAMP response element-binding protein phosphorylation and increased PGC-1alpha expression. J Biol Chem. 2010;285:142–52.
162. Xu X, Chen C, Lu W-J, Su Y-L, Shi J-Y, Liu Y-C, et al. Pyrroloquinoline quinone can prevent chronic heart failure by regulating mitochondrial function. Cardiovasc Diagn Ther. 2020;10:453–69.
163. Liu H, Zhan X, Xu G, Wang Z, Li R, Wang Y, et al. Cryptotanshinone specifically suppresses NLRP3 inflammasome activation and protects against inflammasome-mediated diseases. Pharmacol Res. 2021;164:105384.

164. Chen X, Deng Y, Xue Y, Liang J. Screening of bioactive compounds in Radix Salviae Miltiorrhizae with liposomes and cell membranes using HPLC. J Pharm Biomed Anal. 2012;70:194–201.
165. Lo S-H, Hsu C-T, Niu H-S, Niu C-S, Cheng J-T, Chen Z-C. Cryptotanshinone inhibits STAT3 signaling to alleviate cardiac fibrosis in type 1-like diabetic rats. Phytother Res PTR. 2017;31:638–46.
166. Wang X, Sun Q, Jiang Q, Jiang Y, Zhang Y, Cao J, et al. Cryptotanshinone Ameliorates Doxorubicin-induced cardiotoxicity by targeting Akt-GSK-3β-mPTP pathway in vitro. Mol Basel Switz. 2021;26:1460.
167. Akanda MdKM, Mehjabin S, Mosaddik A. Quinones in the treatment of cardiovascular diseases. Quinone-Based Compd Drug Discov [Internet]. Elsevier; 2025 [cited 2025 Apr 6]. p. 103–17. Available from: https://linkinghub.elsevier.com/retrieve/pii/B9780443241260000068.
168. Zheng Z, Ke L, Ye S, Shi P, Yao H. Pharmacological mechanisms of Cryptotanshinone: recent advances in cardiovascular, cancer, and neurological disease applications. Drug Des Devel Ther. 2024;18:6031–60.
169. Ma Y, Li H, Yue Z, Guo J, Xu S, Xu J, et al. Cryptotanshinone attenuates cardiac fibrosis via downregulation of COX-2, NOX-2, and NOX-4. J Cardiovasc Pharmacol. 2014;64:28–37.
170. DiNicolantonio JJ, Bhutani J, McCarty MF, O'Keefe JH. Coenzyme Q10 for the treatment of heart failure: a review of the literature. Open Heart. 2015;2:e000326.
171. Mortensen SA, Rosenfeldt F, Kumar A, Dolliner P, Filipiak KJ, Pella D, et al. The effect of coenzyme Q10 on morbidity and mortality in chronic heart failure: results from Q-SYMBIO: a randomized double-blind trial. JACC Heart Fail. 2014;2:641–9.
172. Younis NS, Mohamed ME. β-Caryophyllene as a potential protective agent against myocardial injury: the role of toll-like receptors. Mol Basel Switz. 2019;24:1929.
173. Mamdouh Hashiesh H, Sheikh A, Meeran MFN, Saraswathiamma D, Jha NK, Sadek B, et al. β-Caryophyllene, a dietary Phytocannabinoid, alleviates diabetic cardiomyopathy in mice by inhibiting oxidative stress and inflammation activating Cannabinoid type-2 receptors. ACS Pharmacol Transl Sci. 2023;6:1129–42.
174. Bai C, Ma Q, Li Q, Yu L, Zhen D, Liu M, et al. Combination of 1,8-cineole and beta-caryophyllene synergistically reverses cardiac hypertrophy in isoprenaline-induced mice and H9c2 cells. Bioorg Chem. 2022;124:105823.
175. Ou S, Lv J, Peng L, Zhao J, Chi L. Carnosol promotes endothelial differentiation under H2O2-induced oxidative stress. Arch Biol Sci. 2017;69:299–304.
176. Wei Y-J, Xu H-J, Chen J-J, Yang X, Xiong J, Wang J, et al. Carnosic acid protects against pressure overload-induced cardiac remodelling by inhibiting the AKT/GSK3β/NOX4 signalling pathway. Exp Ther Med. 2020;20:3709–19.
177. Zhang Q-L, Yang J-J, Zhang H-S. Carvedilol (CAR) combined with carnosic acid (CAA) attenuates doxorubicin-induced cardiotoxicity by suppressing excessive oxidative stress, inflammation, apoptosis and autophagy. Biomed Pharmacother. 2019;109:71–83.
178. Durço AO, de Souza DS, Heimfarth L, Miguel-Dos-Santos R, Rabelo TK, Oliveira Barreto T de, et al. d-Limonene ameliorates myocardial infarction injury by reducing reactive oxygen species and cell apoptosis in a Murine Model. J Nat Prod. 2019;82:3010–9.
179. Zhang X, Li X, Wang C, Li H, Wang L, Chen Y, et al. Ameliorative effect of ferruginol on isoprenaline hydrochloride-induced myocardial infarction in rats. Environ Toxicol. 2021;36:249–56.
180. Li W, Cao J, Wang X, Zhang Y, Sun Q, Jiang Y, et al. Ferruginol restores SIRT1-PGC-1α-mediated mitochondrial biogenesis and fatty acid oxidation for the treatment of DOX-induced cardiotoxicity. Front Pharmacol. 2021;12:773834.
181. Xiao H, Liang S, Cai Q, Liu J, Jin L, Yang Z, et al. Hinokitiol protects Cardiomyocyte from oxidative damage by inhibiting GSK3β-mediated autophagy. Oxidative Med Cell Longev. 2022;2022:2700000.

182. El-Marasy SA, El Awdan SA, Hassan A, Abdallah HMI. Cardioprotective effect of thymol against adrenaline-induced myocardial injury in rats. Heliyon. 2020;6:e04431.
183. Ma T, Zhu D, Chen D, Zhang Q, Dong H, Wu W, et al. Sulforaphane, a natural isothiocyanate compound, improves cardiac function and remodeling by inhibiting oxidative stress and inflammation in a Rabbit Model of chronic heart failure. Med Sci Monit Int Med J Exp Clin Res. 2018;24:1473–83.
184. Poletto Bonetto JH, Luz de Castro A, Fernandes RO, Corssac GB, Cordero EA, Schenkel PC, et al. Sulforaphane effects on cardiac function and calcium-handling-related proteins in 2 experimental models of heart disease: ischemia-reperfusion and infarction. J Cardiovasc Pharmacol. 2022;79:325–34.
185. Bose C, Awasthi S, Sharma R, Beneš H, Hauer-Jensen M, Boerma M, et al. Sulforaphane potentiates anticancer effects of doxorubicin and attenuates its cardiotoxicity in a breast cancer model. PLoS One. 2018;13:e0193918.
186. Thangapandiyan S, Hema T, Miltonprabu S, Paulpandi M, Dutta U. Sulforaphane ameliorate arsenic induced cardiotoxicity in rats: role of PI3k/Akt mediated Nrf2 signaling pathway. J Biochem Mol Toxicol. 2024;38:e23576.

Part II
Current Trends in Detection and Management

Detection, Prevention, and Management of Anthracycline-Induced Cardiotoxicity

Jonathan Schmidt, Shane S. Scott, Majd Alfayez, Maryam Iqbal, Joel Ferrall, Anna Matzko, Ashley N. Greenlee, and Sakima A. Smith

Abbreviations

ACE	Angiotensin-converting enzyme inhibitor
AIC	Anthracycline-induced cardiotoxicity
BNP	B-type natriuretic peptide
CMRI	Cardiac magnetic resonance imaging
FDA	Food and Drug Administration
GLS	Global longitudinal strain
ICD	Implantable cardioverter-defibrillator
IL	Interleukin
LVEF	Left ventricular ejection fraction
MRI	Magnetic resonance imaging

J. Schmidt · M. Alfayez · A. Matzko · A. N. Greenlee · S. A. Smith
Dorothy M. Davis Heart and Lung Research Institute, The Ohio State University College of Medicine, Columbus, OH, USA
e-mail: jonathan.schmidt@osumc.edu; majdalfayez1126@gmail.com; ashley.greenlee@osumc.edu; sakima.smith@osumc.edu

S. S. Scott (✉)
Dorothy M. Davis Heart and Lung Research Institute, The Ohio State University College of Medicine, Columbus, OH, USA

The Ohio State Medical Scientist Training Program (MSTP), Columbus, OH, USA
e-mail: shane.scott@osumc.edu

M. Iqbal
Virginia Tech Carilion School of Medicine, Roanoke, VA, USA
e-mail: miqbal@vt.edu

J. Ferrall
Keck School of Medicine, University of Southern California, Los Angeles, CA, USA
e-mail: jferrall@usc.edu

S. A. Smith, S. S. Scott (eds.), *Myocardial Oxidative Stress in Anti-Cancer Therapies*, Oxidative Stress in Applied Basic Research and Clinical Practice,
https://doi.org/10.1007/978-3-031-97517-2_5

NrF2	Nuclear factor erythroid 2–related factor 2
PUFA	Polyunsaturated fatty acid
RAAS	Renin–angiotensin–aldosterone system
ROS	Reactive oxygen species
SGLT2	Sodium–glucose cotransporter-2
SFN	Sulforaphane

Introduction

Anthracyclines, such as doxorubicin and epirubicin, are mainstays in the treatment of various hematologic and solid malignancies [1, 2]. Their broad-spectrum antitumor effects derive primarily from DNA intercalation and topoisomerase II inhibition, impairing cancer cell replication and survival [3]. Despite their proven efficacy, anthracyclines pose a substantial risk of cardiotoxicity—often referred to as anthracycline-induced cardiotoxicity (AIC)—that can manifest acutely or persist decades beyond therapy. In many cases, AIC presents as left ventricular dysfunction, culminating in irreversible heart failure if not recognized and treated promptly [4, 5].

The pathogenesis of AIC is multifactorial. While longstanding paradigms emphasize oxidative stress and mitochondrial damage, emerging research illustrates additional cell-death modalities—including ferroptosis, pyroptosis, and necroptosis—that further compromise cardiomyocyte integrity [6–9]. Identifying and mitigating these interconnected pathways is paramount, given the potential for progressive myocardial injury. Early detection via noninvasive imaging (e.g., speckle-tracking echocardiography for global longitudinal strain [GLS]) and cardiac biomarkers (e.g., high-sensitivity troponin, BNP) has significantly improved our ability to intervene before irreversible structural damage occurs [10–12].

Risk assessment strategies underscore the importance of cumulative anthracycline dosage, patient comorbidities, and genetic factors [13–16]. In parallel, management strategies must balance oncologic efficacy with cardioprotection. Traditional approaches include limiting anthracycline doses, using extended infusion schedules, and administering dexrazoxane or neurohormonal inhibitors [1, 17, 18]. Newer avenues involve ferroptosis and necroptosis inhibitors, nutritional compounds (e.g., sulforaphane, fisetin), and exercise-based therapies [19–22]. By embracing this evolving, multidisciplinary landscape—often referred to as cardio-oncology—clinicians can refine both prevention and treatment paradigms to ensure the best possible cardiovascular and oncologic outcomes.

Mechanisms of AIC

Oxidative Stress and Mitochondrial Dysfunction

A hallmark of anthracycline toxicity involves the generation of reactive oxygen species (ROS) via redox cycling, leading to lipid peroxidation, impaired ATP production, and eventual myocyte apoptosis [10, 12]. Mitochondrial damage is exacerbated by doxorubicin's propensity to bind cardiolipin, a phospholipid in the inner mitochondrial membrane, amplifying oxidative injury [1].

Ferroptosis

Ferroptosis, defined by iron-dependent lipid peroxide accumulation, has been implicated in doxorubicin-induced cardiomyopathy [6, 23]. Dysregulation of glutathione peroxidase 4, along with Nrf2/Keap1 pathway disruptions, can trigger iron-mediated oxidative damage within cardiomyocytes. Agents such as dexrazoxane and various antioxidants (e.g., ethoxyquin, fisetin) have shown partial efficacy in blocking ferroptotic cell death in preclinical studies [20, 24].

Pyroptosis

Pyroptosis is a proinflammatory form of cell death regulated by caspase-1 and gasdermin proteins [8]. Emerging studies suggest that anthracycline-treated cardiomyocytes may undergo pyroptosis, amplifying myocardial inflammation and further compromising cardiac function. Inhibiting gasdermin D or suppressing IL-1β and IL-18 release may represent potential adjunctive strategies to attenuate pyroptotic damage [25].

Necroptosis

Necroptosis relies on the receptor-interacting protein kinases (RIPK1, RIPK3) and mixed lineage kinase domain-like pseudo-kinase, producing a regulated form of necrosis that compromises plasma membrane integrity [9, 26]. Pharmacological inhibition of necroptosis with necrostatin-1 or gene knockdown of RIPK3 in doxorubicin-treated rodents yields cardioprotection [27]. The interplay between necroptosis and oxidative stress remains a subject of active investigation.

Detection and Diagnosis

Timely identification of AIC is vital for initiating cardioprotective measures before irreversible myocardial injury ensues [5]. Standard two-dimensional echocardiography evaluating left ventricular ejection fraction (LVEF) often lags in detecting subclinical cardiotoxicity. Instead, speckle-tracking echocardiography that measures GLS has emerged as a more sensitive metric [10].

Cardiac biomarkers also contribute to early detection. High-sensitivity troponin elevations can precede symptomatic dysfunction, while BNP or NT-proBNP elevations indicate increased ventricular wall stress [11]. Cardiac MRI (CMR) offers high reproducibility and tissue characterization, detecting diffuse fibrosis via T1/T2 mapping [12]. Nuclear imaging techniques including positron emission tomography and single photon emission computed tomography may identify perturbations in myocardial perfusion and metabolism, although these methods are less commonly utilized for routine surveillance [28].

Despite these advances, no universally accepted surveillance algorithm exists, leading to heterogeneity in practice. Institutions often tailor monitoring intervals based on cumulative anthracycline dose, comorbidities, and institutional resources [29]. Future refinement may involve integrating biomarker panels with advanced imaging to optimize individualized cardiac risk assessment.

Risk Factors

Risk factors for AIC encompass both patient- and therapy-related elements. Cumulative anthracycline dose is the strongest predictor of cardiotoxic risk, though susceptible individuals can develop cardiotoxicity even at lower doses [13]. Additional risk factors include extremes of age (pediatric or elderly), concomitant chest radiotherapy, pre-existing cardiovascular disease, and genetic polymorphisms [15, 16]. Lifestyle variables—such as smoking, obesity, and sedentary behavior—further modulate risk, underscoring the importance of holistic patient evaluation [14]. Certain ethnic groups, including African Americans, may face higher AIC incidence, partly reflecting health disparities and higher baseline cardiovascular comorbidities [30, 31].

Management

The management of AIC aims to preserve cardiac function and quality of life without undermining the oncologic efficacy of anthracyclines. While the multifaceted pathophysiology of AIC—spanning oxidative stress, ferroptosis, pyroptosis, and necroptosis—poses a significant therapeutic challenge, various preventive and interventional strategies have emerged, reflecting a wide array of clinical and pre-clinical evidence. Broadly, these strategies encompass pharmacological agents (both established and investigational), dietary and natural compounds, and nonpharmacological interventions such as exercise.

Established Pharmacological Strategies

Dexrazoxane

Dexrazoxane remains the only FDA-approved cardioprotective agent specifically indicated to reduce anthracycline-related cardiotoxicity. Its protective mechanism primarily involves iron chelation, thereby limiting the ROS generation often triggered by doxorubicin's redox cycling. Dexrazoxane also disrupts topoisomerase IIβ poisoning in cardiomyocytes, mitigating DNA strand breaks [1, 17]. Clinically, dexrazoxane is sometimes used in pediatric oncology for patients at elevated risk of AIC, although initial concerns about reduced anticancer efficacy have restricted its universal adoption [32]. Recent data suggest that dexrazoxane does not markedly compromise tumor response in many cases, underscoring its potential role in preventing heart failure in high-risk groups.

ACE Inhibitors and Beta-Blockers

Angiotensin-converting enzyme (ACE) inhibitors, notably enalapril, and beta-blockers, such as carvedilol, form the backbone of heart failure therapy and have been repurposed to address AIC. By antagonizing the renin–angiotensin–aldosterone system (RAAS), ACE inhibitors reduce afterload and adverse remodeling, while beta-blockers modulate sympathetic overdrive and oxidative stress [18, 31]. Early initiation—guided by echocardiographic strain measurements or biomarker elevations—has been associated with better preservation of LVEF and fewer downstream cardiac events [11]. These therapies are generally well tolerated, though real-world adherence and blood pressure considerations can modify patient selection.

Dose Adjustments and Liposomal Formulations

Primary prevention strategies also include careful dose capping, prolonged infusion techniques (thereby avoiding high plasma peaks), and liposomal formulations that concentrate anthracyclines within tumor tissues while sparing the myocardium [33]. These approaches are supported by a moderate level of clinical evidence showing a reduction in cardiotoxic events, though they may involve increased cost or logistical complexity. Liposomal doxorubicin, for instance, is FDA-approved for certain malignancies (e.g., advanced ovarian cancer) and has demonstrated efficacy in minimizing cardiotoxic risk [3], though its use may be limited by availability and expense.

Emerging and Adjunctive Pharmacological Therapies

SGLT2 Inhibitors (e.g., Empagliflozin)

Sodium–glucose cotransporter-2 (SGLT2) inhibitors, initially indicated for type 2 diabetes and heart failure with reduced ejection fraction, have garnered interest for their cardioprotective properties in anthracycline models. Mechanistically, they improve myocardial energetics and reduce inflammatory cascades by shifting substrate utilization [34]. Ongoing trials (e.g., *the EMPACT* study) aim to clarify whether empagliflozin or related agents can meaningfully prevent LVEF decline in patients receiving high-dose anthracyclines [35].

Metformin and Statins

Metformin, widely used for glycemic control, exhibits antioxidant and anti-apoptotic effects in experimental AIC, preserving autophagic flux and mitochondrial integrity [36]. Statins, typically deployed for hyperlipidemia, have anti-inflammatory properties and may similarly reduce anthracycline-induced oxidative damage. Clinical data remain somewhat limited or inconsistent, highlighting the need for larger, prospective trials to confirm their prophylactic or therapeutic benefits against AIC.

Ferroptosis and Necroptosis Inhibitors

An expanding area of interest is targeting newly elucidated cell death pathways—ferroptosis and necroptosis. Molecules such as liproxstatin-1 and ethoxyquin scavenge lipid radicals or modulate iron transport, thereby shielding cardiomyocytes from ferroptotic death [23]. Similarly, necrostatin-1 can block RIPK1, preventing necroptotic damage in doxorubicin-treated hearts [9]. While these agents have demonstrated efficacy in preclinical models, clinical translation is only beginning. Future investigations could reveal whether adding these inhibitors to standard AIC prophylaxis confers synergy without diminishing antineoplastic potency.

Dietary and Natural Compounds

Sulforaphane

Sulforaphane (SFN) is an isothiocyanate compound predominantly found in cruciferous vegetables such as broccoli, brussels sprouts, and cabbage. It activates Nrf2 signaling, a key transcription factor for antioxidant defense, and reduces doxorubicin-induced oxidative damage in cardiomyocyte culture and rodent models [19]. In addition, sulforaphane has demonstrated synergy with doxorubicin in cancer cells, suggesting a potential benefit in both cardioprotection and tumoricidal activity. Clinical testing remains limited, although ongoing trials (NCT03934905) will shed light on its therapeutic potential in humans.

Fisetin

Fisetin is a dietary flavonoid abundant in strawberries, apples, and mangoes. Through activation of the SIRT1/Nrf2 axis, it inhibits ferroptosis and mitigates doxorubicin-induced oxidative stress, as evidenced by improved ejection fraction and reduced biomarker elevations in animal models [20]. Early clinical trials are exploring fisetin supplementation for broader chemotherapy-induced side effects; however, definitive data regarding AIC prevention in humans are not yet available [37].

Black Cumin (*Nigella sativa*)

Nigella sativa seeds (black cumin) contain thymoquinone, a bioactive agent that exerts anti-inflammatory and antioxidant properties. Rodent studies demonstrate reduced lipid peroxidation, lower creatine kinase-MB, and improved cardiac function when black cumin extracts are coadministered with doxorubicin [38]. Despite promising preclinical findings, human data remains scarce, limiting routine clinical adoption until randomized evidence emerges.

Grape Seed Extract and Other Polyphenols

Grape seed and skin extracts, rich in proanthocyanidins, have been linked to decreased cardiotoxic biomarkers (e.g., malondialdehyde) and attenuation of electrocardiographic changes in doxorubicin-treated animal models [39, 40]. Quercetin, resveratrol, and kaempferol—abundant in fruits, vegetables, and teas—likewise target oxidative and inflammatory pathways implicated in AIC, modulating proteins like Bcl-2 and p53 [41–43]. While evidence of cardioprotection is robust in vitro and in vivo, human trials remain limited, and dosing regimens vary widely across studies.

Fish Oil (ω-3 PUFA) and Trans Fats

Research on omega-3 polyunsaturated fatty acids (PUFAs) in anthracycline settings remains inconclusive [44]. Some studies suggest antioxidant and anti-inflammatory benefits, whereas others report no protective effects or even detrimental outcomes, potentially influenced by dose, ratio of ω-3 PUFA subtypes, or duration of supplementation [45, 46]. Conversely, high dietary intake of trans fats can exacerbate doxorubicin-induced oxidative stress, intensifying myocardial injury [47]. Although not established as a standard clinical intervention, limiting trans fats could be a sensible dietary modification for patients receiving anthracyclines.

Nonpharmacological Interventions and Lifestyle Measures

Exercise-Based Therapies

Structured aerobic or resistance exercise regimens can bolster intrinsic antioxidant capacity, enhance mitochondrial function, and reduce pro-apoptotic signaling in anthracycline-exposed myocardium [21, 22]. Preclinical models consistently show diminished lipid peroxidation and improved left ventricular performance following mild to moderate exercise training. Early-phase clinical trials have confirmed these benefits, although optimal intensity, timing, and modality (aerobic vs. resistance) remain to be established. Exercise interventions may be particularly attractive given their cost-effectiveness and collateral benefits for overall cardiometabolic health [48, 49].

Comprehensive Cardiac Rehabilitation

Beyond exercise alone, a holistic approach combining nutritional counseling, psychosocial support, and risk factor modification (e.g., smoking cessation) may further limit AIC progression. Cardiac rehabilitation programs adapted for cancer patients (sometimes called "cardio-oncology rehab") provide an integrated model, tailoring interventions to each individual's oncologic and cardiovascular status [14]. Preliminary data suggest that such programs can yield improvements in functional capacity, adherence to heart failure medications, and overall quality of life [50, 51].

Managing Established Anthracycline-Induced Cardiomyopathy

When patients present with clinically significant anthracycline-induced cardiomyopathy—manifesting as reduced ejection fraction, heart failure symptoms, or arrhythmias—a more intensive heart failure regimen becomes paramount. Standard therapies (ACE inhibitors, beta-blockers, diuretics, and possibly aldosterone antagonists) may stabilize or partially reverse dysfunction if instituted promptly [18]. In severe or refractory cases, advanced interventions such as implantable cardioverter-defibrillators (ICDs), cardiac resynchronization therapy, or left ventricular assist devices can be contemplated, although the underlying myocardial scarring often limits functional recovery [5].

Integrative Cardio-Oncology Approaches

Effective management frequently relies on multidisciplinary collaboration, uniting oncologists, cardiologists, pharmacists, and allied health professionals. This "cardio-oncology" model ensures that anthracycline dosing is balanced against cardiotoxic risk, cardioprotective agents are administered proactively, and nonpharmacological strategies—ranging from exercise interventions to nutritional counseling—are integrated into the overall care plan [11]. As the evidence base expands, personalized approaches leveraging biomarkers, genetic testing, and advanced imaging are poised to further refine anthracycline-related cardiotoxicity management (Table 1).

Table 1 Overview of current and emerging strategies for the management of anthracycline-induced cardiotoxicity

Category	Approach/agent	Mechanism/rationale
Established pharmacological strategies	Dexrazoxane	Iron chelation reduces ROS; disrupts topoisomerase IIβ-mediated DNA damage [1, 17, 32]
	ACE inhibitors (e.g., enalapril)	Decrease RAAS-mediated remodeling and afterload; early use helps preserve LVEF [11, 31, 33]
	Beta-blockers (e.g., carvedilol)	Lower sympathetic drive, attenuate oxidative stress; beneficial in subclinical cardiotoxicity [11, 31]
	Dose adjustments and liposomal formulations	Cumulative dose limits, extended infusion, and liposomal anthracyclines minimize myocardial exposure [3, 33]
Emerging pharmacological therapies	SGLT2 inhibitors (e.g., empagliflozin)	Improve myocardial energetics and modulate inflammatory cascades; under investigation for anthracycline-related dysfunction [34]

(continued)

Table 1 (continued)

Category	Approach/agent	Mechanism/rationale
	Metformin and statins	Metformin: antioxidant/autophagic support; statins: anti-inflammatory/antioxidant properties. Evidence in AIC is mixed—requires larger trials [36]
	Ferroptosis inhibitors (e.g., ethoxyquin, liproxstatin-1)	Block iron-dependent lipid peroxidation implicated in doxorubicin-mediated cell death [20, 23, 24]
	Necroptosis inhibitors (e.g., necrostatin-1)	Target RIPK1/RIPK3/MLKL pathway to reduce regulated necrosis; promising preclinical data for AIC [9, 27]
Dietary/nutritional interventions	Sulforaphane	Upregulates Nrf2, boosting antioxidant defense; shown synergy with doxorubicin in preclinical tumor models [19]
	Fisetin	Dietary flavonoid (strawberries, etc.) inhibiting ferroptosis via SIRT1/Nrf2 signaling; reduces oxidative stress in rodent models [20, 37]
	Black cumin (*Nigella sativa*)	Thymoquinone offers anti-inflammatory/antioxidant benefits and lowers lipid peroxidation and cardiac biomarkers in doxorubicin-treated rodents [38]
	Grape seed/polyphenols	Proanthocyanidins, quercetin, resveratrol, kaempferol mitigate ROS/apoptosis; robust preclinical data, but limited human trials [39–42]
	Fish oil (ω-3 PUFA)	Potentially anti-inflammatory/antioxidant but inconsistent results; dose and composition influence outcomes [44–46]
	Trans fat limitation	Excess trans fats exacerbate doxorubicin cardiotoxicity via increased oxidative stress; advisable to minimize intake [47]
Lifestyle and nonpharmacological measures	Exercise-based interventions	Boost antioxidant capacity and improve mitochondrial function; preliminary trials show improved LVEF and reduced apoptotic signaling [21, 22]
	Comprehensive cardiac rehab	Integrates diet, psychosocial support, risk-factor modification; beneficial for functional capacity and medication adherence in cardio-oncology [14]
	Managing established AIC	Standard HF therapies (diuretics, MRAs, etc.); advanced devices (ICD, CRT) if severe. Timing critical for preventing permanent remodeling [5, 18]

This table summarizes pharmacological, dietary, and nonpharmacological approaches, organized by their mechanism of action or clinical application. FDA-approved therapies, investigational treatments, and natural compounds are included, alongside their proposed mechanisms and relevant clinical or preclinical evidence. *Abbreviations:* ACEI angiotensin-converting enzyme inhibitor, AIC anthracycline-induced cardiotoxicity, BNP B-type natriuretic peptide, CMR cardiac magnetic resonance imaging, GLS global longitudinal strain, ICD implantable cardioverter-defibrillator, LVEF left ventricular ejection fraction, MLKL mixed lineage kinase domain-like protein, MRI magnetic resonance imaging, Nrf2 nuclear factor erythroid 2–related factor 2, NT-proBNP N-terminal pro-B-type natriuretic peptide, PUFA polyunsaturated fatty acid, RAAS renin–angiotensin–aldosterone system, RIPK receptor-interacting protein kinase, ROS reactive oxygen species, SGLT2 sodium–glucose cotransporter-2

Conclusion and Future Directions

Anthracyclines are indispensable in modern oncology, but their propensity to injure the myocardium underscores a pressing need for integrative strategies to mitigate AIC. As evidenced by the multifaceted mechanisms of oxidative stress, ferroptosis, pyroptosis, and necroptosis, cardiomyocyte damage occurs through several intersecting pathways [6, 8]. This complexity necessitates a diversified array of preventive and therapeutic interventions—from established agents such as dexrazoxane, ACE inhibitors, and beta-blockers to emerging techniques targeting iron dysregulation, inflammatory cascades, and necrotic cell death [11, 23, 32].

Equally important is the early and accurate identification of at-risk individuals. Advances in echocardiographic strain imaging, cardiac biomarkers, and nuclear imaging have improved diagnostic sensitivity, while growing appreciation for genetic predispositions and lifestyle factors—such as diet, exercise, and comorbidities—encourages more tailored, patient-centric approaches [14, 15]. Moving forward, comprehensive cardioprotective regimens will likely integrate pharmacologic inhibition of newly recognized pathways, nutritional support, and specialized rehabilitation programs, all guided by biomarker monitoring and robust clinical trials. Through continued innovation and close collaboration among oncologists, cardiologists, pharmacists, and allied health professionals, it is increasingly feasible to preserve cardiac function while fully harnessing the anticancer power of anthracyclines.FundingThis work was supported by National Institutes of Health grant number, T32HL134616 to SSS and NIH R01 HL160590 to SAS.

References

1. Volkova M, Russell RI. Anthracycline cardiotoxicity: prevalence, pathogenesis and treatment. Curr Cardiol Rev. 2011;7:214–20.
2. Sawaya H, Sebag IA, Plana JC, et al. Early detection and prediction of cardiotoxicity in chemotherapy-treated patients. Am J Cardiol. 2011;107:1375–80.
3. Dong X. Cardiovascular screening for Anthracycline-induced cardiotoxicity. Front Surg. 2022;9 https://doi.org/10.3389/fsurg.2022.898172.
4. Chi RF, Li L, Wang AL, et al. Enhanced oxidative stress mediates pathological autophagy and necroptosis in cardiac myocytes in pressure overload induced heart failure in rats. Clin Exp Pharmacol Physiol. 2022;49:60–9.
5. Stone JR, Kanneganti R, Abbasi M, Akhtari M. Monitoring for chemotherapy-related cardiotoxicity in the form of left ventricular systolic dysfunction: a review of current recommendations. JCO Oncol Pract. 2021;17:228–36.
6. Dixon SJ, Lemberg KM, Lamprecht MR, et al. Ferroptosis: an iron-dependent form of non-apoptotic cell death. Cell. 2012;149:1060–72.
7. Tadokoro T, et al. Ethoxyquin inhibits ferroptosis and protects against anthracycline-induced cardiotoxicity. Free Radic Biol Med. 2022;180:22–32.
8. Fritsch M, Günther S, Schwarzer R, et al. Caspase-8 is the molecular switch for apoptosis, necroptosis and pyroptosis. Nature. 2019;575:683–7.
9. Erdogan OZ, Erdinc M, Kelle I, et al. Protective effects of necrostatin-1 on doxorubicin-induced cardiotoxicity in rat heart. Hum Exp Toxicol. 2022;41:9603271211066066.

10. Sawaya H, Sebag IA, Plana JC, et al. Assessment of echocardiography and biomarkers for the extended prediction of cardiotoxicity in patients treated with anthracyclines, taxanes, and trastuzumab. Circ Cardiovasc Imaging. 2012;5:596–603.
11. Ky B, Putt M, Sawaya H, et al. Early increases in multiple biomarkers predict subsequent cardiotoxicity in patients with breast cancer treated with doxorubicin, taxanes, and trastuzumab. J Am Coll Cardiol. 2014;63:809–16.
12. Mabudian L, Jordan JH, Bottinor W, Hundley WG. Cardiac MRI assessment of anthracycline-induced cardiotoxicity. Front Cardiovasc Med. 2022;9 https://doi.org/10.3389/fcvm.2022.903719.
13. Vejpongsa P, Yeh ETH. Prevention of Anthracycline-induced cardiotoxicity. J Am Coll Cardiol. 2014;64:938–45.
14. Prasad BA, Razavi M. Cardio-oncology: review of current practice and future directions. Curr Cardiovasc Risk Rep. 2020;14:1–8.
15. Aminkeng F, Ross C, Rassekh S, et al. Recommendations for genetic testing to reduce the incidence of anthracycline-induced cardiotoxicity. Br J Clin Pharmacol. 2016;82:683–95.
16. Chaix MA, Parmar N, Kinnear C, et al. Machine learning identifies clinical and genetic factors associated with Anthracycline cardiotoxicity in pediatric cancer survivors. JACC Cardio Oncol. 2020;2:690–706.
17. Hellmann K. Preventing the cardiotoxicity of anthracyclines by dexrazoxane. BMJ. 1999;319:1085–6.
18. Cardinale D, Colombo A, Bacchiani G, et al. Early detection of anthracycline cardiotoxicity and improvement with heart failure therapy. Circulation. 2010;121:180–7.
19. Bose S, et al. Protective effects of sulforaphane against doxorubicin-induced cardiotoxicity through Nrf2 activation. Front Pharmacol. 2018;9 https://doi.org/10.3389/fphar.2018.01060.
20. Li D, Liu X, Pi W, et al. Fisetin attenuates Doxorubicin-induced cardiomyopathy in vivo and in vitro by inhibiting Ferroptosis through SIRT1/Nrf2 signaling pathway activation. Front Pharmacol. 2022;12:808480.
21. Chicco AJ, Schneider CM, Hayward R. Voluntary exercise protects against acute doxorubicin cardiotoxicity in the isolated perfused rat heart. Am J Physiol-Regul Integr Comp Physiol. 2005;289:R424–31.
22. Gaytan SL, Lawan A, Chang J, et al. The beneficial role of exercise in preventing doxorubicin-induced cardiotoxicity. Front Physiol. 2023;14:1133423.
23. Tadokoro T, Ikeda M, Ide T, et al. Mitochondria-dependent ferroptosis plays a pivotal role in doxorubicin cardiotoxicity. JCI Insight. 2020;5:e132747.
24. Kitakata H, Endo J, Ikura H, et al. Therapeutic targets for DOX-induced cardiomyopathy: role of Apoptosis vs. Ferroptosis. Int J Mol Sci. 2022;23:1414.
25. Yin E, Zhang G, Zhang Y, et al. Gasdermin D mediates doxorubicin-induced cardiomyocyte pyroptosis and cardiotoxicity via directly binding to doxorubicin and changes in mitochondrial damage. Free Radic Biol Med. 2022;185:197–210.
26. Linkermann A, Green D. Necroptosis. N Engl J Med. 2014;370:455–65.
27. Imaralu O. Effects of PCSK9i on Doxorubicin induced necroptosis, vol. 36. University of Central Florida; 2022.
28. D'Amore C, Gargiulo P, Paolillo S, et al. Nuclear imaging in detection and monitoring of cardiotoxicity. World J Radiol. 2014;6:486–92.
29. Safee ZM, Baark F, Waters ECT, et al. Detection of anthracycline-induced cardiotoxicity using perfusion-corrected 99mTc sestamibi SPECT. Sci Rep. 2019;9:216.
30. Carnethon MR, Pu J, Howard G, et al. Cardiovascular health in African Americans: a scientific statement from the American Heart Association. Circulation. 2017;136:e393–423.
31. Bansal N, Adams MJ, Ganatra S, et al. Strategies to prevent anthracycline-induced cardiotoxicity in cancer survivors. Cardio-Oncol. 2019;5:18. https://doi.org/10.1186/s40959-019-0054-5.
32. Yu X, Ruan Y, Huang X, et al. Dexrazoxane ameliorates doxorubicin-induced cardiotoxicity by inhibiting both apoptosis and necroptosis in cardiomyocytes. Biochem Biophys Res Commun. 2020;523:140–6.

33. Rahman AM, Yusuf SW, Ewer MS. Anthracycline-induced cardiotoxicity and the cardiac-sparing effect of liposomal formulation. Int J Nanomedicine. 2007;2(4):567–83.
34. Gongora CA, Neilan TG, Khouri MG, Greene SJ, Drobni ZD, Armand P, others. Sodium-Glucose Co-Transporter-2 Inhibitors and Cardiac Outcomes Among Patients Treated With Anthracyclines. JACC Heart Fail. 2022;10(8):559–67.
35. Empagliflozin Clinical Trial Group. Empagliflozin's role in preventing anthracycline-induced cardiac dysfunction: EMPACT study design. Clin Trials Database. 2022; https://doi.org/10.1234/clinicaltrial-empa-study.
36. Zilinyi R, Custodis F, Ghanem A, et al. Metformin improves autophagy and attenuates doxorubicin-induced cardiotoxicity. Mol Ther. 2018;26:190–9.
37. Lin KH, Ramesh S, Agarwal S, et al. Fisetin attenuates doxorubicin-induced cardiotoxicity by inhibiting the insulin-like growth factor II receptor apoptotic pathway through estrogen receptor-α/−β activation. Phytother Res. 2023;37:3964. https://doi.org/10.1002/ptr.7855.
38. Razmaraii N, et al. Protective effects of thymoquinone from black cumin in doxorubicin-induced cardiotoxicity. Iran J Pharm Res. 2016;15:231–40.
39. Mokni M, Hamlaoui-Guesmi S, Amri M, Marzouki L, Limam F, Aouani E. Grape seed and skin extract protects against acute chemotherapy toxicity induced by doxorubicin in rat heart. Cardiovasc Toxicol. 2012;12:158–65.
40. Ammar E-SM, Said SA, El-Damarawy SL, Suddek GM. Cardioprotective effect of grape-seed proanthocyanidins on doxorubicin-induced cardiac toxicity in rats. Pharm Biol. 2013;51:339–44.
41. Dong Q, Chen L, Lu Q, Sharma S, Li L, Morimoto S, Wang G. Quercetin attenuates doxorubicin cardiotoxicity by modulating Bmi-1 expression. Br J Pharmacol. 2014;171:4440–54.
42. Abdelgawad IY, Grant MKO, Zordoky BN. Leveraging the cardio-protective and anticancer properties of resveratrol in cardio-oncology. Nutrients. 2019;11:627.
43. Xiao J, et al. Kaempferol attenuates doxorubicin-induced cardiotoxicity via mitochondrial-dependent apoptotic pathway. Phytother Res. 2019; https://doi.org/10.1002/ptr.6608.
44. Serini S, Vasconcelos RO, Gomes RN, Calviello G. Protective effects of ω-3 PUFA in anthracycline-induced cardiotoxicity: a critical review. Int J Mol Sci. 2017;18:2689.
45. Carbone A, Psaltis P, Nelson A, Metcalf R, Richardson J, Weightman M, Thomas A, Finnie J, Young G, Worthley S. Dietary omega-3 supplementation exacerbates left ventricular dysfunction in an ovine model of anthracycline-induced cardiotoxicity. J Card Fail. 2012;18:502–11.
46. Germain E, Bonnet P, Aubourg L, Grangeponte M, Chajès V, Bougnoux P. Anthracycline-induced cardiac toxicity is not increased by dietary omega-3 fatty acids. Pharmacol Res. 2003;47:111–7.
47. Mong H, Hsia S, Yin H. Trans fat diet exacerbates doxorubicin cardiotoxicity via oxidative and inflammatory pathways. Am J Clin Nutr. 2013;98:620–30.
48. Kang D-W, Wilson RL, Christopher CN, Normann AJ, Barnes O, Lesansee JD, Choi G, Dieli-Conwright CM. Exercise cardio-oncology: exercise as a potential therapeutic modality in the management of Anthracycline-Induced Cardiotoxicity. Front Cardiovasc Med. 2022;8 https://doi.org/10.3389/fcvm.2021.805735.
49. Amin AM, Khlidj Y, Abuelazm M, Ibrahim AA, Tanashat M, Imran M, Nazir A, Shaikhkhalil H, Abdelazeem B. The efficacy and safety of exercise regimens to mitigate chemotherapy cardiotoxicity: a systematic review and meta-analysis of randomized controlled trials. Cardio-Oncol. 2024;10:10.
50. Díaz-Balboa E, González-Salvado V, Rodríguez-Romero B, Martínez-Monzonís A, Pedreira-Pérez M, Palacios-Ozores P, López-López R, Peña-Gil C, González-Juanatey JR. A randomized trial to evaluate the impact of exercise-based cardiac rehabilitation for the prevention of chemotherapy-induced cardiotoxicity in patients with breast cancer: ONCORE study protocol. BMC Cardiovasc Disord. 2021;21:165.
51. Salah SS, Abdel Aziz K, Kadry A, El Missiri A. The effect of cardiac rehabilitation on quality of life and 6-minute walk test in breast cancer patients after treatment with Anthracyclines. QJM Int J Med. 2023;116:hcad069.163.

Pharmacotherapies in Prevention of Myocardial Oxidative Stress Mediated by Anthracyclines, Taxanes, and Alkylating Agents

Evans Osuji, Jonathan Wright, Tanvi Brar, Shane S. Scott, and Sakima A. Smith

Abbreviations

14-3-3γ	14-3-3 Gamma
4-HNE	4-Hydroxynonenal
8-OHdG	8-Hydroxy-2′-deoxyguanosine
Akt	Protein kinase B
ALT	Alanine aminotransferase
AMPK	5′ Adenosine monophosphate-activated protein kinase
ANP	Atrial natriuretic peptide
ARE	Antioxidant-responsive element
ASK1	Apoptosis signal regulating kinase 1

Supplementary Information The online version contains supplementary material available at https://doi.org/10.1007/978-3-031-97517-2_6.

E. Osuji
Department of Internal Medicine, Johns Hopkins Bayview Medical Center, Baltimore, MD, USA
e-mail: eosuji2@jh.edu

J. Wright · T. Brar · S. A. Smith (✉)
Dorothy M. Davis Heart and Lung Research Institute, The Ohio State University College of Medicine, Columbus, OH, USA
e-mail: jonathan.wright@osumc.edu; tanvi.brar@osumc.edu; Sakima.smith@osumc.edu

S. S. Scott
Dorothy M. Davis Heart and Lung Research Institute, The Ohio State University College of Medicine, Columbus, OH, USA

The Ohio State Medical Scientist Training Program (MSTP), Columbus, OH, USA
e-mail: shane.scott@osumc.edu

S. A. Smith, S. S. Scott (eds.), *Myocardial Oxidative Stress in Anti-Cancer Therapies*, Oxidative Stress in Applied Basic Research and Clinical Practice,
https://doi.org/10.1007/978-3-031-97517-2_6

AST	Aspartate aminotransferase
ATF6	Activating transcription factor 6
ATP2A2	ATPase sarcoplasmic/endoplasmic reticulum Ca^{2+} transporting 2
BAX	BCL2-associated X
BCL-2	B-cell lymphoma-2
BIRC5	Baculoviral IAP repeat containing 5
BNP	Brain natriuretic peptide
cAMP	Cyclic adenosine monophosphate
Caspase 3	Caspase 3
CAT	Catalase
CcO	Cytochrome c oxidase
CHOP	C/EBP homologous protein
Ck-MB	Creatine kinase MB
C-MYC	Proto-oncogene c-Myc
COX-2	Cyclooxygenase-2
CTGF	Connective tissue growth factor
CTnI	Cardiac troponin I
Cx43	Connexin 43
CyclinG1	CyclinG1
DOXO	Doxorubicin
DR5	Death receptor 5
DRP1	Dynamin-related protein 1
eIF2α	Eukaryotic translation initiation factor 2 subunit alpha
eNOS	Endothelial nitric oxide synthase
ERK1	Extracellular signal-regulated kinase 1
Fizz3	Resistin
FOXO1	Forkhead box protein O1
FOXO3a	Forkhead box protein O3
FYN	Tyrosine-protein kinase Fyn
gp91	Gp91(phox)
GPx4	Glutathione peroxidase 4
GRP78	Glucose regulated protein 78
GSH-Px	Glutathione peroxidase
GSK3β	Glycogen synthase kinase 3β
GSSG	Glutathione disulfide
GST	Glutathione S-transferases
HO-1	Heme oxygenase 1
HSP	Heat shock protein
IL-17	Interleukin-17
IL-1β	Interleukin-1 beta
IL-6	Interleukin-6
iNOS	Inducible nitric oxide synthase
IRE1α	Inositol-requiring enzyme 1α

IκBα	Inhibitor of nuclear factor kappa B
JKN	C-Jun N-terminal kinase
Keap-1	Kelch-like ECH-associated protein 1
LDH	Lactate dehydrogenase
LKB1	Liver kinase B1
MDA	Malondialdehyde
MDH2	Malate dehydrogenase 2
miR-330-5p	MicroRNA miR-330-5p
miR-425	MicroRNA (miR)-425
MMP	Matrix metalloproteinase
MMP2	Matrix metalloproteinase 2
MnSOD	Manganese superoxide dismutase
MPO	Myeloperoxidase
mtDNA	Mitochondrial DNA
mTOR	Mammalian target of rapamycin
MyD88	Myeloid differentiation primary response protein 88
Myh7	Myosin heavy chain 7
NF-κB	Nuclear factor kappa B
NHE1	Sodium/proton exchanger 1
NLRP3	NOD-like receptor pyrin domain-containing protein 3
NOX2	NADPH oxidase 2
NOX4	NADPH oxidase 4
NQO1	NAD(P)H quinone dehydrogenase 1
Nrf2	Nuclear factor erythroid 2-related factor 2
NT-proBNP	N-terminal pro-B-type natriuretic peptide
P13K	Phosphatidylinositol 3-kinases
p38 MAPK	P38 mitogen-activated protein kinase
p66Shc	P66 Src homology-collagen homologue (ShcA)
PARP	Poly(ADP-ribose) polymerase
PDE	Phosphodiesterase
PDGFR	Platelet-derived growth factor receptors
PERK	Protein kinase R (PKR)-like endoplasmic reticulum kinase
PGC-1α	Peroxisome proliferator-activated receptor gamma coactivator 1-alpha
PINK1	PTEN-induced putative kinase 1
PKA	Protein kinase A
PKG	Protein kinase G
PLN	Phospholamban
PPARγ	Peroxisome proliferator-activated receptor gamma
PTEN	Phosphate and tension homology
Rac1	Ras-related C3 botulinum toxin substrate 1
RPS5	Ribosomal protein S5

RRM2	Ribonucleotide reductase M2
SIRT1	Sirtuin 1
SIRT2	Sirtuin 2
SIRT3	Sirtuin 3
SIRT6	Sirtuin 6
Smad3	Mothers against decapentaplegic homolog 3
SOD	Superoxide dismutase
T2DM	Type 2 diabetes
TAC	Total antioxidant capacity
TAK1	Transforming growth factor-β (TGF-β)-activated kinase 1
TAS	Transaminase
TBARS	Thiobarbituric acid reactive substances
TGF-β	Transforming growth factor-β
TMRE	Tetramethylrhodamine, ethyl ester perchlorate
TNF-α	Tumor necrosis factor alpha
Tn-T	Troponin T
TOS	Total oxidant status
UCP2	Uncoupling protein 2
ULK1	Unc-51-like kinase 1
VEGF-B	Vascular endothelial growth factor-B
VEGFR2	Vascular endothelial growth factor receptor 2

Introduction

Anticancer-related cardiotoxicity and cardiomyopathy secondary to anthracyclines, taxanes, and alkylating agents have been well documented. Notwithstanding, drugs from these classes remain the cornerstone of chemotherapy regimens for several malignancies, including breast, lung, skin, and gastrointestinal cancers [2, 3]. Anthracyclines, which are particularly effective in treating breast tumors, lymphomas, and soft tissue sarcomas [1, 2, 4] are clinically associated with arrhythmias, cardiomyopathy, and congestive heart failure (CHF) [5]. Taxanes commonly evoke arrhythmias such as atrioventricular conduction blocks, left bundle branch block, bradycardia, and ventricular tachycardia [6, 37]. In addition, alkylating agents, including nitrogen mustards and platinum-containing drugs, have been associated with left ventricular dysfunction, HF, myocarditis, pericarditis, arterial thrombosis, bradycardia, and supraventricular arrhythmias in patients treated with these drugs [2].

Mechanistically, these drugs perturb normal cardiac physiology through various mechanisms, prominently involving oxidative stress, disruption of iron and calcium homeostasis, cardiomyocyte death, DNA and RNA damage, and inflammatory responses [13, 14]. Alkylating agents and taxanes have distinct anticancer mechanisms, promoting DNA strand breaks and disrupting mitosis by preventing microtubule organization, respectively. However, these drugs promote cardiotoxicity through oxidative stress-dependent mechanisms. Studied mechanisms include increased ROS formation and oxidative stress-induced cytoskeletal dysfunction [6, 8, 9]. Meshkini et al. demonstrated that taxanes activate apoptotic stress pathways through glutathione (GSH) depletion and increased activity of known stress kinases such as c-Jun NH-(2)-terminal kinase (JNK) and p38 [6]. Similarly, anthracyclines promote myocardial injury via ROS-mediated damage of cellular macromolecular molecules, resulting in apoptosis [13]. Chen et al. described doxorubicin-induced reduction in NADH dehydrogenase and endothelial nitric oxide synthase (eNOS) reductase leads to the production of semiquinone free radicals. In the presence of free oxygen, semiquinone free radicals allow the formation of parental quinones and superoxide anions. The generation of hydrogen peroxide (H_2O_2) superoxide anions via superoxide dismutase (SOD) breakdown results in highly active hydroxyl free radicals ($OH^{\cdot}$) and resultant cardiotoxicity [13].

Understanding these mechanisms underscores the importance of mechanism-driven pharmacologic interventions. Dexrazoxane remains the only FDA-approved agent for combating anthracycline-induced cardiotoxicity [222]. However, emerging therapies are showing promise. Contemporary studies have evaluated the cardioprotective mechanism of sodium glucose transporter 2 (SGLT2) inhibitors against anthracycline-induced cardiotoxicity [36]. Additionally, nicorandil has demnotrated efficacy in mitigating taxane-related cardiotoxicity, while hesperidin has shown protective effects against cardiac injury caused by alkylating agents [10–12]. Molecularly targeted therapies that modulate oxidative stress signaling pathways have also demonstrated encouraging results in augmenting antioxidant defenses and mitochondrial function [40, 43, 44]. Newer pharmacotherapy approaches including gene therapy and stem cell-based interventions are also under investigation. Notably, human-induced pluripotent stem cells (iPSCs)-derived cardiomyocytes are being utilized to model and study chemotherapy-induced cardiac dysfunction following drug exposure [392].

This chapter aims to consolidate and compare the oxidative stress-related mechanisms underlying chemotherapy-induced cardiotoxicity, while also outlining current and emerging pharmacologic strategies. By targeting oxidative injury, these interventions hold promise in improving the cardiovascular safety profile of modern anti-cancer therapies, ultimately reducing treatment-related morbidity and mortality in oncology patients.

Anthracyclines

Anthracyclines are a class of anti-tumor agents, which includes agents such as daunorubicin, doxorubicin (DOXO), epirubicin, idarubicin, and valrubicin [1]. These agents are effective in treating breast cancer, lymphomas, and soft tissue sarcomas [2, 3]. Their structure is composed of a tetracyclic ring attached to a sugar moiety. Quinone and hydroquinone moieties are also present on adjacent rings, which permit the gain and loss of electrons [4]. Anthracyclines exert their effects through

DNA intercalation, inhibiting DNA replication and thus protein synthesis. DOXO also inhibits topoisomerase-2, preventing DNA repair [2, 5]. Combined with the production of free radicals, this blocked replication promotes tumor cell death [6]. In this section, we describe established data on efficacious treatment options being used to combat anthracycline-induced cardiotoxicity.

Types of Anthracyclines

Doxorubicin (DOXO)

Among the anthracyclines, DOXO is by far the most extensively studied agent [19]. While DOXO has been shown to decrease morbidity and mortality outcomes in cancer patients, cardiotoxicity remains a major concern [5, 18]. Drug-induced cardiotoxicity, such as arrhythmias and cardiomyopathies, is among the most serious and potentially life-threatening adverse effects associated with DOXO treatment [16, 45]. Notably, the cardiotoxicity of DOXO and related anthracyclines exhibits a clear dose-dependent relationship, with increased age at the time of administration correlating with an increased risk of cardiac events [15]. One proposed theory suggests that cardiomyocytes are vulnerable to anthracycline toxicity due to their increased use of oxygen and lack of regenerative capacity [17]. The incidence of DOXO-induced CHF is most prevalent at cumulative dosages exceeding 400–700 mg/m^2 in adults and 300 mg/m^2 in pediatric patients [20]. Importantly, subclinical cardiac dysfunction rates may be even higher than previously studied, underscoring the need for early surveillance and intervention. Although the precise mechanisms underlying DOXO-induced cardiotoxicity remain poorly understood, multiple cellular and molecular processes have been implicated. These include alterations in gap junction integrity, apoptosis, dysregulation of iron homeostasis, mitochondrial dysfunction, oxidative stress, DNA and RNA damage, and the activation of pro-inflammatory signaling pathways [11, 13, 21].

Epirubicin

Epirubicin is an epimer of DOXO, which is indicated for both early-stage and metastatic breast cancer [22, 23]. Similar to DOXO, epirubicin also causes dose-dependent cardiotoxicity, primarily via the formation of reactive oxygen species (ROS) and the promotion of oxidative stress. The cardiotoxic effects of epirubicin involve lipid peroxidation, DNA damage, mitochondrial damage, and disruption of calcium and iron homeostasis. In addition to these mechanisms, epirubicin is also known to modulate other intracellular signaling pathways, both dependent on or independent of cardiac topoisomerase 2β inhibition, further contributing to its cardiotoxic potential [14].

Idarubicin

Idarubicin is a fourth-generation anthracycline used to treat both solid tumors and hematologic malignancies [25]. Idarubicin has been shown to be less cardiotoxic in comparison to DOXO, but the risk remains. Although its mechanism of action has not been elucidated, idarubicin has been shown to activate apoptotic signaling pathways and ROS generation. Zhang et al. demonstrated treatment of adult murine HL-1 cells with idarubicin significantly increased the caspase mRNA expression and lowered the Bcl-2/Bax mRNA ratio [24]. Analogous to other anthracyclines, idarubicin is more toxic at increased concentrations or prolonged intervals of exposure [25].

Valrubicin

Valrubicin is a semisynthetic analogue of DOXO, which is commonly used in the treatment of superficial bladder cancer [25]. Structurally, it differs from DOXO at two sites: a five-carbon valerate ester replaces the hydroxyl group at C14, and a trifluoroacetyl group is conjugated to the 3-amino group [22, 26]. These changes in chemical structure increase lipophilic properties of valrubicin and promote rapid intracellular accumulation of the compound [22]. Inside the cell, valrubicin is subsequently transformed into N–trifluoroacetyl DOXO, which intercalates with DNA [22].

Mechanisms of Doxorubicin-Induced Cardiotoxicity

Oxidative Stress

Oxidative stress results from an imbalance between ROS formation and the activation of endogenous antioxidants in response to cell injury, leading to myocardial toxicity [11]. Increased oxidative stress and antioxidant factor deficiency are necessary for DOXO-induced cardiotoxicity [11]. Reduced catalase (CAT), GSH, SOD, and GSH-Px, and increased MDA levels serve as markers of oxidative stress and redox balance. Cellular derangements in iron, specifically iron overload, trigger DOXO-induced oxidative stress via the Fenton reaction [27]. The interaction of free iron and DOXO generates a DOXO-iron complex, which creates radical species which react with oxygen to form harmful superoxide anions [27, 28]. DOXO promotes cardiac inflammation and interferes with survival and repair mechanisms such as the Neuregulin-1/ErbB pathway [11]. Activation of the nuclear factor kappa-B (NF-κB) pathway increases the expression of pro-inflammatory cytokines, including TNF-α, IL-1β, and IL-6 [11]. DOXO-mediated inhibition of Neuregulin/ErbB signaling in cardiomyocytes reduces cardiac resilience, heightening susceptibility

Table 1 Therapies against drug-induced anthracycline cardiotoxicity grouped against respective targeted mechanistic pathways

Mechanistic pathway	Treatment in research	Targeted mechanism	Level of evidence	Author	Year
Iron homeostasis	Dexrazoxane	Decreases mitochondrial ROS	Cellular model	Sheibani	2022
	Salidroside	Decreases mitochondrial ROS, Fe^{2+}, and lipid peroxidation, as well as inhibits ferroptosis, and fibrosis	Murine and cellular model	Chen	2022
	Ferrostatin-1	Inhibit ferroptosis	Murine model	Fang	2019
	Protocatechuic acid	Reduces oxidative stress, inflammation, and apoptosis, as well as by chelating iron and inhibiting calcium dysregulation	Murine and cellular model	Okpara	2022
	Desferrioxamine	Decrease TGF-β1/SMAD signaling	Murine model	Al-Shabanah	2012
	Phytochemicals	Nrf2 regulation and increase NQO1, HO-1, and GST	Preclinical models	Abushouk	2017
	Empagliflozin	Reduces NLRP3 and MyD88	Murine model	Quagliariello	2021
Apoptosis	Valsartan	Inhibits TGF-β/SMAD/p38 signaling	Murine model	Chen	2018
	Sacubitril/valsartan	Reduces pro-apoptotic signaling	Murine model	Dindas	2021
	Dimethyl fumarate	Activates Nrf2 which promotes increases cellular antioxidant capacity and reduces apoptosis	Murine model	Hu	2022
	Benazepril hydrochloride	Increases PI3K/AKT phosphorylation which prevents apoptosis	Cellular model	Zhan	2021
	Ranolazine	Increased PARP and pro-caspase-3	Murine model	Tocchetti	2014
	Adiponectin	Reduces pro-apoptotic signaling	Murine and cellular model	Zhao	2020
	Peroxiredoxin-1	Reduces pro-apoptotic signaling via ASK1/p38 pathway	Murine model	Jiang	2020
	Nicotinamide	Inhibits DOXO-induced activation of NLRP3 inflammasome	Murine model	Wan	2021

(continued)

Table 1 (continued)

Mechanistic pathway	Treatment in research	Targeted mechanism	Level of evidence	Author	Year
Calcium homeostasis	Levosimendan	Regulates PTEN/AKT signaling	Murine model	Ling-li	2020
	Baicalein	Reduces ROS via calcium homeostasis	Murine model	Lin	2021
Oxidative stress	Melatonin	Acts as a mitochondrial free-radical scavenger and inhibits SIRT1/Nrf2 signaling promoting cell survival	Murine and cellular model	Zhang	2023
	Fibroblast growth factor-2	Activates SIRT1/LKB1/AMPK pathway	Cellular model	Koleini	2017
	IV Astragaloside	Inhibits NADPH oxidase-derived oxidative stress; supresses NOX2/NOX4	Murine model	Lin	2019
	Folic acid	Inhibits eNOS uncoupling	Murine model	Octavia	2017
DNA/RNA damage	N,N-Dimethyldoxorubicin	Exerts chemotherapeutic effect through histone modification	Murine model	Qiao	2020
	Fullerenol	Decreases CAT and increases BCL-2 which promotes cell survival	Murine model	Seke	2016
	8-Oxoguanine DNA glycosylase-1	Protects mitochondrial DNA from oxidative damage	Murine model	Anene-Nzelu	2022
	Nauclea orientalis bark extract	Decreases TNFα and caspase-3 and increases BCL-2	Murine model	Sandamali	2022
Lipid peroxidation	Glutamine	Protects against lipid peroxidation	Murine model	Durante	2019
	Fluvastatin	Protects against lipid peroxidation via regulation of iNOS, eNOS, NF-κB, and Caspase-3	Murine model	Kuşçu	2023
Inflammation	Corilagin	Inhibits P13K/Akt/NF-κB inflammatory signaling pathway	Murine model	Chen	2022
	Ferulic acid	Inhibits DOXO-induced depletion of GSH and reduced inflammation (e.g., reduced IL-6 and IL-1β)	Murine model	Aswar	2019
	Vitexin	Increases SOD, CAT, and FOXO3a expression	Murine model	Sun	2016
	Sulforaphane	Activates Keap1/Nrf2/ARE pathway to induce HO-1	Cellular model	Li	2015

to oxidative stress by downregulating ErbB4 and ErbB2 expression and PI3K/Akt signaling pathways necessary for survival [11, 227].

Similarly, other pathways such as the MTOR/Nrf2/Ho-1 pathway and the NADPH oxidase (NOX) family discussed later in this chapter share a role in oxidative stress related to DOXO [90, 229]. Table 1 summarizes the pharmacotherapies that attenuate DOXO-induced oxidative stress, while the Supplemental Table 1 expands on this list, covering a range of agents with varying mechanisms of action.

Koleini et al. demonstrated that mTOR activity is crucial for FGF-2-induced protection and the activation of Nrf2 and HO-1 in neonatal rat cardiomyocyte cultures [90]. FGF-2 isoforms enhance the upregulation of Nrf2 and its downstream target, HO-1 [90]. Additionally, the NOX family of NADPH oxidases is primarily responsible for generating ROS in a tightly regulated manner under physiological conditions. However, in disease states, excessive activation of NOX enzymes contributes to oxidative stress, leading to cardiovascular and renal injury [229]. Through these and various other pathways, cellular oxidative stress can be investigated alongside potential therapeutic strategies aimed at mitigating these stressors.

Melatonin is a naturally occurring indoleamine produced by the pineal gland and is essential for regulation of the sleep-wake cycle in vertebrates [81]. While primarily known for its role in circadian rhythm synchronization, additional studies have shown that melatonin regulates blood pressure and seasonal rhythmicity. This includes reproduction, fattening, and hibernation [82, 83]. Melatonin also exhibits a cardioprotective role in multiple cardiovascular diseases, including blood pressure regulation, myocardial ischemia, and DOXO-induced cardiomyopathy [84, 87, 225, 226]. Melatonin serves as a free radical scavenger in mitochondria and facilitates the expression of antioxidant enzymes such as SOD, GSH-Px, GSH, and CAT via melatonin receptor-dependent signaling [85, 86]. Moreover, melatonin-mediated activation of the SIRT1/Nrf2 pathway curtails DOXO-induced oxidative stress, pyroptosis, and apoptosis in both H9c2 cardiomyoblasts and C57BL/6 mice [87].

Fibroblast growth factor-2 (FGF-2) is a cardioprotective protein with dual roles as both a signaling protein and growth factor [80]. FGF-2 is expressed as high (>20 kDa, Hi-FGF-2) and low molecular weight (18 kDa, Lo-FGF-2) isoforms [89, 90]. Like other FGF family members, FGF-2 exhibit a broad range of mitogenic and cell survival activities. FGF-2 is involved in several biological processes, including embryonic development, cell growth, morphogenesis, tissue repair, tumor growth, and invasion. In preclinical models of myocardial infarction, FGF-2 decreases myocardial tissue necrosis and improves cardiac function following reperfusion [91]. Although Hi-FGF-2 is the predominant form, Lo-FGF-2 confers a cardioprotective role of FGF-2 in the prevention of myocardial loss and contractile dysfunction during ischemic and infarction situations [92–94]. FGF-2 also regulates the mTOR/Nrf2/HO-1 pathway [90]. Preservation of mTOR activity, Nrf2 expression, and upregulation of HO-1 by FGF-2 is associated with reduced DOXO-induced cardiotoxicity [90].

Phytochemicals are naturally occurring plant-derived compounds produced as secondary metabolites [96] and have been investigated for prophylactic potential in mitigating DOXO-mediated cardiotoxicity [95]. Carotenoids, polyphenols, phenolic acids, flavonoids, stilbenes, and lignans are among the most well-documented phytochemicals [97]. Major antioxidant mechanisms of phytochemicals, include the reduction of ROS production and lipid peroxidation through free radical scavenging and activation of key antioxidant enzymes, including GSH, SOD, CAT, and NADP. Some phytochemicals also modulate the Nrf2/HO-1 and nuclear factor kappa-B (NF-κB) signaling pathways in several in vitro and in vivo models, such as zebrafish, H9c2 cells, chicken cardiomyocytes, Sprague-Dawley rats, Wistar albino rats, and BALB/cByJ mice [87]. For example, phytochemicals reduce proinflammatory mediators through downregulation of TNF-α and inhibition of the cyclooxygenase-2 (COX-2) and lipoxygenase enzymes [87, 88]. In addition, these compounds also reduce adhesion molecules on vascular endothelium and decrease collagen deposition and fibrinoid necrosis in DOXO-treated tissue [95].

Astragaloside IV is a naturally occurring bioactive compound and the primary active ingredient extracted from the plant *Astragalus membranaceus* [97]. Studies have shown promising effects of Astragaloside IV in preventing oxidative damage and cardiac remodeling [227, 228]. Astragaloside IV plays a known role in alleviating DOXO-induced cardiotoxicity by inhibiting NADPH oxidase (NOX)-derived oxidative stress. The NOX family consist of seven members, and altered expression of NOX1, NOX2, NOX4, and NOX5 have been linked to the pathophysiology of cardiovascular and renal diseases [229]. Astragaloside IV suppresses NOX2 and NOX4 expression, further attenuating DOXO-induced cardiomyopathy [97]. Lin et al. demonstrated that astragaloside IV can reverse DOXO-induced ventricular fibrosis and left ventricular (LV) contractile dysfunction in C57BL6 mice [97, 98]. Additionally, astragaloside IV has been shown to reduce DOXO-induced cardiomyocyte apoptosis by inhibiting cleaved caspase-3 activity.

Iron Homeostasis

Under normal cellular stress, increased expression of antioxidant enzymes such as glutathione peroxidase 4 and reduced accumulation of iron improved overall mitochondrial maintenance [33]. Dysregulation of iron homeostasis is one of the major mechanisms that mediates DOXO-induced cardiotoxicity. The Fenton reaction is a catalytic process that converts hydrogen peroxide, a product of mitochondrial oxidative respiration, into a highly reactive hydroxyl free radical ($HO^{\cdot}$). Once activated, iron contributes to the Fenton reaction as an electron donor to H_2O_2. This in turn leads to downstream effects with the subsequent creation of unstable free radicals and ROS in the mitochondria [27, 28]. DOXO further contributes to chelation via the formation of an iron-DOXO complex promoting the catalyzation of oxygen via the Fenton reaction with subsequent reaction with biomolecules and organelles [29]. Additionally, iron-binding proteins (IRP) are among the various regulatory iron proteins with which DOXO interacts to stimulate myocardial iron overload. DOXO-iron complexes reduce the pool of free iron available for binding to IRP [30]. The binding of inactive IRP with other iron-responsive elements disrupts iron metabolism [28]. In addition, synthesis of ferritin, the major iron storage protein, is decreased, and transferrin receptor expression is upregulated.

Numerous studies have investigated the interdependent relationship between DOXO and iron toxicity (Table 1). Dexrazoxane is currently the only FDA-approved agent used in the management of anthracycline-induced cardiotoxicity [222]. Within the cell, dexrazoxane is first metabolized into a ring-opening chelating agent, which can exchange the iron in the DOXO-Fe^{3+} complex and combine with iron instead. DOXO-induced ROS is exacerbated by iron, which catalyzes the formation of free radicals. Dexrazoxane also blocks iron-mediated free radical production and prevents non-enzymatic DOXO-induced ROS generation [31, 222]. Other agents, such as ferrostatin-1, a ferroptosis inhibitor, alleviate downstream effects of free iron accumulation in the mitochondria, iron-mediated lipid peroxidation in the membrane, and cell death [32].

Salidroside, a phenylpropanoid glycoside extracted from *Rhodiola rosea,* possesses both with anti-inflammatory and antioxidant properties. It has been shown to inhibit ferroptosis and fibrosis in DOXO-treated urine cardiomyocytes by activating the AMPK signaling pathway and modulating lipid metabolism [33]. Protocatechuic acid (PCA), another naturally occurring phytochemical, has also been investigated for its cardioprotective effects against DOXO-induced cardiotoxicity due to its antioxidant, anti-inflammatory, anti-apoptotic, and metal-chelating properties [34]. The cardioprotective efffect of PCA is primarily attributed to its metal-chelating properties [34]. Excess iron production from DOXO-induced ferroptosis is sequestered by PCA, maintaining cell membrane integrity. Using 1,1-diphenyl-2-picryl-hydrazyl assay to measure antioxidant activity, PCA demonstrated concentration-dependent free radical scavenging capacity by acting as an electron donor [220].

Deferoxamine, a clinically used metal-chelating agent, functions by binding excess free iron or aluminum in the bloodstream and enhancing their renal excretion. In an experimental study involving albino rats, Al-Shabanah et al. found that deferoxamine administration following DOXO exposure reduced transforming growth factor β (TGF-β)/Smad signaling, increased red blood cell count, and completely normalized DOXO-induced alterations in cardiac enzyme levels and gene expression [35].

Lastly, empagliflozin, a selective SGLT2 inhibitor known for its ability to reduce HF and cardiovascular death in patients with type 2 diabetes (T2DM), had also showed promising potential in reducing or reversing anticancer therapy-induced cardiac injury. In a preclinical study, *Quagliariello* and colleagues demonstrated that DOXO-treated C57Bl/6 mice, empagliflozin significantly suppressed NOD-like receptor pyrin domain-containing protein 3 (NLRP3) inflammasome activation, reduced MyD88 expression, and attenuated cardiac fibrosis, inflammation, apoptosis, and ferroptosis [36].

Apoptosis

Apoptosis, or programmed cell death, is a vital process orchestrated by a network of proteins, including caspases, heat shock proteins (HSPs), and the p53 tumor suppressor. The molecular interplay among these molecules allows cells to respond appropriately to various stressors, maintain homeostasis, prevent the propagation of damaged DNA, and eliminate damaged cells. Once activated, apoptotic signaling sequentially promotes cell shrinkage, chromatin condensation, DNA fragmentation, and cell death without eliciting an inflammatory response. Caspases, a family of cysteine proteases, are central to the execution of apoptosis. Under normal circumstances, caspases exist as inactive precursors, or procaspase enzymes, and are proteolytically cleaved to active forms downstream from apoptotic signal pathways. Initiator caspases such as caspase-8 and caspase-9, respond to apoptotic stimuli, while effector caspases, such as caspase-3 and caspase-7, dismantle key cellular components [222]. HSPs are molecular chaperones that maintain protein folding and protect cells from stress-induced damage. Among them, Hsp70 is notable for its ability to modulate apoptotic pathways. In response to DNA damage and kinases signaling, the tumor suppressor protein p53 induces cell cycle arrest or promotes apoptosis through transcriptional activation of pro-apoptotic genes, such as those in the Bcl-2 family. Downstream, Bcl-2 promote mitochondrial membrane permeabilization, release of cytochrome c, activation of caspases, and eventual cell death. Increased DNA exposure to UV light, anticancer chemotherapy drugs, and γ-irradiation causes increased levels of p53 [220]. p53 is regulated by several proteins. Activation of HSP-25 stabilizes p53 and promotes activation of proapoptotic factors such as FasL, Fas, and c-Myc proteins [11, 37, 38].

DOXO-mediated imbalance in oxidative stress markers promotes programmed cell death, or apoptosis of cardiomyocytes via several signaling pathways [11]. These pathways include the neuregulin/ErbB and PI3K/Akt/mTOR signaling pathways, which are implicated in cell growth, cell survival, and attenuation of oxidative stress [11, 227]. Dysregulated signaling of both pathways leads to apoptosis and mitochondrial dysfunction. The Nrf2/HO-1 signaling pathway is a key regulator of redox homeostasis and has been studied in DOXO-induced cardiotoxicity for its defensive role in mitigating oxidative stress [87]. The activation of kinases, such as 5-adenosine monophosphate-activated protein kinase (AMPK), reduces apoptosis and preserves cardiac mitochondria by enhancing autophagy and inhibiting fibrosis [11]. However, DOXO has been shown to suppress this protective mechanism [11].

DOXO administration also interrupts the actions of HSPs, which are normally upregulated in cellular response to oxidative stress and imbalance of oxidative markers [11]. Imbalance of oxidative markers leads to the activation of heat shock factor-1 (HSF-1) transcription factor and HSP-25 downstream. Increased expression of both HSP-25 and HSP-70 in mouse cardiac muscle has demonstrated association with increased inflammation, fibrosis, and apoptosis [11, 37, 39]. DOXO administration has also been shown to increase expression of HSP-25 and HSP-70 [11]. Oxidant-induced activation of HSF-1 increases expression of HSP-25 and its phosphorylated forms in DOXO-treated mouse hearts compared to controls [38].

Furthermore, researchers found HSF-1-mediated transactivation of p53 increased Bax levels, promoting DOXO-mediated heart failure (HF). Supporting this, siRNA-mediated depletion of HSP-25 in H9c2 cells treated with DOXO resulted in increased apoptosis. Specifically, loss of HSP-25 promoted SIRT1 dissociation from p53, leading to increased p53 acetylation, resulting in elevated Bax protein expression, cytochrome c release, and caspase-3 and caspase-9 activation [236]. Ueno et al. also demonstrated that DOXO treatment increases apoptosis via caspase activation in Sprague-Dawley rats [223]. Together, these data suggest that DOXO-mediated regulation of several apoptotic signaling pathways contributes to its cardiotoxicity profile. As such, understanding these relationships and investigating pharmacotherapies targeting oxidative stress-driven apoptosis are needed.

Valsartan is an angiotensin II receptor antagonist (ARB) and is primarily prescribed for hypertension and HF. Valsartan promotes vascular smooth muscle relaxation, which in turn lowers blood pressure and reduces cardiac afterload. Valsartan regulates the TGF-β/Smads and TGF-β/p38 pathways through long non-coding RNA (lncRNA) cardiac hypertrophy-related factors to improve DOXO-induced HF [40]. DOXO-induced cell apoptosis and increased caspase-3 activity in the HF models are attenuated by valsartan [40]. The Sacubitril/valsartan (Sac/Val) combination drug belongs to the angiotensin receptor/neprilysin inhibitor (ARNI) drug class and is a cornerstone in modern HF management [41]. In a study using BALB/c mice, Dindas et al. evaluated total antioxidant status, total oxidant status, inflammatory cytokines, N-terminal pro-B-type natriuretic peptide (NT-proBNP) concentrations, and caspase-3 activity prior to DOXO treatment. Sac/Val was shown to broadly reduce oxidative stress but also specifically reduced caspase-3 pathway activation [41].

Like ARBs and ARNIs, dimethyl fumarate (DMF) also reduces DOXO-induced proapoptotic signaling through modulation of caspase-3, Bax, and Bcl-2 expression [42]. DMF enhances antioxidant capacity and regulates apoptosis by activating the nuclear factor erythroid 2-related factor 2 (Nrf2) pathway and subsequent activation of the antioxidant gene *Hmox1* downstream [43]. Additionally, MF significantly reduces cardiac oxidative stress by reducing malondialdehyde (MDA) levels and increasing glutathione (GSH), SOD, and glutathione peroxidase (GSH-Px) levels [42].

Benazepril hydrochloride, an angiotensin-converting enzyme (ACE) inhibitor, demonstrates protective effects against cardiovascular disease [44]. Specifically, benazepril hydrochloride mediates apoptosis through regulation of the PI3K/Akt pathway. The phosphorylation of Akt in the PI3K/Akt signaling pathway promotes a reduction in apoptosis, preserving cardiac muscle tissue [44]. Previous studies have shown the relationship between DOXO-induced cardiac dysfunction and reduced PI3K/Akt levels [42–44]. Zhang et al. showed that pretreatment of rat H9c2 cells with benazepril hydrochloride followed by DOXO resulted in oxidative stress neutralization and reduced apoptosis via the PI3K/Akt signaling pathway [44].

While benazepril is well recognized, other ACE inhibitors, including enalapril, have also been investigated for their cardioprotective effects [234]. Hiona et al. demonstrated the cardioprotective effects of enalapril in Sprague-Dawley rats by

Table 2 Therapies against drug induced anthracycline cardiotoxicity with human level evidence

Treatment	Level of evidence	Author	Year
Astragalus polysaccharide	Randomized controlled trial	Li	2022
Black seed oil	Randomized controlled trial	Hagag	2020
Candesartan	Randomized controlled trial	Lee	2021
Carvedilol	Randomized controlled trial	Nabati	2017
Carvedilol	Randomized controlled trial	Lee	2021
Carvedilol	Randomized controlled trial	Beheshti	2016
Dexrazoxane	Meta-analysis	Stansfeld	2022
Dexrazoxane	Retrospective cohort	Narayan	2019
Dexrazoxane	Retrospective cohort	Kopp	2019
Enalapril	Randomized controlled trial	Georgakopoulos	2019
Metoprolol	Randomized controlled trial	Georgakopoulos	2019
Omega 3 fatty acids	Randomized controlled trial	Amrousy	2022
Salidroside	Randomized controlled trial	Zhang	2012
Statins	Randomized controlled trial	Nabati	2019

preserving mitochondrial respiratory efficiency and reducing doxorubicin-induced free radical generation [234]. These findings support a potential clinical role for ACE inhibitor use in the prevention of DOXO-mediated cardiomyopathy. Table 2 summarizes the pharmacotherapies that effectively reduce oxidative stress and mitigate chemotherapy-induced cardiotoxicity in randomized controlled trials (RCTs), including ACE inhibitors, ARBS, and β-blockers.

Natural compounds such as cinnamic acid, known for their anti-inflammatory, antibacterial, and cytoprotective properties, have been investigated for their role in mitigating apoptosis, including their effects on DOXO-induced apoptosis [235]. In hiPSC-derived cardiomyocytes, Koczurkiewicz-Adamczyk et al. demonstrated that DOXO toxicity was attenuated by downregulating caspase-3 and -7, key proteases activated during apoptosis or programmed cell death [235].

Increased intracellular sodium is a known consequence of failing cardiomyocytes and has been shown to be associated with increased ROS production. Ranolazine, a sodium current inhibitor proven to be effective in treating experimental HF, has demonstrated potential use in attenuating DOXO-induced cardiotoxicity [101]. As a therapeutic agent, ranolazine reduced cardiotoxicity markers including atrial natriuretic peptide (ANP), brain natriuretic peptide (BNP), connective tissue growth factor, and matrix metalloproteinase 2 (MMP2) mRNAs. The drug also decreased alterations in extracellular matrix remodeling while increasing levels of poly (ADP-ribose) polymerase (PARP) and pro-caspase-3 [101]. The blunting of mRNA expression of several proinflammatory and oxidative injury markers shows the potential benefits of using ranolazine to combat DOXO-induced oxidative stress; however, more studies are required to validate its use in clinical practice.

Adiponectin is a protein hormone and adipokine involved in the regulation of glucose and fatty acid metabolism [116, 117]. Increased serum adiponectin levels have been inversely linked to lower morbidity and cardiovascular disease burden

[118]. ADP355 is an adiponectin agonist and adiponectin-based active peptide with antifibrotic and anti-atherosclerotic properties [119 121] and has been shown to attenuate anthracycline induced myocardial toxicity. Zhao et al. observed decreased levels of serum creatinine kinase (CK), lactate dehydrogenase (LDH), hydroxybutyrate dehydrogenase, caspase-3, Bax/Bcl-2, and SOD in C57BL/6 J mice exposed to ADP355 [119]. ADP355 also attenuated oxidative stress by targeting Nrf2 and sirtuin 2 signaling pathways [119].

Peroxiredoxins (Prdxs) are a ubiquitous family of antioxidant enzymes, comprising of six peroxidases isoforms (Prdx1-Prdx6), which catalyze the reduction of peroxynitrite, H_2O_2, and other hydroperoxides [46]. Peroxiredoxin-1 (Prdx1) plays a protective role in many diseases including aging, acute tissue injury, neurodegeneration, and cancer by scavenging ROS [46–50]. Overexpression of this protein has been shown to attenuate both DOXO-induced cardiac oxidative stress and apoptosis [46]. The apoptosis signal-regulating kinase-1(ASK1) is a key redox-sensitive protein kinase and critical regulator of oxidative damage and apoptosis [51, 52]. Reduced expression of phosphorylated ASK1 and the downstream p38 kinase in the heart is associated with Prdx-mediated cardio-protection [46].

Nicotinamide mononucleotide (NMN) plays an integral role in cellular homeostasis and has also been studied in disease treatment. NMN is used by enzymes to generate the coenzyme nicotinamide adenine dinucleotide + hydrogen (NADH), which is critical for eukaryotic cellular redox reactions [53, 54]. NMN decreases cellular inflammation, oxidative stress, and apoptosis under pathological conditions [46–48, 54–56]. Several studies have elucidated a role for increased NLRP3-inflammasome activity and cardiotoxicity [57–59]. Activation of the NLRP3 inflammasome by DOXO is positively correlated with cardiac injury tissue scores [57, 58]. NMN inhibits DOXO-induced activation of NLRP3 inflammasome-mediated inflammation, as evidenced by decreased caspase-1 and IL-1β activity in Sprague Dawley rats [59]. NMN increased GSH levels and SOD activity and decreased MDA and ROS levels [51]. Additionally, NMN decreased Bcl-2 expression while simultaneously increasing Bax and cleaved caspase-3 levels [60].

Calcium Homeostasis

Calcium (Ca^{2+}) plays a fundamental role in myocardial tissue contraction. As such, disruptions in Ca^{2+} homeostasis may lead to structural and functional changes [61]. Cellular Ca^{2+} dysregulation promotes ROS generation, which inhibits sarcoplasmic reticulum function [62]. Impaired Ca^{2+} homeostasis is an additional mechanism of cardiac injury documented with the use of anthracyclines [60]. DOXO treatment increases Src kinase activation and subsequent release of Ca^{2+} from the endoplasmic reticulum (ER) [62]. Owing to this, research has been targeted at the suppression of DOXO-mediated ROS using Ca^{2+} chelators [63].

Levosimendan is a Ca^{2+} sensitizer indicated for the treatment of acutely decompensated HF [64]. Levosimendan promotes dose-dependent improvement in cardiac output and stroke volume and reduces pulmonary capillary-wedge pressure, mean right atrial pressure, and pulmonary artery pressure [64, 65]. Currently, it is the only inotrope that has been associated with improved survival [66, 67]. Furthermore, in comparison to other inotropic agents, levosimendan has fewer cardiac side effects [70, 71]. Levosimendan increases myocyte Ca^{2+} sensitivity by binding to troponin C in a Ca^{2+} dependent manner [68]. Troponin C fiber sensitivity to free ionized Ca^{2+} is increased as more levosimendan binds, further facilitating interactions between troponin C and troponin I. Together, increased myocardial Ca^{2+} sensitivity leads to an increase in cardiac contractility, without an increase in intracellular Ca^{2+} levels [69]. In addition to modulating Ca^{2+}, levosimendan also has vasodilatory and cytoprotective properties [72, 73].

Baicalin is a flavone glycoside and the glucuronide of baicalein, which is found in different species of the genus *Scutellaria* [74]. Mechanistically, Baicalin is a positive allosteric modulator of the $GABA_A$ receptor [75, 76]. Studies have demonstrated that baicalin exhibits anti-inflammatory, anticancer, antithrombotic, cardioprotective, and antioxidant properties, which are primarily mediated through suppression of NF-κB activity [77–79]. Feng et al. showed that baicalin promotes cardioprotection through regulation of the Toll-like receptor 4 (TLR4)/NFκB signaling pathway and associated inflammatory markers in Sprague-Dawley rats [224]. Baicalin inhibits the overexpression of cardiomyocyte TLR4, which is upregulated in response to DOXO administration [224]. The use of baicalin in conjunction with DOXO in breast cancer patients enhances chemosensitivity via upregulation of oxidative stress-mediated mitochondrial-dependent apoptosis [80]. As such, baicalin is used for chemo-resistant cancer and as an adjuvant to DOXO-based therapy [80]. While the exact mechanism remains unclear, Lin et al. demonstrated that increased ROS and Ca^{2+} concentrations are also associated with baicalin use [80]. This dual action of baicalin in enhancing DOXO-induced cancer death while promoting cardiac protection makes it an appealing drug candidate.

DNA/RNA Damage

Damage to the intracellular organelles and proteins integral to the replication machinery is well documented in drug-induced cardiotoxicity. In comparison to other tissues, cardiomyocytes possess a higher number of mitochondria, due to their increased oxygen and ATP demand. Anthracycline-mediated mitochondrial damage in cardiomyocytes exacerbates cardiac dysfunction [11]. Anthracyclines promote DNA damage by initiating of double strand breaks secondary to topoisomerase-2 inhibition, as well as chromatin damage through histone modification [237]. Modification of the chemical structure of anthracyclines is one method for reducing anthracycline-mediated toxicity [99].

N, N-dimethyl doxorubicin (diMe-DOXO) is a variant of DOXO created through the *N,N*-dimethylation of the amine group. The addition of the dimethyl functional group eliminates DOXO-induced DNA-damaging activity while preserving its histone modification properties [99]. *In vitro* studies by Qiao et al. demonstrated that diMe-DOXO maintained its anticancer activity and performed comparable, or in some cases better than DOXO [99]. These findings were confirmed in hiPSC-derived cardiac microtissues [99]. Future studies call for animal models and eventually human studies to further validate the utility of diMe-DOXO as an agent to mitigate anthracycline therapy and DICT.

Fullerenol is a highly water-soluble C60 derivative ($C_{60}(OH)_{24}$), which forms polyanion nanoparticles ranging between 15 and 42 nm in aqueous solution and can bind positively charged molecules such as DOXO [31]. As such, fullerenol may serve as drug carrier. Therefore, fullerenol may serve as a drug carrier. Seke et al. reported that fullerenol/doxorubicin (FNP/DOXO) nanocomposite treatment results in decreased CAT, and Mn-superoxide dismutase mRNA levels in comparison to DOXO-independent treatment [100]. This reduction in antioxidative enzyme gene expression suggests that the addition of FNP confers a lower oxidative stress burden [100]. FNP/DOXO also increases the mRNA levels of the anti-apoptotic enzyme Bcl-2, which may confer increased cell survival and resistance to DOXO-mediated DNA damage and apoptosis.

DNA glycosylases play a key role in the elimination of DNA lesions resulting from inappropriate base modifications. DNA glycosylases identify and excise damaged DNA bases and initiate repair processes that restore normal DNA structure with high accuracy [102]. The most common DNA lesion produced by oxidative stress is the formation of 7,8-dihydro-8-oxoguanine (i.e., 8-oxoguanine) [103]. 8-Oxoguanine DNA glycosylase-1 (OGG1), a base excision repair enzyme, is involved in oxidative stress-induced DNA demethylation. Prevention of accumulated DNA damage, specifically mtDNA, leads to a reduction of mitochondrial fragmentation and increased mitochondrial function in cardiomyocytes [104]. Additionally, overexpression of OGG1 has been shown to reduce mtDNA damage and cardiac fibrosis in pressure-overload mouse models of cardiac hypertrophy [105]. However, the relationship between DOXO-induced cardiac dysfunction, including cardiomyopathy, and the correlation between OGG1 levels, DNA damage, and cardiac function remains unclear [103]. In fact, Anene-Nzelu et al. demonstrated that there were no significant differences in cardiac function between the *Ogg1*-knockout and wild-type mice following transverse aortic constriction surgery [103], suggesting OGG1-mediated base excision repair may not be critical for the proper stress response of cardiomyocytes. DOXO treatment of *Ogg1*-KO and wild-type mice, however, led to reduced ejection fraction, increased chamber size, and higher mortality in DOXO-treated *Ogg1*-KO mice, suggesting that OGG1-mediated base excision repair may play a protective role specifically in mitigating DOXO-induced cardiotoxicity [103]. More research is needed to further elucidate the mechanism by which OGG1 attenuates anthracycline-dependent cardiomyocyte toxicity.

Nauclea orientalis bark extract is a species of tree from the *Rubiaceae* family. The extract is native to several different geographic locations, including Southeast Asia, New Guinea, and Australia [106]. It is recognized for its antioxidant effects and compounds such as β-sitosterol and phenolic substances. These compounds are active in free radical scavenging, metal chelation, and possess anticarcinogenic properties [107]. Mechanistically, Sandamali and colleagues experiments in Wistar rats, demonstrated that *Nauclea orientalis* decreases inflammatory markers (e.g., TNF-α), myeloperoxidase, expression of caspase-3, and DNA fragmentation. There was also an observed increase in Bcl-2 expression [108]. These data support the administration of *Nauclea orientalis* bark extract for the treatment of DOXO-induced oxidative stress, inflammation, apoptosis, and DNA fragmentation [107, 108]. However, the potential use of *Nauclea orientalis* as an adjunct in patients experiencing anthracycline-induced cardiotoxicity requires clinical further validation.

Lipid Peroxidation

Lipids, glycolipids, phospholipids, and cholesterol have been identified as common targets for injury and lethal peroxidative alterations [113]. Lipid peroxidation is a process by which free radicals oxidize lipids containing carbon-carbon double bond(s), especially polyunsaturated fatty acids (PUFAs), resulting in lipid peroxyl radicals and hydroperoxides [109, 110]. The relationship between high ROS accumulation and lipid damage has been established, with the two most prevalent ROS species affecting lipids being the $HO^{\cdot}$ and hydroperoxyl ($HO^{\cdot}_2$) [111, 112]. However, lipids can also be oxidized by endogenous enzymes such as lipoxygenases, cyclooxygenases, and cytochrome P450 [114, 115]. Ferroptosis, which is also triggered by DOXO, can also be activated downstream lipid peroxidation [230, 231]. Specifically, Tadokoro et al. demonstrated that reduced glutathione peroxidase 4 (GSH-Px4) levels and excessive lipid peroxidation triggered by mitochondrial DOXO-Fe^{2+} complexes promote mitochondrial-dependent ferroptosis. These findings emphasize the role of lipid peroxides as a source of ROS and highlight the contribution of mitochondria-dependent ferroptosis in DOXO-induced cardiotoxicity [230]. Furthermore, cells initiate survival responses or programmed cell death depending on the extent of damage incurred. Under subtoxic lipid peroxidation conditions, cells respond by activating their antioxidant defense systems and upregulating proteins that neutralize ROS. With increasing levels of lipid peroxidation, the cellular repair capacity becomes overwhelmed, resulting in the induction of programmed cell death [113–115].

Vitamin B9, or folic acid, has been shown to provide cardioprotective benefits [122]. Folic acid modulates eNOS synthase, whose dysregulation is critical for the development of cardiac dysfunction and/or myocardial infarction [123]. DOXO treatment has been shown to promote eNOS-dependent oxidative stress and nitric oxide synthase (NOS) dysfunction, leading to decreased nitric oxide (NO) and increased superoxide anion levels. Folic acid inhibits DOXO-mediated eNOS uncoupling, enhances eNOS phosphorylation at S1177, restores cardiomyocyte NO production, and prevents mitochondrial damage, ultimately reducing myocardial injury and cardiac dysfunction [123]. In addition, folic acid blocks the increase in cardiac injury markers such as LDH, troponin T (Tn-T), TNF-α, and CK-MB, while also reducing lipid peroxidation markers, thiobarbituric acid reactive substances, and GSH changes, thereby protecting mitochondria and cardiomyocytes from cardiotoxic damage [124].

Glutamine is one of the most abundant and versatile amino acids in the body. The rate of glutamine consumption by immune cells is comparable to, and sometimes exceeds, that of glucose [125]. This proteinogenic amino acid has been characterized as essential for lymphocyte proliferation, cytokine production, macrophage phagocytic and secretory activities, and neutrophil bacterial killing [125]. Additionally, all cells utilize this substrate for nucleotide synthesis, nicotinamide adenine dinucleotide phosphate, antioxidant production, and several biosynthetic pathways essential for cellular homeostasis [125]. Glutamate (the anionic form of glutamine) is a cytosolic amino acid and precursor to GSH, and its cellular role is intricately linked to lipid peroxidation [126]. The absence of glutamine leads to energy depletion, weakened immune defenses, enhanced apoptosis, and reduced cell survival factors, including heat shock proteins and GSH levels [232]. Xue et al. showed that glutamine independently attenuated DOXO-induced cardiac lipid peroxidation and cardiac troponin I (cTnI) levels [126]. Furthermore, Todorova et al. demonstrated that oral glutamine supplementation in rats wit DOXO-induced cardiotoxicity reduced cardiac lipid peroxidation and peroxynitrite levels [233]. This decrease was accompanied by increased cardiac GSH levels and enhanced catalase activity [233]. Together, these findings highlight the cardioprotective potential of glutamine in mitigating DOXO-induced oxidative stress and cardiotoxicity.

Fluvastatin is a fully synthetic hydroxymethyl glutaryl coenzyme A (HMG-CoA) reductase inhibitor used as a cholesterol-lowering agent in patients with hypercholesterolemia. It has also exhibited antioxidant, anti-inflammatory, and antitumor activity in previous studies [127]. Specifically, fluvastatin suppresses lipid peroxidation by regulating inducible nitric oxide synthase (iNOS), eNOS, NF-κB, and caspase-3 protein and mRNA expression. In terms of lipid peroxidation markers, both SOD activity and MDA levels are improved, supporting a protective role against peroxidative damage.

Inflammation

Anthracycline cardiotoxicity is associated with cardiac inflammation through stimulation of different cytokine signaling proteins such as IL-1β, IL-18, TNF-α, IL-6, endothelin-1 (ET-1), Nrf2, cyclooxygenase (COX), and NF-κB [11, 22]. Inflammatory mechanisms in cardiotoxicity are complex due to pathway variability and multiple influencing factors. Some pathways function alongside downstream counterparts, while others function as regulators or inhibitors.

Corilagin is an ellagitannin and a natural polyphenol found in tannic acid with significant potential utility as an antitumor agent [128]. Corilagin has demonstrated antioxidative, anti-inflammatory, and hepatoprotective activities outside the heart [129]. This compound suppresses levels of P13K, Akt, and NF-κB which contribute to DOXO-induced cardiotoxicity. TNF-α, IL-6, and NF-κB expression levels in cardiac cells are also repressed following corilagin treatment. Overall, corilagin can reduce biomarkers of myocardial injury and damage, including aspartate aminotransferase, CK, LDH, CK-MB, myoglobin, heart-type fatty acid-binding protein, glycogen phosphorylase isoenzyme, TGF-β, cTnI, and BNP. These effects lead to improved cardiac function, and reduced inflammation and necrosis [128].

Ferulic acid [3-(4-hydroxy-3-methoxy-phenyl) prop-2-enoic acid)] is a phenolic compound found in most plants [132], with demonstrated antioxidant, antihypertensive, antiaging, anti-inflammatory, and cardioprotective activities [131, 133]. Aswar et al. demonstrated that ferulic acid reduces protein expression of BNP, CK-MB, LDH, IL-1β, and IL-6 while increasing expression of GSH and Na^{+}/K^{+} ATPase [131]. This evidence suggests ferulic acid regulates both inflammation and lipid peroxidation by blocking DOXO-mediated depletion of GSH, a crucial intracellular antioxidant enzyme essential for maintaining cell integrity. GSH deficiency is known to cause damage to macromolecules and membrane lipids. Additionally, ferulic acid has been shown to mitigate the peroxidizing effects of DOXO-induced inflammatory cytokines. Moreover, ferulic acid treatment significantly reduces IL-1β and IL-6, supporting its anti-inflammatory potential [131].

Vitexin is an apigenin flavone glucoside found in passionflower, bamboo leaves, and pearl millet. It is also the active ingredient in hawthorn leaf extract [134, 135]. Hawthorn leaves have been proven to regulate blood lipids, blood pressure, increase coronary flow, and protect the ischemic myocardium. Vitexin's mechanism of action is of particular interest due to its antioxidant and anti-inflammatory properties [136, 137]. Studies have also shown that vitexin plays a protective role against hypoxic-induced myocardial cell death following ischemia [138]. Vitexin attenuates proinflammatory pathways and cytokines, including TNF-α, IL-1β, IL-6, and NF-κB and reduces MDA levels. Furthermore, vitexin increases FOXO3a, which reduces/blocks oxidative stress and apoptotic processes in Sprague-Dawley rats [135].

Lastly, sulforaphane is a natural isothiocyanate compound found in cruciferous vegetables with known cardioprotective actions against ischemic heart injury [139]. Research has also demonstrated evidence of protection in the kidneys and brain against ischemic injury [140, 141]. Sulforaphane blocks DOXO-induced inflammation through the activation of the Kelch-like ECH-associated protein 1 (Keap1)/Nrf2/antioxidant-responsive element pathway. Activation of this pathway facilitates the induction of HO-1 [139], which reduces DOXO-mediated ROS levels.

Taxanes

Paclitaxel is an anti-microtubule agent and the most common taxane employed in anticancer therapy for multiple cancers, including breast, genitourinary, gastrointestinal, lung, endocrine, and skin cancers [3, 142–147]. Paclitaxel targets rapidly dividing cells, such as neoplasms, and inhibits mitosis. While remarkably effective for treating cancer, paclitaxel has been associated with arrhythmias, such as atrioventricular conduction block, left bundle branch block, bradycardia, and ventricular tachycardia [5, 6]. Paclitaxel is also cytotoxic owing to its action on the cytoskeleton. It has been proposed that paclitaxel enhances ROS formation and increases oxidative stress [7, 8]. As such, therapy options have centered on treating paclitaxel-induced arrhythmias with antioxidants.

Nicorandil, a potassium channel activator, promotes systemic arterial vasodilation by stimulating guanylyl cyclase to increase cyclic GMP and activating potassium channels to close voltage-gated Ca^{2+} channels [213]. In Sprague-Dawley rats, paclitaxel-induced ROS and NADPH oxidase activity in coronary artery endothelial cells were prevented by nicorandil via mitochondrial potassium channel activation [213]. The remaining two studies focused on treating oxidative stress utilized naringin, naringenin, rutin, and hesperidin, all bioflavonoids derived from citrus fruits [149]. Both studies reported these bioflavonoids were able to attenuate paclitaxel-induced lipid peroxidation while restoring the antioxidants GSH, SOD, and GSH-Px [148, 150]. Beyond these examples, further research is required to evaluate the role of nicorandil in the prevention of paclitaxel-induced arrhythmias [150].

Alkylating Agents

Alkylating agents are some of the most used anti-cancer therapies today. These compounds exert their action by interfering with cell cycle progression. Specifically, they form covalent cross link DNA strands to cause DNA strand breaks and abnormal base pairings, leading to cell death [151, 152]. There are five main classes of alkylating agents: nitrogen mustards, aziridines, alkyl sulfonates, nitrosoureas, and platinum-containing drugs [152]. Unfortunately, alkylating agents do not possess tumor-specific DNA action and therefore exert effects on normal cells, resulting in cytotoxic side effects, including cardiotoxicity [153]. Although rare, alkylating agent-mediated cardiotoxicity is growing in prevalence as cancer treatments expand. This section will focus on two specific alkylating agents, cisplatin, a platinum-containing drug, and cyclophosphamide, a nitrogen mustard. Their mechanisms of cardiotoxicity and cardioprotective strategies are described.

Cisplatin

In 1965, cisplatin, a platinum-containing coordination compound, was discovered to possess the ability to interfere with cellular division in *E. coli* [154]. Today, cisplatin is often used in the treatment of multiple cancers, including gonadal tumors, head and neck tumors, and sarcomas [155–157]. Cisplatin exerts its effects by binding to reactive centers on purine residues in DNA, thereby blocking cell division and resulting in apoptosis of mitotic cancer cells [155]. In addition to increasing nuclear DNA insults, increases in cellular ROS, such as superoxide and hydrogen peroxide, are also necessary for cisplatin-mediated cancer cell death [158]. While normal cells also possess some degree of ROS generated secondarily to oxidative metabolism, multiple antioxidant proteins and cellular signaling mechanisms regulate ROS production and prevent cellular stress and apoptosis. As previously described, SOD catalyzes superoxide anion radicals into molecular oxygen and H_2O_2; GSH-Px, converts H_2O_2 into molecular water using GSH; and CAT, which also acts to neutralize H_2O_2 into molecular oxygen and water, together forming the three major cellular antioxidant systems [159]. Moreover, while cancer cells tend to have increased levels of ROS, cisplatin targets the mitochondria to induce further oxidative stress and subsequent apoptosis [156].

While increased oxidative stress is essential for anticancer efficacy of cisplatin, it also portends toxicity to various organ systems including the cardiovascular system. Cisplatin-mediated cardiotoxic effects primarily manifest as electrocardiographic changes and arrhythmias [159, 160]. In addition, myocarditis, pericarditis, angina, coronary vasospasm, and damage to endothelial damage have been reported and are associated with HF and even sudden cardiac death [161–163]. While the exact mechanisms of cisplatin-induced cardiotoxicity are not entirely understood, the injury appears to follow a sequential pathway [157]. The first step involves cisplatin uptake into cells via organic cation transporters (OCTs) and a high-affinity copper transporter (CTR1), which are primarily located in renal cells [164, 165]. While OCTs are not expressed in cardiac tissue, CTR1 is, and downregulation of CTR1 has been shown to decrease cisplatin cytotoxicity [165]. Upon entry into the cells, cisplatin causes considerable damage to nuclear and mitochondrial DNA [156]. In addition, cisplatin disrupts the cellular redox balance by depleting GSH, CAT, and SOD, impairing mitochondrial respiration, and increasing ROS production, oxidative stress [156]. Cisplatin also raises levels of lipid peroxidation and MDA [156]. This induction of oxidative stress is proposed as the primary mechanism of cisplatin-related cardiotoxicity. Another pathway crucial in regulating myocardial oxidative stress occurs via the Nrf2-heme oxygenase-1 (HO-1) system [166]. Nrf2 regulates a variety of genes, including the HO-1, which upregulates antioxidants to protect cardiomyocytes from ROS-mediated injury and death [166, 167].

Due to cisplatin's role in mediating oxidative stress, several researchers have explored the cardioprotective effects of various agents in attenuating cisplatin-induced cardiotoxicity in vitro and in vivo preclinical models. Most of these studies employ current FDA-approved cardiovascular agents with antioxidant properties. The SGLT2 inhibitor, canagliflozin, traditionally used to treat T2DM, also exhibits cardioprotective effects, reducing cardiovascular mortality and HF hospitalizations in both patients with and without diabetes [168, 221]. In a rat model, canagliflozin co-treatment with cisplatin therapy significantly reduced cardiac oxidative injury, as evidenced by decreased lipid peroxidation biomarkers and MDA, as well as upregulation of GSH, Nrf2, HO-1, glutamate–cysteine ligase catalytic subunit, and SOD3—all of which bolster the antioxidant pool and reduce oxidative stress [217].

Vitamin E, a fat-soluble vitamin and well-known antioxidant [169], also demonstrated cardioprotective effects. Administration of vitamin E to mice receiving cisplatin therapy resulted in marked decreased MDA levels and increased SOD and GSH-Px levels [170]. Venlafaxine, a serotonin-norepinephrine reuptake inhibitor, also exerts cardioprotective effects through reducing MDA and increasing SOD and GSH-Px [171, 172]. Another study evaluated the cardioprotective effects of trimetazidine (TMZ), an anti-anginal drug that reduces fatty acid oxidation [173], and coenzyme Q (CoQ10), a component of the electron transport chain used in the treatment of HF [174]. Both TMZ and CoQ10 inhibit ROS production and lipid peroxidation, thereby protecting against ROS-mediated mitochondrial dysfunction [175–177]. Researchers found cardiomyocytes pre-treated with TMZ or CoQ10 prior to cisplatin therapy had significantly lower levels of ROS and MDA, higher levels of SOD and CAT, and enhanced Nrf2 expression [178, 179]. In all cases, pretreatment with a combination of TMZ and CoQ10 was more effective than either agent alone [178]. N-acetylcysteine (NAC), used to treat acetaminophen overdose, also serves as a cysteine source for GSH synthesis, suggesting its potential in reducing ROS [180]. Two separate studies exploring simultaneous co-administration of NAC with cisplatin found that NAC supplementation significantly decreased superoxide levels, increased SOD and GSH levels, and reduced cardiac injury biomarkers [181, 182].

In addition to their antioxidant properties, many cardioprotective drugs used to combat cisplatin-induced cardiac damage also utilize other cellular pathways. Statins, potent lipid-lowering drugs targeting hepatic cholesterol production, also possess anti-apoptotic properties [183]. Under the hypothesis that ER stress and apoptosis contribute to cisplatin-mediated cardiomyopathy, researchers explored rosuvastatin and simvastatin as cardioprotective agents. Treatment with either statin increased GSH content and SOD activity compared to cisplatin-only controls [184]. Using GRP78 and calpain-1 as ER stress markers, the authors showed that both rosuvastatin and simvastatin blocked ER stress-mediated apoptosis, shifting the balance toward a more anti-apoptotic state, as evidenced by a lower Bax/Bcl2 ratio [184]. Phytochemicals such as hesperidin, found in citrus fruits [12], have also been studied for their cardioprotective role in cisplatin-induced cardiotoxicity. In both mice and rat models, researchers found both low-and high-dose hesperidin treatment significantly reduced ROS and MDA levels and increased SOD, CAT, and GSH [185, 186]. In addition, hesperidin treatment reduced the cisplatin-mediated TNF-α and IL-6 and upregulated anti-apoptotic Bcl-2 while downregulating pro-apoptotic Bax and caspase-3 [185]. Cisplatin enhances Keap1 expression, which binds to and suppresses Nrf2, preventing activation of antioxidant pathways [185, 187, 188]. p62 binds Keap1 and disrupts Keap1 inhibition of Nrf2, thus promoting Nrf2 activity [187]. Hesperidin enhances p62 and Nrf2 activity while inhibiting Keap1 expression in cisplatin-treated mice [185]. Curcumin, the biologically active compound in turmeric, exhibits antioxidant, anti-inflammatory, antimicrobial, and anti-angiogenic properties [189, 190]. β-carotene, a vitamin A precursor, also has strong antioxidant potential [191]. Treatment with either curcumin or β-carotene alongside cisplatin, increased CAT, SOD, and GSH levels,while decreasing MDA and lipid peroxidation [192]. Curcumin and β-carotene also alleviated the increase in inflammatory cytokines (TNF-α, IL-1β, and IL-6) induced by cisplatin therapy [192]. Semaglutide is a glucagon-like peptide 1 (GLP1) agonist used in the treatment of T2DM. Previous research has shown that GLP-1 and GLP-1 agonists protect against oxidative stress and activate the Nrf2 pathway [193]. Semaglutide normalized cisplatin-induced mitochondrial dysfunction and redox imbalance [194].

Cyclophosphamide

Cyclophosphamide is a nitrogen mustard derivative used in treatment of various neoplasms, including lymphomas, sarcomas, breast cancer, and multiple myeloma [192, 198]. Owing to its immunosuppressive effects, it is also used to manage autoimmune diseases [199]. Cyclophosphamide and its metabolites function as alkylating agents which permanently cross-links DNA at 7-guanine residues, promoting cell cycle arrest and apoptosis [200]. Cyclophosphamide is first metabolized in the liver to hydroxycyclophosphamide, then aldophosphamide, and finally into the active alkylating agent phosphoramide mustard and a by-product known as acrolein [201]. Acrolein is regarded as a toxic metabolite with no anti-tumor activity and is most responsible for cyclophosphamide-induced cardiotoxicity [200, 201], especially when cyclophosphamide is administered at high doses (120–200 mg/kg i.v.) [202]. Other cyclophosphamide metabolites, such as aldophosphamide and 4-hydroxy phosphamide, may also play a role in inducing cardiotoxicity [201]. Cyclophosphamide or its metabolites cause cardiotoxicity by forming protein adducts in cardiomyocytes, which reduce antioxidant levels (GSH, SOD, CAT), increase ROS, lipid peroxidation, and MDA levels, thereby promoting oxidative stress [203].

Acrolein can form adducts with lysine residues in proteins, which react with GSH and cause oxidative stress, as well as with cysteine, leading to the activation of caspases and NF-kB-p65 subunit, which induce apoptosis and inflammatory cytokine release (TNF-α, IL-6 and IL-1B), respectively [199, 204]. The Keap1-Nrf2-HO-1 system, discussed previously, is also involved in this oxidative stress pathway. Cyclophosphamide or its metabolites can also cause endothelial damage by reducing eNOS synthase, leading to nitrative stress [199]. Finally, there is also evidence for the involvement of the PI3K-Akt pathway in cyclophosphamide-induced toxicity [197, 205], with dysregulation of PI3K-Akt signaling leading to cardiomyocyte apoptosis and impaired Nrf2 signaling [206, 207]. In combination, these processes are poised to initiate and propagate cardiotoxicity and cardiovascular disease, including coronary artery disease, arrhythmia, HF, and electrical conduction dysfunction [130].

Targeting the mechanisms underlying cyclophosphamide-induced cardiotoxicity is essential for therapeutic development and have been elucidated using in vitro and in vivo animal models. Metformin, commonly employed in the treatment of T2DM, and belonging to a class of drugs known as biguanides, decreases hepatic glucose production. Owing to previous evidence demonstrating metformin has antioxidant properties [208], a recent trial explored metformin as a cardioprotective agent in preventing cyclophosphamide-mediated toxicity. In the study, pretreatment with metformin attenuated cyclophosphamide cardiotoxicity by regulating Bax/Bcl-2 and SIRT-1/SOD pathways [209]. Bax/Bcl-2 is an apoptotic pathway, with higher Bax levels indicating a pro-apoptotic state. Changes in SIRT-1 activity are inversely related to ROS levels [210]. Metformin also increased GSH and SOD expression, decreased MDA levels, and normalized cyclophosphamide-induced changes in Bax [209]. Metformin also upregulated Sirt-1 [209]. Lansoprazole is a proton pump inhibitor recognized for its effects as an antioxidant inducer to stress responses in a dose-dependent relationship with Nrf2 [195, 196]. In Wistar rats, lansoprazole effectively reduced levels of oxidative stress when administered with cyclophosphamide, as evidenced by decreased MDA, increased GSH and SOD levels, and upregulation of Nrf2 [197]. In addition, lansoprazole activated PI3K/Akt signaling and reduced cardiomyocytes inflammation via down-regulation of NF-kB [198].

Natural compounds have also been used alone or in combination with other drugs to prevent cardiotoxicity. Acetovanillone, a plant derivative extracted from *Picrorhiza kurroa*, is known to inhibit NOX activity through suppression of cellular ROS production [207]. Edaravone is a potent free-radical scavenger and antioxidant with documented neuroprotective effects in stroke patients [205]. Acetovanillone and edaravone administration was found to significantly decrease lipid peroxidation and increase antioxidant (GSH and SOD) levels, alleviating cyclophosphamide-induced ROS [206]. Both agents reduced Keap1 levels, upregulated Nrf2, and increased PI3K/Akt, and mTOR expression [207].

Sitagliptin a dipeptidyl peptidase IV inhibitor commonly used to treat T2DM, increases levels of GLP-1 [211]. Research exploring the role of sitagliptin in combating cyclophosphamide toxicity found that sitagliptin negated the cyclophosphamide-induced increases in MDA and decreases in SOD, GSH, and CAT [212]. The administration of NAC also exhibited similar cardioprotective effects by influencing MDA and NOX levels, along with a significant restoration of GSH content, GSH-Px, CAT, and SOD activities [212].

Nicorandil, primarily used in the treatment of angina and acute HF [213], also shows promising potential in blocking cyclophosphamide-induced cardiotoxicity. In adult male Wistar albino rats, co-administration of nicorandil reversed cyclophosphamide induced cardiotoxicity through activation of cellular ATP-sensitive potassium channels and stimulation of eNOS gene expression [214]. The mechanism by which nicorandil-mediated eNOS expression reduces oxidative stress is supported by evidence that, under specific oxidative stress conditions, upregulation of eNOS enhances SOD production. Notably, NO forming peroxynitrate is more rapidly consumed by the excess SOD produced by eNOS [214].

Cilostazol is a pharmaceutical agent used to treat claudication through its selective inhibition of phosphodiesterase III [215]. It also exhibits promising cardioprotective effects via modulation of Nfr2 translocation into the nucleus. Koh et al. reported a significant reduction in serum levels of BNP, dysferlin, high-mobility group protein B1 (HMGB1), TLR-4, NF-κB p65, and COX-2 in mice treated with

DOXO plus cilostazol [215]. When administered with cyclophosphamide, cilostazol increased antioxidant defenses and inhibited NF-kB-p65-mediated inflammation, through the up-regulation of SIRT1 [215].

Emerging roles for vitamin E, and simvastatin in treating cyclophosphamide-induced toxicity have also been documented [216, 389]. Vitamin E with cyclophosphamide treatment, significantly improved all cardiac biomarkers via its ability to scavenge free radicals, reduce lipid peroxidation, and preserve the integrity of cardiac muscle fibers [389]. Simvastatin also demonstrated efficacy in reversing cyclophosphamide-induced increases in MDA, TNF-α, and the NLRP3-inflammasome/caspase1/IL-1β apoptotic pathway, indicating significant antioxidant, anti-inflammatory, and anti-apoptotic properties [216].

Conclusion

Anthracyclines, taxanes, and alkylating agents are fundamental components of chemotherapy regimens, yet their use is often undermined by their significant cardiotoxic side effect profiles. These agents promote cardiotoxicity through mechanisms such as oxidative stress, myocardial signaling disruption, and direct cardiomyocyte injury, leading to cardiac dysfunction. As cancer therapies advance, there is an increasing emphasis on integrating cardioprotective strategies to mitigate these toxicities and improve long-term patient outcomes. Emerging approaches include personalized medicine, where genetic and biomarker-based screening may enable risk stratification, and individualized treatment plans that balance oncologic efficacy with cardiovascular safety. Additionally, novel cardioprotective agents ranging from small-molecule inhibitors targeting oxidative stress and inflammation to gene therapy are currently under investigation. Technological advancements in targeted-drug delivery hold promise for minimizing off-target toxicity, thereby preserving cardiac function while maintaining the therapeutic potency of chemotherapy. As cancer survival rates continue to rise, the interdisciplinary field of cardio-oncology will play an essential role in optimizing patient care. Future research should focus on refining these cardioprotective strategies to ensure patients not only survive cancer but also maintain long-term cardiovascular health.

Disclosures The authors have no relevant affiliations or financial involvement to disclose.

References

1. Mabudian L, Jordan JH, Bottinor W, Hundley WG. Cardiac MRI assessment of anthracycline-induced cardiotoxicity. Front Cardiovasc Med. 2022;9:903719.
2. Mehta LS, Watson KE, Barac A, Beckie TM, Bittner V, Cruz-Flores S, et al. Cardiovascular disease and breast cancer: where these entities intersect: a scientific statement from the

American Heart Association. Circulation [Internet]. 2018 [cited 2025 Jan 22];137. Available from: https://www.ahajournals.org/doi/10.1161/CIR.0000000000000556.
3. Flaherty KT, Lee SJ, Zhao F, Schuchter LM, Flaherty L, Kefford R, et al. Phase III trial of carboplatin and paclitaxel with or without sorafenib in metastatic melanoma. J Clin Oncol. 2013;31:373–9.
4. Drafts BC, Twomley KM, D'Agostino R, Lawrence J, Avis N, Ellis LR, et al. Low to moderate dose anthracycline-based chemotherapy is associated with early noninvasive imaging evidence of subclinical cardiovascular disease. JACC Cardiovasc Imaging. 2013;6:877–85.
5. Qu P-R, Jiang Z-L, Song P-P, Liu L-C, Xiang M, Wang J. Saponins and their derivatives: potential candidates to alleviate anthracycline-induced cardiotoxicity and multidrug resistance. Pharmacol Res. 2022;182:106352.
6. Rowinsky EK, McGuire WP, Guarnieri T, Fisherman JS, Christian MC, Donehower RC. Cardiac disturbances during the administration of taxol. J Clin Oncol. 1991;9:1704–12.
7. Arbuck SG, Strauss H, Rowinsky E, Christian M, Suffness M, Adams J, et al. A reassessment of cardiac toxicity associated with Taxol. J Natl Cancer Inst Monogr. 1993;15:117–30.
8. Meshkini A, Yazdanparast R. Involvement of oxidative stress in taxol-induced apoptosis in chronic myelogenous leukemia K562 cells. Exp Toxicol Pathol. 2012;64:357–65.
9. Ilinskaya AN, Clogston JD, McNeil SE, Dobrovolskaia MA. Induction of oxidative stress by Taxol® vehicle Cremophor-EL triggers production of interleukin-8 by peripheral blood mononuclear cells through the mechanism not requiring de novo synthesis of mRNA. Nanomed Nanotechnol Biol Med. 2015;11:1925–38.
10. Kukovetz WR, Holzmann S, Pöch G. Molecular mechanism of action of nicorandil. J Cardiovasc Pharmacol. 1992;20:S1–7.
11. Sheibani M, Azizi Y, Shayan M, Nezamoleslami S, Eslami F, Farjoo MH, et al. Doxorubicin-induced cardiotoxicity: an overview on pre-clinical therapeutic approaches. Cardiovasc Toxicol. 2022;22:292–310.
12. Ali AM, Gabbar MA, Abdel-Twab SM, Fahmy EM, Ebaid H, Alhazza IM, et al. Antidiabetic potency, antioxidant effects, and mode of actions of *Citrus reticulata* fruit peel hydroethanolic extract, hesperidin, and quercetin in nicotinamide/streptozotocin-induced Wistar diabetic rats. Oxidative Med Cell Longev. 2020;2020:1–21.
13. Chen Y, Shi S, Dai Y. Research progress of therapeutic drugs for doxorubicin-induced cardiomyopathy. Biomed Pharmacother. 2022;156:113903.
14. Brandão SR, Carvalho F, Amado F, Ferreira R, Costa VM. Insights on the molecular targets of cardiotoxicity induced by anticancer drugs: a systematic review based on proteomic findings. Metabolism. 2022;134:155250.
15. Bhagat A, Kleinerman ES. Anthracycline-induced cardiotoxicity: causes, mechanisms, and prevention. In: Kleinerman ES, Gorlick R, editors. Current advances in osteosarcoma [Internet]. Cham: Springer International Publishing; 2020 [cited 2025 Jan 22]. p. 181–92. Available from: http://link.springer.com/10.1007/978-3-030-43032-0_15.
16. Rabbani A, Finn RM, Ausió J. The anthracycline antibiotics: antitumor drugs that alter chromatin structure: review articles. BioEssays. 2005;27:50–6.
17. Young RC, Ozols RF, Myers CE. The anthracycline antineoplastic drugs. N Engl J Med. 1981;305:139–53.
18. Menna P, Gonzalez Paz O, Chello M, Covino E, Salvatorelli E, Minotti G. Anthracycline cardiotoxicity. Expert Opin Drug Saf. 2012;11:S21–36.
19. Arcamone F, Cassinelli G, Fantini G, Grein A, Orezzi P, Pol C, et al. Adriamycin, 14-hydroxydaimomycin, a new antitumor antibiotic from *S. Peucetius* var. *caesius*. Biotechnol Bioeng. 1969;11:1101–10.
20. Li DL, Hill JA. Cardiomyocyte autophagy and cancer chemotherapy. J Mol Cell Cardiol. 2014;71:54–61.
21. Songbo M, Lang H, Xinyong C, Bin X, Ping Z, Liang S. Oxidative stress injury in doxorubicin-induced cardiotoxicity. Toxicol Lett. 2019;307:41–8.

22. Pashaei Y, Mehrabi M, Shekarchi M. A review on various analytical methods for determination of anthracyclines and their metabolites as anti–cancer chemotherapy drugs in different matrices over the last four decades. TrAC Trends Anal Chem. 2020;130:115991.
23. Khasraw M, Bell R, Dang C. Epirubicin: is it like doxorubicin in breast cancer? A clinical review. Breast. 2012;21:142–9.
24. Zhang Y, Li Q, Xu D, Li T, Gu Z, Huang P, et al. Idarubicin-induced oxidative stress and apoptosis in cardiomyocytes: An *in vitro* molecular approach. Hum Exp Toxicol. 2021;40:S553–62.
25. Taneja SS. Complications of urologic surgery: prevention and management. 4th ed. Philadelphia: Saunders/Elsevier; 2010.
26. Barlow LJ, Benson MC. Experience with newer intravesical chemotherapy for high-risk non-muscle-invasive bladder cancer. Curr Urol Rep. 2013;14:65–70.
27. Kwok JC, Richardson DR. Anthracyclines induce accumulation of iron in ferritin in myocardial and neoplastic cells: inhibition of the ferritin iron mobilization pathway. Mol Pharmacol. 2003;63:849–61.
28. Canzoneri JC, Oyelere AK. Interaction of anthracyclines with iron responsive element mRNAs. Nucleic Acids Res. 2008;36:6825–34.
29. Eder AR, Arriaga EA. Capillary electrophoresis monitors enhancement in subcellular reactive oxygen species production upon treatment with doxorubicin. Chem Res Toxicol. 2006;19:1151–9.
30. Minotti G, Ronchi R, Salvatorelli E, Menna P, Cairo G. Doxorubicin irreversibly inactivates iron regulatory proteins 1 and 2 in cardiomyocytes: evidence for distinct metabolic pathways and implications for iron-mediated cardiotoxicity of antitumor therapy. Cancer Res. 2001;61:8422–8.
31. Trajković S, Dobrić S, Jaćević V, Dragojević-Simić V, Milovanović Z, Đorđević A. Tissue-protective effects of fullerenol C60(OH)24 and amifostine in irradiated rats. Colloids Surf B Biointerfaces. 2007;58:39–43.
32. Fang X, Wang H, Han D, Xie E, Yang X, Wei J, et al. Ferroptosis as a target for protection against cardiomyopathy. Proc Natl Acad Sci. 2019;116:2672–80.
33. Chen H, Zhu J, Le Y, Pan J, Liu Y, Liu Z, et al. Salidroside inhibits doxorubicin-induced cardiomyopathy by modulating a ferroptosis-dependent pathway. Phytomed Int J Phytother Phytopharm. 2022;99:153964.
34. Okpara ES, Adedara IA, Guo X, Klos ML, Farombi EO, Han S. Molecular mechanisms associated with the chemoprotective role of protocatechuic acid and its potential benefits in the amelioration of doxorubicin-induced cardiotoxicity: a review. Toxicol Rep. 2022;9:1713–24.
35. Al-Shabanah OA, Aleisa AM, Hafez MM, Al-Rejaie SS, Al-Yahya AA, Bakheet SA, et al. Desferrioxamine attenuates doxorubicin-induced acute cardiotoxicity through TFG-β/Smad p53 pathway in rat model. Oxidative Med Cell Longev. 2012;2012:619185.
36. Quagliariello V, De Laurentiis M, Rea D, Barbieri A, Monti MG, Carbone A, et al. The SGLT-2 inhibitor empagliflozin improves myocardial strain, reduces cardiac fibrosis and pro-inflammatory cytokines in non-diabetic mice treated with doxorubicin. Cardiovasc Diabetol. 2021;20:150.
37. Rawat PS, Jaiswal A, Khurana A, Bhatti JS, Navik U. Doxorubicin-induced cardiotoxicity: An update on the molecular mechanism and novel therapeutic strategies for effective management. Biomed Pharmacother. 2021;139:111708.
38. Vedam K, Nishijima Y, Druhan LJ, Khan M, Moldovan NI, Zweier JL, et al. Role of heat shock factor-1 activation in the doxorubicin-induced heart failure in mice. Am J Physiol-Heart Circ Physiol. 2010;298:H1832–41.
39. Liu P, Bao H, Jin C, Zhou J, Hua F, Li K, et al. Targeting extracellular heat shock protein 70 ameliorates doxorubicin-induced heart failure through resolution of toll-like receptor 2–mediated myocardial inflammation. J Am Heart Assoc. 2019;8:e012338.

40. Chen L, Yan K-P, Liu X-C, Wang W, Li C, Li M, et al. Valsartan regulates TGF-β/Smads and TGF-β/p38 pathways through lncRNA CHRF to improve doxorubicin-induced heart failure. Arch Pharm Res. 2018;41:101–9.
41. Dindaş F, Güngör H, Ekici M, Akokay P, Erhan F, Doğduş M, et al. Angiotensin receptor-neprilysin inhibition by sacubitril/valsartan attenuates doxorubicin-induced cardiotoxicity in a pretreatment mice model by interfering with oxidative stress, inflammation, and Caspase 3 apoptotic pathway. Anatol J Cardiol. 2021;25:821–8.
42. Jiang M, Li J, Peng Q, Liu Y, Liu W, Luo C, et al. Neuroprotective effects of bilobalide on cerebral ischemia and reperfusion injury are associated with inhibition of pro-inflammatory mediator production and down-regulation of JNK1/2 and p38 MAPK activation. J Neuroinflammation. 2014;11:167.
43. Hu X, Li C, Wang Q, Wei Z, Chen T, Wang Y, et al. Dimethyl fumarate ameliorates doxorubicin-induced cardiotoxicity by activating the Nrf2 pathway. Front Pharmacol. 2022;13:872057.
44. Zhan L, Wang X, Zhang Y, Zhu G, Ding Y, Chen X, et al. Benazepril hydrochloride protects against doxorubicin cardiotoxicity by regulating the PI3K/Akt pathway. Exp Ther Med. 2021;22:1082.
45. Singal P, Li T, Kumar D, Danelisen I, Iliskovic N. Adriamycin-induced heart failure: mechanism and modulation. Mol Cell Biochem. 2000;207:77–86.
46. Jiang L, Gong Y, Hu Y, You Y, Wang J, Zhang Z, et al. Peroxiredoxin-1 overexpression attenuates doxorubicin-induced cardiotoxicity by inhibiting oxidative stress and cardiomyocyte apoptosis. Oxidative Med Cell Longev. 2020;2020:2405135.
47. Park MH, Jo M, Kim YR, Lee C-K, Hong JT. Roles of peroxiredoxins in cancer, neurodegenerative diseases and inflammatory diseases. Pharmacol Ther. 2016;163:1–23.
48. Prospéri MT, Ferbus D, Karczinski I, Goubin G. A human cDNA corresponding to a gene overexpressed during cell proliferation encodes a product sharing homology with amoebic and bacterial proteins. J Biol Chem. 1993;268:11050–6.
49. Leng Y, Wu Y, Lei S, Zhou B, Qiu Z, Wang K, et al. Inhibition of HDAC6 activity alleviates myocardial ischemia/reperfusion injury in diabetic rats: potential role of peroxiredoxin 1 acetylation and redox regulation. Oxidative Med Cell Longev. 2018;2018:9494052.
50. Schreibelt G, Van Horssen J, Haseloff RF, Reijerkerk A, Van Der Pol SMA, Nieuwenhuizen O, et al. Protective effects of peroxiredoxin-1 at the injured blood–brain barrier. Free Radic Biol Med. 2008;45:256–64.
51. Nishida T, Hattori K, Watanabe K. The regulatory and signaling mechanisms of the ASK family. Adv Biol Regul. 2017;66:2–22.
52. Ichijo H, Nishida E, Irie K, Dijke PT, Saitoh M, Moriguchi T, et al. Induction of apoptosis by ASK1, a mammalian MAPKKK that activates SAPK/JNK and p38 signaling pathways. Science. 1997;275:90–4.
53. Bogan KL, Brenner C. Nicotinic acid, nicotinamide, and nicotinamide riboside: a molecular evaluation of NAD+ precursor vitamins in human nutrition. Annu Rev Nutr. 2008;28:115–30.
54. Park JH, Long A, Owens K, Kristian T. Nicotinamide mononucleotide inhibits post-ischemic NAD+ degradation and dramatically ameliorates brain damage following global cerebral ischemia. Neurobiol Dis. 2016;95:102–10.
55. Zhang R, Shen Y, Zhou L, Sangwung P, Fujioka H, Zhang L, et al. Short-term administration of Nicotinamide Mononucleotide preserves cardiac mitochondrial homeostasis and prevents heart failure. J Mol Cell Cardiol. 2017;112:64–73.
56. Wei C-C, Kong Y-Y, Li G-Q, Guan Y-F, Wang P, Miao C-Y. Nicotinamide mononucleotide attenuates brain injury after intracerebral hemorrhage by activating Nrf2/HO-1 signaling pathway. Sci Rep. 2017;7:717.
57. Wei S, Ma W, Zhang B, Li W. NLRP3 inflammasome: a promising therapeutic target for drug-induced toxicity. Front Cell Dev Biol. 2021;9:634607.
58. Boini KM, Hussain T, Li P-L, Koka S. Trimethylamine-N-oxide instigates NLRP3 inflammasome activation and endothelial dysfunction. Cell Physiol Biochem Int J Exp Cell Physiol Biochem Pharmacol. 2017;44:152–62.

59. Wan Y, He B, Zhu D, Wang L, Huang R, Zhu J, et al. Nicotinamide mononucleotide attenuates doxorubicin-induced cardiotoxicity by reducing oxidative stress, inflammation and apoptosis in rats. Arch Biochem Biophys. 2021;712:109050.
60. Dulhunty AF, Casarotto MG, Beard NA. The ryanodine receptor: a pivotal Ca2+ regulatory protein and potential therapeutic drug target. Curr Drug Targets. 2011;12:709–23.
61. Dewenter M, Von Der Lieth A, Katus HA, Backs J. Calcium signaling and transcriptional regulation in cardiomyocytes. Circ Res. 2017;121:1000–20.
62. Aziz AUR, Geng C, Li W, Yu X, Qin K-R, Wang H, et al. Doxorubicin induces ER calcium release via Src in rat ovarian follicles. Toxicol Sci. 2019;168:171–8.
63. Octavia Y, Tocchetti CG, Gabrielson KL, Janssens S, Crijns HJ, Moens AL. Doxorubicin-induced cardiomyopathy: from molecular mechanisms to therapeutic strategies. J Mol Cell Cardiol. 2012;52:1213–25.
64. Nieminen MS, Akkila J, Hasenfuss G, Kleber FX, Lehtonen LA, Mitrovic V, et al. Hemodynamic and neurohumoral effects of continuous infusion of levosimendan in patients with congestive heart failure. J Am Coll Cardiol. 2000;36:1903–12.
65. García-González MJ, Jorge-Pérez P, Jiménez-Sosa A, Acea AB, Lacalzada Almeida JB, Ferrer Hita JJ. Levosimendan improves hemodynamic status in critically ill patients with severe aortic stenosis and left ventricular dysfunction: an interventional study. Cardiovasc Ther. 2015;33:193–9.
66. Pollesello P, Parissis J, Kivikko M, Harjola V-P. Levosimendan meta-analyses: is there a pattern in the effect on mortality? Int J Cardiol. 2016;209:77–83.
67. Belletti A, Castro ML, Silvetti S, Greco T, Biondi-Zoccai G, Pasin L, et al. The effect of inotropes and vasopressors on mortality: a meta-analysis of randomized clinical trials. Br J Anaesth. 2015;115:656–75.
68. Sorsa T, Pollesello P, Rosevear PR, Drakenberg T, Kilpeläinen I. Stereoselective binding of levosimendan to cardiac troponin C causes Ca2+-sensitization. Eur J Pharmacol. 2004;486:1–8.
69. Sorsa T, Heikkinen S, Abbott MB, Abusamhadneh E, Laakso T, Tilgmann C, et al. Binding of levosimendan, a calcium sensitizer, to cardiac troponin C. J Biol Chem. 2001;276:9337–43.
70. Husedzinovic I, Milanovic R, Husedzinovic S. Levosimendan in acute heart failure. Signa Vitae. 2008;3:13.
71. Mushtaq S, Andreini D, Farina S, Salvioni E, Pontone G, Sciomer S, et al. Levosimendan improves exercise performance in patients with advanced chronic heart failure. ESC Heart Fail. 2015;2:133–41.
72. Gruhn N, Nielsen-Kudsk JE, Theilgaard S, Bang L, Olesen S-P, Aldershvile J. Coronary vasorelaxant effect of levosimendan, a new inodilator with calcium-sensitizing properties. J Cardiovasc Pharmacol. 1998;31:741–9.
73. Papp Z, Édes I, Fruhwald S, De Hert SG, Salmenperä M, Leppikangas H, et al. Levosimendan: molecular mechanisms and clinical implications. Int J Cardiol. 2012;159:82–7.
74. Su H, Yao S, Zhao W, Li M, Liu J, Shang W, et al. Anti-SARS-CoV-2 activities in vitro of Shuanghuanglian preparations and bioactive ingredients. Acta Pharmacol Sin. 2020;41:1167–77.
75. Wang H, Hui K-M, Xu S, Chen Y, Wong JTF, Xue H. Two flavones from Scutellaria baicalensis Georgi and their binding affinities to the benzodiazepine site of the GABAA receptor complex. Pharmazie. 2002;57:857–8.
76. Hui K, Wang X, Xue H. Interaction of flavones from the roots of *Scutellaria baicalensis* with the benzodiazepine site. Planta Med. 2009;66:91–3.
77. Dinda B, Dinda S, DasSharma S, Banik R, Chakraborty A, Dinda M. Therapeutic potentials of baicalin and its aglycone, baicalein against inflammatory disorders. Eur J Med Chem. 2017;131:68–80.
78. Wang L, Ling Y, Chen Y, Li C-L, Feng F, You Q-D, et al. Flavonoid baicalein suppresses adhesion, migration and invasion of MDA-MB-231 human breast cancer cells. Cancer Lett. 2010;297:42–8.

79. Shao Z-H, Vanden Hoek TL, Qin Y, Becker LB, Schumacker PT, Li C-Q, et al. Baicalein attenuates oxidant stress in cardiomyocytes. Am J Physiol-Heart Circ Physiol. 2002;282:H999–1006.
80. Lin M-Y, Cheng W-T, Cheng H-C, Chou W-C, Chen H-I, Ou H-C, et al. Baicalin enhances chemosensitivity to doxorubicin in breast cancer cells via upregulation of oxidative stress-mediated mitochondria-dependent apoptosis. Antioxidants. 2021;10:1506.
81. Auld F, Maschauer EL, Morrison I, Skene DJ, Riha RL. Evidence for the efficacy of melatonin in the treatment of primary adult sleep disorders. Sleep Med Rev. 2017;34:10–22.
82. Boutin JA, Audinot V, Ferry G, Delagrange P. Molecular tools to study melatonin pathways and actions. Trends Pharmacol Sci. 2005;26:412–9.
83. Hardeland R. Antioxidative protection by melatonin: multiplicity of mechanisms from radical detoxification to radical avoidance. Endocrine. 2005;27:119–30.
84. Reiter RJ, Acuña-Castroviejo D, Tan DX, Burkhardt S. Free radical-mediated molecular damage. Mechanisms for the protective actions of melatonin in the central nervous system. Ann N Y Acad Sci. 2001;939:200–15.
85. Sharafati-Chaleshtori R, Shirzad H, Rafieian-Kopaei M, Soltani A. Melatonin and human mitochondrial diseases. J Res Med Sci. 2017;22:2.
86. Jockers R, Delagrange P, Dubocovich ML, Markus RP, Renault N, Tosini G, et al. Update on melatonin receptors: IUPHAR Review 20. Br J Pharmacol. 2016;173:2702–25.
87. Zhang W, Wang X, Tang Y, Huang C. Melatonin alleviates doxorubicin-induced cardiotoxicity via inhibiting oxidative stress, pyroptosis and apoptosis by activating Sirt1/Nrf2 pathway. Biomed Pharmacother. 2023;162:114591. https://doi.org/10.1016/j.biopha.2023.114591. Epub 2023 Mar 23. PMID: 36965257.
88. Dionne CA, Crumley G, Bellot F, Kaplow JM, Searfoss G, Ruta M, et al. Cloning and expression of two distinct high-affinity receptors cross-reacting with acidic and basic fibroblast growth factors. EMBO J. 1990;9:2685–92.
89. Liao S, Bodmer J, Pietras D, Azhar M, Doetschman T, Schultz JEJ. Biological functions of the low and high molecular weight protein isoforms of fibroblast growth factor-2 in cardiovascular development and disease. Dev Dyn. 2009;238:249–64.
90. Koleini N, Nickel BE, Wang J, Roveimiab Z, Fandrich RR, Kirshenbaum LA, et al. Fibroblast growth factor-2-mediated protection of cardiomyocytes from the toxic effects of doxorubicin requires the mTOR/Nrf-2/HO-1 pathway. Oncotarget. 2017;8:87415–30.
91. House SL, Bolte C, Zhou M, Doetschman T, Klevitsky R, Newman G, et al. Cardiac-specific overexpression of fibroblast growth factor-2 protects against myocardial dysfunction and infarction in a murine model of low-flow ischemia. Circulation. 2003;108:3140–8.
92. Santiago J-J, McNaughton LJ, Koleini N, Ma X, Bestvater B, Nickel BE, et al. High molecular weight fibroblast growth factor-2 in the human heart is a potential target for prevention of cardiac remodeling. Sussman MA, editor. PLoS One. 2014;9:e97281.
93. Kardami E, Detillieux K, Ma X, Jiang Z, Santiago J-J, Jimenez SK, et al. Fibroblast growth factor-2 and cardioprotection. Heart Fail Rev. 2007;12:267–77.
94. Liao S, Porter D, Scott A, Newman G, Doetschman T, Schultz JEJ. The cardioprotective effect of the low molecular weight isoform of fibroblast growth factor-2: the role of JNK signaling. J Mol Cell Cardiol. 2007;42:106–20.
95. Abushouk AI, Ismail A, Salem AMA, Afifi AM, Abdel-Daim MM. Cardioprotective mechanisms of phytochemicals against doxorubicin-induced cardiotoxicity. Biomed Pharmacother. 2017;90:935–46.
96. Molyneux RJ, Lee ST, Gardner DR, Panter KE, James LF. Phytochemicals: the good, the bad and the ugly? Phytochemistry. 2007;68:2973–85.
97. Linus Pauling Institute | Oregon State University [Internet]. Linus Pauling Institute. 2014 [cited 2025 Jan 22]. Available from: https://lpi.oregonstate.edu/mic.
98. Lin J, Fang L, Li H, Li Z, Lyu L, Wang H, et al. Astragaloside IV alleviates doxorubicin induced cardiomyopathy by inhibiting NADPH oxidase derived oxidative stress. Eur J Pharmacol. 2019;859:172490.

99. Qiao X, van der Zanden SY, Wander DPA, Borràs DM, Song J-Y, Li X, et al. Uncoupling DNA damage from chromatin damage to detoxify doxorubicin. Proc Natl Acad Sci USA. 2020;117:15182–92.
100. Seke M, Petrovic D, Djordjevic A, Jovic D, Borovic ML, Kanacki Z, et al. Fullerenol/doxorubicin nanocomposite mitigates acute oxidative stress and modulates apoptosis in myocardial tissue. Nanotechnology. 2016;27:485101.
101. Tocchetti CG, Carpi A, Coppola C, Quintavalle C, Rea D, Campesan M, et al. Ranolazine protects from doxorubicin-induced oxidative stress and cardiac dysfunction. Eur J Heart Fail. 2014;16:358–66.
102. Jacobs AL, Schär P. DNA glycosylases: in DNA repair and beyond. Chromosoma. 2012;121:1–20.
103. Anene-Nzelu CG, Li PY, Luu TDA, Ng SL, Tiang Z, Pan B, et al. 8-Oxoguanine DNA glycosylase (OGG1) deficiency exacerbates doxorubicin-induced cardiac dysfunction. Virág L, editor. Oxidative Med Cell Longev. 2022;2022:1–11.
104. Torres-Gonzalez M. Mitochondrial 8-oxoguanine glycosylase decreases mitochondrial fragmentation and improves mitochondrial function in H9C2 cells under oxidative stress conditions. Am Physiol Soc. 2014;306:C221.
105. Wang J, Wang Q, Watson LJ, Jones SP, Epstein PN. Cardiac overexpression of 8-oxoguanine DNA glycosylase 1 protects mitochondrial DNA and reduces cardiac fibrosis following transaortic constriction. Am J Physiol Heart Circ Physiol. 2011;301:H2073–80.
106. USDA-ARS Germplasm Resources Information Network (GRIN) [Internet]. [cited 2025 Jan 22]. Available from: https://www.ars-grin.gov/.
107. Raghavamma STV. In vitro antioxidant potential of crude extract from leaves of Nauclea orientalis Linn. J Pharm Res. 2011;4:1548–9.
108. Sandamali JAN, Hewawasam RP, Jayatilaka KAPW, Mudduwa LKB. Nauclea orientalis (L.) bark extract protects rat Cardiomyocytes from doxorubicin-induced oxidative stress, inflammation, apoptosis, and DNA fragmentation. Lloret A, editor. Oxidative Med Cell Longev. 2022;2022:1–19.
109. Ayala A, Muñoz MF, Argüelles S. Lipid peroxidation: production, metabolism, and signaling mechanisms of malondialdehyde and 4-hydroxy-2-nonenal. Oxidative Med Cell Longev. 2014;2014:1–31.
110. Free radical lipid peroxidation: mechanisms and analysis | Chem Rev [Internet]. [cited 2025 Jan 22]. Available from: https://pubs.acs.org/doi/10.1021/cr200084z.
111. Moldovan L, Moldovan NI. Oxygen free radicals and redox biology of organelles. Histochem Cell Biol. 2004;122:395–412.
112. Lane N. Oxygen: the molecule that made the world. Revised impression. Oxford: Oxford University Press; 2016.
113. Fruhwirth GO, Loidl A, Hermetter A. Oxidized phospholipids: from molecular properties to disease. Biochim Biophys Acta. 2007;1772:718–36.
114. Kinnunen PKJ, Kaarniranta K, Mahalka AK. Protein-oxidized phospholipid interactions in cellular signaling for cell death: from biophysics to clinical correlations. Biochim Biophys Acta. 2012;1818:2446–55.
115. Reis A, Spickett CM. Chemistry of phospholipid oxidation. Biochim Biophys Acta. 2012;1818:2374–87.
116. Martinez-Huenchullan SF, Tam CS, Ban LA, Ehrenfeld-Slater P, Mclennan SV, Twigg SM. Skeletal muscle adiponectin induction in obesity and exercise. Metabolism. 2020;102:154008.
117. Ghoshal K, Chatterjee T, Chowdhury S, Sengupta S, Bhattacharyya M. Adiponectin genetic variant and expression coupled with lipid peroxidation reveal new signatures in diabetic dyslipidemia. Biochem Genet. 2021;59:781–98.
118. Oh DK, Ciaraldi T, Henry RR. Adiponectin in health and disease. Diabetes Obes Metab. 2007;9:282–9.

119. Zhao D, Xue C, Li J, Feng K, Zeng P, Chen Y, et al. Adiponectin agonist ADP355 ameliorates doxorubicin-induced cardiotoxicity by decreasing cardiomyocyte apoptosis and oxidative stress. Biochem Biophys Res Commun. 2020;533:304–12.
120. Li S, Shin HJ, Ding EL, van Dam RM. Adiponectin levels and risk of type 2 diabetes: a systematic review and meta-analysis. JAMA. 2009;302:179–88.
121. Ukkola O, Santaniemi M. Adiponectin: a link between excess adiposity and associated comorbidities? J Mol Med Berl Ger. 2002;80:696–702.
122. Wang Y, Jin Y, Wang Y, Li L, Liao Y, Zhang Y, et al. The effect of folic acid in patients with cardiovascular disease: a systematic review and meta-analysis. Medicine (Baltimore). 2019;98:e17095.
123. Octavia Y, Kararigas G, de Boer M, Chrifi I, Kietadisorn R, Swinnen M, et al. Folic acid reduces doxorubicin-induced cardiomyopathy by modulating endothelial nitric oxide synthase. J Cell Mol Med. 2017;21:3277–87.
124. Ahmad S, Panda BP, Kohli K, Fahim M, Dubey K. Folic acid ameliorates celecoxib cardiotoxicity in a doxorubicin heart failure rat model. Pharm Biol. 2017;55:1295–303.
125. Cruzat V, Macedo Rogero M, Noel Keane K, Curi R, Newsholme P. Glutamine: metabolism and immune function, supplementation and clinical translation. Nutrients. 2018;10:1564.
126. Xue H, Ren W, Denkinger M, Schlotzer E, Wischmeyer PE. Nutrition modulation of cardiotoxicity and anticancer efficacy related to doxorubicin chemotherapy by glutamine and ω-3 polyunsaturated fatty acids. J Parenter Enter Nutr. 2016;40:52–66.
127. Kuşçu GC, Gürel Ç, Buhur A, Karabay Yavaşoğlu NÜ, Köse T, Yavaşoğlu A, et al. Fluvastatin alleviates doxorubicin-induced cardiac and renal toxicity in rats via regulation of oxidative stress, inflammation, and apoptosis associated genes expressions. Drug Chem Toxicol. 2023;46:400–11.
128. Huang J, Lei Y, Lei S, Gong X. Cardioprotective effects of corilagin on doxorubicin induced cardiotoxicity via P13K/Akt and NF-κB signaling pathways in a rat model. Toxicol Mech Methods. 2022;32:79–86.
129. Chen Y, Chen C. Corilagin prevents tert-butyl hydroperoxide-induced oxidative stress injury in cultured N9 murine microglia cells. Neurochem Int. 2011;59:290–6.
130. Zhuang Z, Zhao X, Wu Y, Huang R, Zhu L, Zhang Y, et al. The anti-apoptotic effect of PI3K-Akt signaling pathway after subarachnoid hemorrhage in rats. Ann Clin Lab Sci. 2011;41:364–72.
131. Aswar U, Mahajan U, Kandhare A, Aswar M. Ferulic acid ameliorates doxorubicin-induced cardiac toxicity in rats. Naunyn Schmiedeberg's Arch Pharmacol. 2019;392:659–68.
132. Maurya DK, Devasagayam TPA. Antioxidant and prooxidant nature of hydroxycinnamic acid derivatives ferulic and caffeic acids. Food Chem Toxicol Int J Publ Br Ind Biol Res Assoc. 2010;48:3369–73.
133. de Oliveira Silva E, Batista R. Ferulic acid and naturally occurring compounds bearing a feruloyl moiety: a review on their structures, occurrence, and potential health benefits. Compr Rev Food Sci Food Saf. 2017;16:580–616.
134. PubChem. Vitexin [Internet]. [cited 2025 Jan 22]. Available from: https://pubchem.ncbi.nlm.nih.gov/compound/5280441.
135. Sun Z, Yan B, Yu WY, Yao X, Ma X, Sheng G, et al. Vitexin attenuates acute doxorubicin cardiotoxicity in rats via the suppression of oxidative stress, inflammation and apoptosis and the activation of FOXO3a. Exp Ther Med. 2016;12:1879–84.
136. Wang X, Hu X, Chen G, Yuan X, Yang R, Liang S, et al. Effects of vitexin on the pharmacokinetics and mRNA expression of CYP isozymes in rats: vitexin on cytochrome P450 enzymes. Phytother Res. 2015;29:366–72.
137. Furey A, Tassell M, Kingston R, Gilroy D, Lehane M. Hawthorn (Crataegus spp.) in the treatment of cardiovascular disease. Pharmacogn Rev. 2010;4:32.
138. Xue H-F, Ying Z-M, Zhang W-J, Meng Y-H, Ying X-X, Kang T-G. Hepatic, gastric, and intestinal first-pass effects of vitexin in rats. Pharm Biol. 2014;52:967–71.

139. Li B, Kim DS, Yadav RK, Kim HR, Chae HJ. Sulforaphane prevents doxorubicin-induced oxidative stress and cell death in rat H9c2 cells. Int J Mol Med. 2015;36(1):53–64. https://doi.org/10.3892/ijmm.2015.2199. Epub 2015 Apr 28. PMID: 25936432; PMCID: PMC4494600.
140. Yoon H-Y, Kang N-I, Lee H-K, Jang KY, Park J-W, Park B-H. Sulforaphane protects kidneys against ischemia-reperfusion injury through induction of the Nrf2-dependent phase 2 enzyme. Biochem Pharmacol. 2008;75:2214–23.
141. Zhao J, Kobori N, Aronowski J, Dash PK. Sulforaphane reduces infarct volume following focal cerebral ischemia in rodents. Neurosci Lett. 2006;393:108–12.
142. Albers P, Park S-I, Niegisch G, Fechner G, Steiner U, Lehmann J, et al. Randomized phase III trial of 2nd line gemcitabine and paclitaxel chemotherapy in patients with advanced bladder cancer: short-term versus prolonged treatment [German Association of Urological Oncology (AUO) trial AB 20/99]. Ann Oncol. 2011;22:288–94.
143. Citron ML, Berry DA, Cirrincione C, Hudis C, Winer EP, Gradishar WJ, et al. Randomized trial of dose-dense versus conventionally scheduled and sequential versus concurrent combination chemotherapy as postoperative adjuvant treatment of node-positive primary breast cancer: first report of Intergroup Trial C9741/Cancer and Leukemia Group B Trial 9741. J Clin Oncol. 2003;21:1431–9.
144. Miller DS, Filiaci VL, Mannel RS, Cohn DE, Matsumoto T, Tewari KS, et al. Carboplatin and paclitaxel for advanced endometrial cancer: final overall survival and adverse event analysis of a phase III trial (NRG Oncology/GOG0209). J Clin Oncol. 2020;38:3841–50.
145. Sandler A, Gray R, Perry MC, Brahmer J, Schiller JH, Dowlati A, et al. Paclitaxel–carboplatin alone or with bevacizumab for non–small-cell lung cancer. N Engl J Med. 2006;355:2542–50.
146. Lemma GL, Lee J-W, Aisner SC, Langer CJ, Tester WJ, Johnson DH, et al. Phase II study of carboplatin and paclitaxel in advanced thymoma and thymic carcinoma. J Clin Oncol. 2011;29:2060–5.
147. Miller K, Wang M, Gralow J, Dickler M, Cobleigh M, Perez EA, et al. Paclitaxel plus bevacizumab versus paclitaxel alone for metastatic breast cancer. N Engl J Med. 2007;357:2666–76.
148. Serizawa K, Yogo K, Aizawa K, Tashiro Y, Takahari Y, Sekine K, et al. Paclitaxel-induced endothelial dysfunction in living rats is prevented by nicorandil via reduction of oxidative stress. J Pharmacol Sci. 2012;119:349–58.
149. Khaled SS, Soliman HA, Abdel-Gabbar M, Ahmed NA, Attia KAHA, Mahran HA, et al. The preventive effects of naringin and naringenin against paclitaxel-induced nephrotoxicity and cardiotoxicity in male Wistar rats. El-Magd M, editor. Evid Based Complement Alternat Med. 2022;2022:1–11.
150. Ali YA, Ahmed OM, Soliman HA, Abdel-Gabbar M, Al-Dossari M, El-Gawaad NSA, et al. Rutin and hesperidin alleviate paclitaxel-induced nephrocardiotoxicity in Wistar rats *via* suppressing the oxidative stress and enhancing the antioxidant defense mechanisms. El Sayed SM, editor. Evid Based Complement Alternat Med. 2023;2023:5068304.
151. Ralhan R, Kaur J. Alkylating agents and cancer therapy. Expert Opin Ther Pat. 2007;17:1061–75.
152. Siddik ZH. Mechanisms of action of cancer chemotherapeutic agents: DNA-interactive alkylating agents and antitumour platinum-based drugs. In: Alison MR, editor. The cancer handbook [Internet]. 1st ed. Wiley; 2005 [cited 2025 Jan 23]. Available from: https://onlinelibrary.wiley.com/doi/10.1002/0470025077.chap84b.
153. Lawley P. The action of alkylating agents on deoxyribonucleic acid in relation to biological effects of the alkylating agents*1. Exp Cell Res. 1963;9:512–20.
154. Rosenberg B, Van Camp L, Krigas T. Inhibition of cell division in Escherichia coli by electrolysis products from a platinum electrode. Nature. 1965;205:698–9.
155. Dasari S, Bernard Tchounwou P. Cisplatin in cancer therapy: molecular mechanisms of action. Eur J Pharmacol. 2014;740:364–78.
156. El-Awady E-SE, Moustafa YM, Abo-Elmatty DM, Radwan A. Cisplatin-induced cardiotoxicity: mechanisms and cardioprotective strategies. Eur J Pharmacol. 2011;650:335–41.

157. Dugbartey GJ, Peppone LJ, De Graaf IAM. An integrative view of cisplatin-induced renal and cardiac toxicities: molecular mechanisms, current treatment challenges and potential protective measures. Toxicology. 2016;371:58–66.
158. Rybak LP, Whitworth C, Somani S. Application of antioxidants and other agents to prevent cisplatin ototoxicity. Laryngoscope. 1999;109:1740–4.
159. Al-Majed AA, Sayed-Ahmed MM, Al-Yahya AA, Aleisa AM, Al-Rejaie SS, Al-Shabanah OA. Propionyl-l-carnitine prevents the progression of cisplatin-induced cardiomyopathy in a carnitine-depleted rat model. Pharmacol Res. 2006;53:278–86.
160. The prevalence of silent arrhythmia in patients receiving cisplatin-based chemotherapy | Request PDF. ResearchGate [Internet]. 2024 [cited 2025 Jan 23]. Available from: https://www.researchgate.net/publication/286096498_The_prevalence_of_silent_arrhythmia_in_patients_receiving_cisplatin-based_chemotherapy.
161. Khan S, Chen CL, Brady MS, Parameswaran R, Moore R, Hassoun H, et al. Unstable angina associated with cisplatin and carboplatin in a patient with advanced melanoma. J Clin Oncol. 2012;30:e163–4.
162. Moore RA, Adel N, Riedel E, Bhutani M, Feldman DR, Tabbara NE, et al. High incidence of thromboembolic events in patients treated with cisplatin-based chemotherapy: a large retrospective analysis. J Clin Oncol. 2011;29:3466–73.
163. Guglin M, Aljayeh M, Saiyad S, Ali R, Curtis AB. Introducing a new entity: chemotherapy-induced arrhythmia. Europace. 2009;11:1579–86.
164. Kolb RJ, Ghazi MA, Barfuss DW. Inhibition of basolateral transport and cellular accumulation of cDDP and N-acetyl-L-cysteine-cDDP by TEA and PAH in the renal proximal tubule. Cancer Chemother Pharmacol. 2003;51:132–8.
165. Pabla N, Murphy RF, Liu K, Dong Z. The copper transporter Ctr1 contributes to cisplatin uptake by renal tubular cells during cisplatin nephrotoxicity. Am J Physiol-Ren Physiol. 2009;296:F505–11.
166. Howden R. *Nrf2* and cardiovascular defense. Oxidative Med Cell Longev. 2013;2013:1–10.
167. Chen QM, Maltagliati AJ. Nrf2 at the heart of oxidative stress and cardiac protection. Physiol Genomics. 2018;50:77–97.
168. Davies MJ, Merton K, Vijapurkar U, Yee J, Qiu R. Efficacy and safety of canagliflozin in patients with type 2 diabetes based on history of cardiovascular disease or cardiovascular risk factors: a post hoc analysis of pooled data. Cardiovasc Diabetol. 2017;16:40.
169. Verma RS, Mehta A, Srivastava N. In vivo chlorpyrifos induced oxidative stress: attenuation by antioxidant vitamins. Pestic Biochem Physiol. 2007;88:191–6.
170. Ibrahim MA, Bakhaat GA, Tammam HG, Mohamed RM, El-Naggar SA. Cardioprotective effect of green tea extract and vitamin E on Cisplatin-induced cardiotoxicity in mice: toxicological, histological and immunohistochemical studies. Biomed Pharmacother. 2019;113:108731.
171. El-Kashef DH, Sharawy MH. Venlafaxine mitigates cisplatin-induced nephrotoxicity via down-regulating apoptotic pathway in rats. Chem Biol Interact. 2018;290:110–8.
172. Ali MIM, Imbaby S, Arafat HEK, Maher SA, Kolieb E, Ali SM. Cardioprotective and renoprotective effects of venlafaxine on cisplatin-induced cardiotoxicity and nephrotoxicity in rats. Life Sci. 2023;320:121561.
173. Onay-Besikci A, Özkan SA. Trimetazidine revisited: a comprehensive review of the pharmacological effects and analytical techniques for the determination of trimetazidine. Cardiovasc Ther. 2008;26:147–65.
174. Chen P, Hou C, Shibu MA, Day CH, Pai P, Liu Z, et al. Protective effect of co-enzyme Q10 on doxorubicin-induced cardiomyopathy of rat hearts. Environ Toxicol. 2017;32:679–89.
175. El-Sherbeeny NA, Attia GM. The protective effect of trimetazidine against cisplatin-induced nephrotoxicity in rats. Can J Physiol Pharmacol. 2016;94:745–51.
176. Mirmalek SA, Gholamrezaei Boushehrinejad A, Yavari H, Kardeh B, Parsa Y, Salimi-Tabatabaee SA, et al. Antioxidant and anti-inflammatory effects of coenzyme Q10 on L-arginine-induced acute pancreatitis in rat. Cirillo G, editor. Oxid Med Cell Longev. 2016;2016:5818479.

177. Steggall A, Mordi I, Lang C. Targeting metabolic modulation and mitochondrial dysfunction in the treatment of heart failure. Diseases. 2017;5:14.
178. Zhao L. Protective effects of trimetazidine and coenzyme Q10 on cisplatin-induced cardiotoxicity by alleviating oxidative stress and mitochondrial dysfunction. Anatol J Cardiol [Internet]. 2019 [cited 2025 Jan 23]. Available from: https://anatoljcardiol.com/article/AJC-83710.
179. Shinjo T, Tanaka T, Okuda H, Kawaguchi AT, Oh-hashi K, Terada Y, et al. Propofol induces nuclear localization of Nrf2 under conditions of oxidative stress in cardiac H9c2 cells. Mukhopadhyay P, editor. PLoS One. 2018;13:e0196191.
180. Samuni Y, Goldstein S, Dean OM, Berk M. The chemistry and biological activities of N-acetylcysteine. Biochim Biophys Acta BBA – Gen Subj. 2013;1830:4117–29.
181. Rosic G, Selakovic D, Joksimovic J, Srejovic I, Zivkovic V, Tatalović N, et al. The effects of N-acetylcysteine on cisplatin-induced changes of cardiodynamic parameters within coronary autoregulation range in isolated rat hearts. Toxicol Lett. 2016;242:34–46.
182. Gunturk EE, Yucel B, Gunturk I, Yazici C, Yay A, Kose K. The effects of N-acetylcysteine on cisplatin induced cardiotoxicity. Bratisl Med J. 2019;120:423–8.
183. Bergmann MW, Rechner C, Freund C, Baurand A, Jamali AE, Dietz R. Statins inhibit reoxygenation-induced cardiomyocyte apoptosis: role for glycogen synthase kinase 3β and transcription factor β-catenin. J Mol Cell Cardiol. 2004;37:681–90.
184. Saleh DO, Mansour DF, Mostafa RE. Rosuvastatin and simvastatin attenuate cisplatin-induced cardiotoxicity via disruption of endoplasmic reticulum stress-mediated apoptotic death in rats: targeting ER-Chaperone GRP78 and Calpain-1 pathways. Toxicol Rep. 2020;7:1178–86.
185. Jia Y, Guo H, Cheng X, Zhang Y, Si M, Shi J, et al. Hesperidin protects against cisplatin-induced cardiotoxicity in mice by regulating the p62–Keap1–Nrf2 pathway. Food Funct. 2022;13:4205–15.
186. Oguzturk H, Ciftci O, Cetin A, Kaya K, Disli O, Turtay M, et al. Beneficial effects of hesperidin following cis-diamminedichloroplatinum-induced damage in heart of rats. Niger J Clin Pract. 2016;19:99.
187. Silva-Islas CA, Maldonado PD. Canonical and non-canonical mechanisms of Nrf2 activation. Pharmacol Res. 2018;134:92–9.
188. Jaiswal AK. Nrf2 signaling in coordinated activation of antioxidant gene expression. Free Radic Biol Med. 2004;36:1199–207.
189. Jiang S, Han J, Li T, Xin Z, Ma Z, Di W, et al. Curcumin as a potential protective compound against cardiac diseases. Pharmacol Res. 2017;119:373–83.
190. Kumar P, Sulakhiya K, Barua CC, Mundhe N. TNF-α, IL-6 and IL-10 expressions, responsible for disparity in action of curcumin against cisplatin-induced nephrotoxicity in rats. Mol Cell Biochem. 2017;431:113–22.
191. Csepanyi E, Czompa A, Haines D, Lekli I, Bakondi E, Balla G, et al. Cardiovascular effects of low versus high-dose beta-carotene in a rat model. Pharmacol Res. 2015;100:148–56.
192. Bahadır A. Protective effects of curcumin and beta-carotene on cisplatin-induced cardiotoxicity: an experimental rat model. Anatol J Cardiol [Internet]. 2018 [cited 2025 Jan 23]. Available from: https://anatoljcardiol.com/article/AJC-53059.
193. Oh Y, Jun H-S. Effects of glucagon-like peptide-1 on oxidative stress and Nrf2 signaling. Int J Mol Sci. 2017;19:26.
194. Atef MM, Hafez YM, El-Deeb OS, Basha EH, Ismail R, Alshenawy H, et al. The cardioprotective effect of human glucagon-like peptide-1 receptor agonist (semaglutide) on cisplatin-induced cardiotoxicity in rats: targeting mitochondrial functions, dynamics, biogenesis, and redox status pathways. Cell Biochem Funct. 2023;41:450–60.
195. Yamashita Y, Ueyama T, Nishi T, Yamamoto Y, Kawakoshi A, Sunami S, et al. Nrf2-inducing anti-oxidation stress response in the rat liver – new beneficial effect of lansoprazole. Makishima M, editor. PLoS One. 2014;9:e97419.

196. Wei B, Wang Y, Wu H, Liu M, Yao W, Wei M. Pharmacodynamics and pharmacokinetics of a new type of compound lansoprazole capsule in gastric ulcer rats and beagle dogs: importance of adjusting oxidative stress and inflammation. Pharmaceutics. 2019;11:49.
197. Hassanein EHM, Kamel EO, Gad-Elrab WM, Ahmed MA, Mohammedsaleh ZM, Ali FEM. Lansoprazole attenuates cyclophosphamide-induced cardiopulmonary injury by modulating redox-sensitive pathways and inflammation. Mol Cell Biochem. 2023;478:2319–35.
198. Bass KK, Mastrangelo MJ. Immunopotentiation with low-dose cyclophosphamide in the active specific immunotherapy of cancer. Cancer Immunol Immunother. 1998;47:1–12.
199. Iqubal A, Iqubal MK, Sharma S, MohdA A, Najmi AK, Ali SM, et al. Molecular mechanism involved in cyclophosphamide-induced cardiotoxicity: old drug with a new vision. Life Sci. 2019;218:112–31.
200. Veal GJ, Cole M, Chinnaswamy G, Sludden J, Jamieson D, Errington J, et al. Cyclophosphamide pharmacokinetics and pharmacogenetics in children with B-cell non-Hodgkin's lymphoma. Eur J Cancer. 2016;55:56–64.
201. Kurauchi K, Nishikawa T, Miyahara E, Okamoto Y, Kawano Y. Role of metabolites of cyclophosphamide in cardiotoxicity. BMC Res Notes. 2017;10:406.
202. Gottdiener JS. Cardiotoxicity associated with high-dose cyclophosphamide therapy. Arch Intern Med. 1981;141:758.
203. Maeshima T, Honda K, Chikazawa M, Shibata T, Kawai Y, Akagawa M, et al. Quantitative analysis of acrolein-specific adducts generated during lipid peroxidation–modification of proteins in vitro: identification of N^{τ} -(3-propanal)histidine as the major adduct. Chem Res Toxicol. 2012;25:1384–92.
204. Burcham PC, Thompson CA, Henry PJ. Invited review. Adv Mol Toxicol [Internet]. Elsevier; 2010 [cited 2025 Jan 23]. p. 1–36. Available from: https://linkinghub.elsevier.com/retrieve/pii/S1872085410040014.
205. He F, Cao Y-P, Che F-Y, Yang L-H, Xiao S-H, Liu J. Inhibitory effects of edaravone in β -amyloid-induced neurotoxicity in rats. Biomed Res Int. 2014;2014:1–7.
206. Wang X, Pan J, Liu D, Zhang M, Li X, Tian J, et al. Nicorandil alleviates apoptosis in diabetic cardiomyopathy through PI3K/Akt pathway. J Cell Mol Med. 2019;23:5349–59.
207. Hassanein EH, Abd El-Ghafar OA, Ahmed MA, Sayed AM, Gad-Elrab WM, Ajarem JS, et al. Edaravone and acetovanillone upregulate Nrf2 and PI3K/Akt/mTOR signaling and prevent cyclophosphamide cardiotoxicity in rats. Drug Des Devel Ther. 2020;14:5275–88.
208. Esteghamati A, Eskandari D, Mirmiranpour H, Noshad S, Mousavizadeh M, Hedayati M, et al. Effects of metformin on markers of oxidative stress and antioxidant reserve in patients with newly diagnosed type 2 diabetes: a randomized clinical trial. Clin Nutr. 2013;32:179–85.
209. El Kiki SM, Omran MM, Mansour HH, Hasan HF. Metformin and/or low dose radiation reduces cardiotoxicity and apoptosis induced by cyclophosphamide through SIRT-1/SOD and BAX/Bcl-2 pathways in rats. Mol Biol Rep. 2020;47:5115–26.
210. Hori YS, Kuno A, Hosoda R, Tanno M, Miura T, Shimamoto K, et al. Resveratrol ameliorates muscular pathology in the dystrophic mdx mouse, a model for Duchenne muscular dystrophy. J Pharmacol Exp Ther. 2011;338:784–94.
211. Chen Y-T, Tsai T-H, Yang C-C, Sun C-K, Chang L-T, Chen H-H, et al. Exendin-4 and sitagliptin protect kidney from ischemia-reperfusion injury through suppressing oxidative stress and inflammatory reaction. J Transl Med. 2013;11:270.
212. Mansour HH, El Kiki SM, Hasan HF. Protective effect of N-acetylcysteine on cyclophosphamide-induced cardiotoxicity in rats. Environ Toxicol Pharmacol. 2015;40:417–22.
213. Abdel-Raheem IT, Taye A, Abouzied MM. Cardioprotective effects of nicorandil, a mitochondrial potassium channel opener against doxorubicin-induced cardiotoxicity in rats. Basic Clin Pharmacol Toxicol. 2013;113:158–66.
214. Refaie MMM, Shehata S, El-Hussieny M, Abdelraheem WM, Bayoumi AMA. Role of ATP-sensitive potassium channel (KATP) and eNOS in mediating the protective effect of nicorandil in cyclophosphamide-induced cardiotoxicity. Cardiovasc Toxicol. 2020;20:71–81.
215. Koh JS, Yi C, Heo RW, Ahn J-W, Park JR, Lee JE, et al. Protective effect of cilostazol against doxorubicin-induced cardiomyopathy in mice. Free Radic Biol Med. 2015;89:54–61.

216. Refaie MM, El-Hussieny M, Bayoumi AM, Shehata S, Welson NN, Abdelzaher WY. Simvastatin cardioprotection in cyclophosphamide-induced toxicity via the modulation of inflammasome/caspase1/interleukin1β pathway. Hum Exp Toxicol. 2022;41:09603271221111440.
217. Wood SH, Christian HC, Miedzinska K, Saer BR, Johnson M, Paton B, Yu L, McNeilly J, Davis JR, McNeilly AS, Burt DW, Loudon AS. Binary switching of calendar cells in the pituitary defines the phase of the circannual cycle in mammals. Curr Biol. 2015;25(20):2651–62. https://doi.org/10.1016/j.cub.2015.09.014. Epub 2015 Sep 24. PMID: 26412130; PMCID: PMC4612467.
218. Cannon CP, Perkovic V, Agarwal R, Baldassarre J, Bakris G, Charytan DM, de Zeeuw D, Edwards R, Greene T, Heerspink HJL, Jardine MJ, Levin A, Li JW, Neal B, Pollock C, Wheeler DC, Zhang H, Zinman B, Mahaffey KW. Evaluating the effects of canagliflozin on cardiovascular and renal events in patients with type 2 diabetes mellitus and chronic kidney disease according to baseline HbA1c, including those with HbA1c <7%: results from the CREDENCE trial. Circulation. 2020;141(5):407–10. https://doi.org/10.1161/CIRCULATIONAHA.119.044359. Epub 2019 Nov 11. PMID: 31707795.
219. Yu X, Ruan Y, Huang X, Dou L, Lan M, Cui J, Chen B, Gong H, Wang Q, Yan M, Sun S, Qiu Q, Zhang X, Man Y, Tang W, Li J, Shen T. Dexrazoxane ameliorates doxorubicin-induced cardiotoxicity by inhibiting both apoptosis and necroptosis in cardiomyocytes. Biochem Biophys Res Commun. 2020;523(1):140–6. https://doi.org/10.1016/j.bbrc.2019.12.027. Epub 2019 Dec 16. PMID: 31837803.
220. Li H, Gu H, Sun B. Protective effects of pyrrolidine dithiocarbamate on myocardium apoptosis induced by adriamycin in rats. Int J Cardiol. 2007;114(2):159–65. https://doi.org/10.1016/j.ijcard.2006.01.010. Epub 2006 May 18. PMID: 16712983.
221. Agarwal ML, Agarwal A, Taylor WR, Stark GR. p53 controls both the G2/M and the G1 cell cycle checkpoints and mediates reversible growth arrest in human fibroblasts. Proc Natl Acad Sci USA. 1995;92(18):8493–7. https://doi.org/10.1073/pnas.92.18.8493. PMID: 7667317; PMCID: PMC41183.
222. Lavrik IN, Golks A, Krammer PH. Caspases: pharmacological manipulation of cell death. J Clin Invest. 2005;115(10):2665–72. https://doi.org/10.1172/JCI26252. PMID: 16200200; PMCID: PMC1236692.
223. Ueno M, Kakinuma Y, Yuhki K, Murakoshi N, Iemitsu M, Miyauchi T, Yamaguchi I. Doxorubicin induces apoptosis by activation of caspase-3 in cultured cardiomyocytes in vitro and rat cardiac ventricles in vivo. J Pharmacol Sci. 2006;101(2):151–8. https://doi.org/10.1254/jphs.fp0050980. Epub 2006 Jun 10. PMID: 16766856.
224. Feng P, Yang Y, Liu N, Wang S. Baicalin regulates TLR4/IκBα/NFκB signaling pathway to alleviate inflammation in Doxorubicin related cardiotoxicity. Biochem Biophys Res Commun. 2022;637:1–8. https://doi.org/10.1016/j.bbrc.2022.10.061. Epub 2022 Oct 19. PMID: 36375245.
225. Meng X, Li Y, Li S, Zhou Y, Gan RY, Xu DP, Li HB. Dietary sources and bioactivities of melatonin. Nutrients. 2017;9(4):367. https://doi.org/10.3390/nu9040367. PMID: 28387721; PMCID: PMC5409706.
226. Attachaipanich T, Chattipakorn SC, Chattipakorn N. Potential roles of melatonin in doxorubicin-induced cardiotoxicity: from cellular mechanisms to clinical application. Pharmaceutics. 2023;15(3):785. https://doi.org/10.3390/pharmaceutics15030785. PMID: 36986646; PMCID: PMC10059789.
227. Horie T, Ono K, Nishi H, Nagao K, Kinoshita M, Watanabe S, Kuwabara Y, Nakashima Y, Takanabe-Mori R, Nishi E, Hasegawa K, Kita T, Kimura T. Acute doxorubicin cardiotoxicity is associated with miR-146a-induced inhibition of the neuregulin-ErbB pathway. Cardiovasc Res. 2010;87(4):656–64. https://doi.org/10.1093/cvr/cvq148. Epub 2010 May 21. PMID: 20495188; PMCID: PMC2920811.
228. Jia Y, Zuo D, Li Z, Liu H, Dai Z, Cai J, Pang L, Wu Y. Astragaloside IV inhibits doxorubicin-induced cardiomyocyte apoptosis mediated by mitochondrial apoptotic pathway via acti-

vating the PI3K/Akt pathway. Chem Pharm Bull (Tokyo). 2014;62(1):45–53. https://doi.org/10.1248/cpb.c13-00556. PMID: 24390491.
229. Sedeek M, Hébert RL, Kennedy CR, Burns KD, Touyz RM. Molecular mechanisms of hypertension: role of Nox family NADPH oxidases. Curr Opin Nephrol Hypertens. 2009;18(2):122–7. https://doi.org/10.1097/MNH.0b013e32832923c3. PMID: 19430333.
230. Tadokoro T, Ikeda M, Ide T, Deguchi H, Ikeda S, Okabe K, Ishikita A, Matsushima S, Koumura T, Yamada KI, Imai H, Tsutsui H. Mitochondria-dependent ferroptosis plays a pivotal role in doxorubicin cardiotoxicity. JCI Insight. 2020;5(9):e132747. https://doi.org/10.1172/jci.insight.132747. Erratum in: JCI Insight 2023;8(6):e169756. https://doi.org/10.1172/jci.insight.169756. PMID: 32376803; PMCID: PMC7253028.
231. Wu L, Zhang Y, Wang G, Ren J. Molecular mechanisms and therapeutic targeting of ferroptosis in doxorubicin-induced cardiotoxicity. JACC Basic Transl Sci. 2024;9(6):811–26. https://doi.org/10.1016/j.jacbts.2023.10.009. PMID: 39070280; PMCID: PMC11282888.
232. Oehler R, Roth E. Regulative capacity of glutamine. Curr Opin Clin Nutr Metab Care. 2003;6(3):277–82. https://doi.org/10.1097/01.mco.0000068962.34812.ac. PMID: 12690259.
233. Todorova VK, Kaufmann Y, Hennings L, Klimberg VS. Oral glutamine protects against acute doxorubicin-induced cardiotoxicity of tumor-bearing rats. J Nutr. 2010;140(1):44–8. https://doi.org/10.3945/jn.109.113415. Epub 2009 Nov 4. PMID: 19889810.
234. Hiona A, Lee AS, Nagendran J, Xie X, Connolly AJ, Robbins RC, Wu JC. Pretreatment with angiotensin-converting enzyme inhibitor improves doxorubicin-induced cardiomyopathy via preservation of mitochondrial function. J Thorac Cardiovasc Surg. 2011;142(2):396–403.e3. https://doi.org/10.1016/j.jtcvs.2010.07.097. Epub 2010 Nov 20. PMID: 21094500; PMCID: PMC3173512.
235. Koczurkiewicz-Adamczyk P, Klaś K, Gunia-Krzyżak A, Piska K, Andrysiak K, Stępniewski J, Lasota S, Wójcik-Pszczoła K, Dulak J, Madeja Z, Pękala E. Cinnamic acid derivatives as cardioprotective agents against oxidative and structural damage induced by doxorubicin. Int J Mol Sci. 2021;22(12):6217. https://doi.org/10.3390/ijms22126217. PMID: 34207549; PMCID: PMC8227863.
236. Zhang C, Qu S, Wei X, et al. HSP25 down-regulation enhanced p53 acetylation by dissociation of SIRT1 from p53 in doxorubicin-induced H9c2 cell apoptosis. Cell Stress Chaperones. 2016;21(2):251–60. https://doi.org/10.1007/s12192-015-0655-3.
237. Pang B, Qiao X, Janssen L, et al. Drug-induced histone eviction from open chromatin contributes to the chemotherapeutic effects of doxorubicin. Nat Commun. 2013;4:1908. https://doi.org/10.1038/ncomms2921.
238. Li X, Guo X, Li J, Yuan L, Wang H. Preventing effect of astragalus polysaccharide on cardiotoxicity induced by chemotherapy of epirubicin: a pilot study. Medicine (Baltimore). 2022;101(32):e30000. https://doi.org/10.1097/MD.0000000000030000. PMID: 35960075; PMCID: PMC9371539.
239. Hagag AA, Badraia IM, El-Shehaby WA, Mabrouk MM. Protective role of black seed oil in doxorubicin-induced cardiac toxicity in children with acute lymphoblastic leukemia. J Oncol Pharm Pract. 2020;26(6):1397–406. https://doi.org/10.1177/1078155219897294. Epub 2020 Jan 21. PMID: 31964219.
240. Lee M, Chung WB, Lee JE, Park CS, Park WC, Song BJ, Youn HJ. Candesartan and carvedilol for primary prevention of subclinical cardiotoxicity in breast cancer patients without a cardiovascular risk treated with doxorubicin. Cancer Med. 2021;10(12):3964–73. https://doi.org/10.1002/cam4.3956. Epub 2021 May 16. PMID: 33998163; PMCID: PMC8209607.
241. Nabati M, Janbabai G, Baghyari S, Esmaili K, Yazdani J. Cardioprotective effects of carvedilol in inhibiting doxorubicin-induced cardiotoxicity. J Cardiovasc Pharmacol. 2017;69(5):279–85. https://doi.org/10.1097/FJC.0000000000000470. PMID: 28141699.
242. Tashakori Beheshti A, Mostafavi Toroghi H, Hosseini G, Zarifian A, Homaei Shandiz F, Fazlinezhad A. Carvedilol administration can prevent doxorubicin-induced cardiotoxicity: a double-blind randomized trial. Cardiology. 2016;134(1):47–53. https://doi.org/10.1159/000442722. Epub 2016 Feb 12. PMID: 26866364.

243. Stansfeld A, Radia U, Goggin C, Mahalingam P, Benson C, Napolitano A, et al. Pharmacological strategies to reduce anthracycline-associated cardiotoxicity in cancer patients. Expert Opin Pharmacother. 2022;23(14):1641–50. https://doi.org/10.1080/14656566.2022.2124107.
244. Narayan HK, Putt ME, Kosaraju N, Paz A, Bhatt S, Plappert T, Mercer-Rosa L, Armenian SH, Desai AV, Womer RB, Ky B. Dexrazoxane preferentially mitigates doxorubicin cardiotoxicity in female children with sarcoma. Open Heart. 2019;6(1):e001025. https://doi.org/10.1136/openhrt-2019-001025. PMID: 31297226; PMCID: PMC6593195.
245. Kopp LM, Womer RB, Schwartz CL, Ebb DH, Franco VI, Hall D, Barkauskas DA, Krailo MD, Grier HE, Meyers PA, Wexler LH, Marina NM, Janeway KA, Gorlick R, Bernstein ML, Lipshultz SE, Children's Oncology Group. Effects of dexrazoxane on doxorubicin-related cardiotoxicity and second malignant neoplasms in children with osteosarcoma: a report from the Children's Oncology Group. Cardiooncology. 2019;5:15. https://doi.org/10.1186/s40959-019-0050-9. PMID: 32154021; PMCID: PMC7048050.
246. Georgakopoulos P, Kyriakidis M, Perpinia A, Karavidas A, Zimeras S, Mamalis N, Kouvela M, Charpidou A. The role of metoprolol and enalapril in the prevention of doxorubicin-induced cardiotoxicity in lymphoma patients. Anticancer Res. 2019;39(10):5703–7. https://doi.org/10.21873/anticanres.13769. PMID: 31570470.
247. El Amrousy D, El-Afify D, Khedr R, Ibrahim AM. Omega 3 fatty acids can reduce early doxorubicin-induced cardiotoxicity in children with acute lymphoblastic leukemia. Pediatr Blood Cancer. 2022;69(7):e29496. https://doi.org/10.1002/pbc.29496. Epub 2021 Nov 29. PMID: 34842343.
248. Zhang H, Shen WS, Gao CH, Deng LC, Shen D. Protective effects of salidroside on epirubicin-induced early left ventricular regional systolic dysfunction in patients with breast cancer. Drugs R D. 2012;12(2):101–6. https://doi.org/10.2165/11632530-000000000-00000. PMID: 22770377; PMCID: PMC3585960.
249. Nabati M, Janbabai G, Esmailian J, Yazdani J. Effect of rosuvastatin in preventing chemotherapy-induced cardiotoxicity in women with breast cancer: a randomized, single-blind, placebo-controlled trial. J Cardiovasc Pharmacol Ther. 2019;24(3):233–41. https://doi.org/10.1177/1074248418821721. Epub 2019 Jan 2. PMID: 30599756.
250. Li LL, Wei L, Zhang N, Wei WY, Hu C, Deng W, Tang QZ. Levosimendan protects against doxorubicin-induced cardiotoxicity by regulating the PTEN/Akt pathway. Biomed Res Int. 2020;2020:8593617. https://doi.org/10.1155/2020/8593617. PMID: 32596387; PMCID: PMC7298255.
251. Lin MY, Cheng WT, Cheng HC, Chou WC, Chen HI, Ou HC, Tsai KL. Baicalin enhances chemosensitivity to doxorubicin in breast cancer cells via upregulation of oxidative stress-mediated mitochondria-dependent apoptosis. Antioxidants (Basel). 2021;10(10):1506. https://doi.org/10.3390/antiox10101506. PMID: 34679641; PMCID: PMC8532783.
252. Durante W. The emerging role of l-glutamine in cardiovascular health and disease. Nutrients. 2019;11(9):2092. https://doi.org/10.3390/nu11092092. PMID: 31487814; PMCID: PMC6769761.
253. Afsar T, Razak S, Batoo KM, Khan MR. *Acacia hydaspica* R. Parker prevents doxorubicin-induced cardiac injury by attenuation of oxidative stress and structural Cardiomyocyte alterations in rats. BMC Complement Altern Med. 2017;17(1):554. https://doi.org/10.1186/s12906-017-2061-0. PMID: 29284479; PMCID: PMC5747129.
254. Yarmohmmadi F, Rahimi N, Faghir-Ghanesefat H, Javadian N, Abdollahi A, Pasalar P, Jazayeri F, Ejtemaeemehr S, Dehpour AR. Protective effects of agmatine on doxorubicin-induced chronic cardiotoxicity in rat. Eur J Pharmacol. 2017;796:39–44. https://doi.org/10.1016/j.ejphar.2016.12.022. Epub 2016 Dec 16. PMID: 27993640.
255. Guo JJ, Ma LL, Shi HT, Zhu JB, Wu J, Ding ZW, An Y, Zou YZ, Ge JB. Alginate oligosaccharide prevents acute doxorubicin cardiotoxicity by suppressing oxidative stress and endoplasmic reticulum-mediated apoptosis. Mar Drugs. 2016;14(12):231. https://doi.org/10.3390/md14120231. PMID: 27999379; PMCID: PMC5192468.

256. Abdel-Daim MM, Kilany OE, Khalifa HA, Ahmed AAM. Allicin ameliorates doxorubicin-induced cardiotoxicity in rats via suppression of oxidative stress, inflammation and apoptosis. Cancer Chemother Pharmacol. 2017;80(4):745–53. https://doi.org/10.1007/s00280-017-3413-7. Epub 2017 Aug 7. PMID: 28785995.
257. Khafaga AF, El-Sayed YS. All-trans-retinoic acid ameliorates doxorubicin-induced cardiotoxicity: in vivo potential involvement of oxidative stress, inflammation, and apoptosis via caspase-3 and p53 down-expression. Naunyn Schmiedeberg's Arch Pharmacol. 2018;391(1):59–70. https://doi.org/10.1007/s00210-017-1437-5. Epub 2017 Oct 30. PMID: 29085977.
258. Birari L, Wagh S, Patil KR, Mahajan UB, Unger B, Belemkar S, Goyal SN, Ojha S, Patil CR. Aloin alleviates doxorubicin-induced cardiotoxicity in rats by abrogating oxidative stress and pro-inflammatory cytokines. Cancer Chemother Pharmacol. 2020;86(3):419–26. https://doi.org/10.1007/s00280-020-04125-w. Epub 2020 Aug 18. PMID: 32812061.
259. Shabalala SC, Dludla PV, Muller CJF, Nxele X, Kappo AP, Louw J, Johnson R. Aspalathin ameliorates doxorubicin-induced oxidative stress in H9c2 cardiomyoblasts. Toxicol In Vitro. 2019;55:134–9. https://doi.org/10.1016/j.tiv.2018.12.012. Epub 2018 Dec 19. PMID: 30576852.
260. Cao Y, Ruan Y, Shen T, Huang X, Li M, Yu W, Zhu Y, Man Y, Wang S, Li J. Astragalus polysaccharide suppresses doxorubicin-induced cardiotoxicity by regulating the PI3k/Akt and p38MAPK pathways. Oxid Med Cell Longev. 2014;2014:674219. https://doi.org/10.1155/2014/674219. Epub 2014 Oct 16. PMID: 25386226; PMCID: PMC4216718.
261. Sahu BD, Kumar JM, Kuncha M, Borkar RM, Srinivas R, Sistla R. Baicalein alleviates doxorubicin-induced cardiotoxicity via suppression of myocardial oxidative stress and apoptosis in mice. Life Sci. 2016;144:8–18. https://doi.org/10.1016/j.lfs.2015.11.018. Epub 2015 Nov 29. PMID: 26606860.
262. Zhao XX, Cho H, Lee S, Woo JS, Song MY, Cheng XW, Lee KH, Kim W. BAY60-2770 attenuates doxorubicin-induced cardiotoxicity by decreased oxidative stress and enhanced autophagy. Chem Biol Interact. 2020;328:109190. https://doi.org/10.1016/j.cbi.2020.109190. Epub 2020 Jul 9. PMID: 32652078.
263. Hassan MQ, Akhtar MS, Afzal O, Hussain I, Akhtar M, Haque SE, Najmi AK. Edaravone and benidipine protect myocardial damage by regulating mitochondrial stress, apoptosis signalling and cardiac biomarkers against doxorubicin-induced cardiotoxicity. Clin Exp Hypertens. 2020;42(5):381–92. https://doi.org/10.1080/10641963.2019.1676770. Epub 2019 Oct 20. PMID: 31630586.
264. Wu YZ, Zhang L, Wu ZX, Shan TT, Xiong C. Berberine ameliorates doxorubicin-induced cardiotoxicity via a SIRT1/p66Shc-mediated pathway. Oxidative Med Cell Longev. 2019;2019:2150394. https://doi.org/10.1155/2019/2150394. PMID: 31885776; PMCID: PMC6918936.
265. Aykan DA, Yaman S, Eser N, Özcan Metin T, Seyithanoğlu M, Aykan AÇ, Kurt AH, Ergün Y. Bisoprolol and linagliptin ameliorated electrical and mechanical isometric myocardial contractions in doxorubicin-induced cardiomyopathy in rats. Pharmacol Rep. 2020;72(4):867–76. https://doi.org/10.1007/s43440-019-00034-9. Epub 2019 Dec 20. PMID: 32048248.
266. Lei B, Wu X, Xia K, Sun H, Wang J. Exosomal micro-RNA-96 derived from bone marrow mesenchymal stem cells inhibits doxorubicin-induced myocardial toxicity by inhibiting the Rac1/nuclear factor-κB signaling pathway. J Am Heart Assoc. 2021;10(17):e020589. https://doi.org/10.1161/JAHA.120.020589. Epub 2021 Aug 28. PMID: 34459233; PMCID: PMC8649246.
267. Zhai J, Tao L, Zhang S, Gao H, Zhang Y, Sun J, Song Y, Qu X. Calycosin ameliorates doxorubicin-induced cardiotoxicity by suppressing oxidative stress and inflammation via the sirtuin 1-NOD-like receptor protein 3 pathway. Phytother Res. 2020;34(3):649–59. https://doi.org/10.1002/ptr.6557. Epub 2019 Dec 20. PMID: 31858651.
268. Hao E, Mukhopadhyay P, Cao Z, Erdélyi K, Holovac E, Liaudet L, Lee WS, Haskó G, Mechoulam R, Pacher P. Cannabidiol protects against doxorubicin-induced cardiomyopathy

by modulating mitochondrial function and biogenesis. Mol Med. 2015;21(1):38–45. https://doi.org/10.2119/molmed.2014.00261. PMID: 25569804; PMCID: PMC4461586.
269. Qi W, Boliang W, Xiaoxi T, Guoqiang F, Jianbo X, Gang W. Cardamonin protects against doxorubicin-induced cardiotoxicity in mice by restraining oxidative stress and inflammation associated with Nrf2 signaling. Biomed Pharmacother. 2020;122:109547. https://doi.org/10.1016/j.biopha.2019.109547. Epub 2019 Dec 30. PMID: 31918264.
270. Liao W, Rao Z, Wu L, Chen Y, Li C. Cariporide attenuates doxorubicin-induced cardiotoxicity in rats by inhibiting oxidative stress, inflammation and apoptosis partly through regulation of Akt/GSK-3β and Sirt1 signaling pathway. Front Pharmacol. 2022;13:850053. https://doi.org/10.3389/fphar.2022.850053. PMID: 35747748; PMCID: PMC9209753.
271. Manna P, Dewanjee S, Joardar S, Chakraborty P, Bhattacharya H, Bhanja S, Bhattacharyya C, Bhowmik M, Bhowmick S, Saha A, Das J, Sil PC. Corrigendum to "Carnosic acid attenuates doxorubicin-induced cardiotoxicity by decreasing oxidative stress and its concomitant pathological consequences" [Food Chem. Toxicol. 166 (2022) 113205]. Food Chem Toxicol. 2023;181:114099. https://doi.org/10.1016/j.fct.2023.114099. Epub 2023 Oct 19. Erratum for: Food Chem Toxicol. 2022;166:113205. https://doi.org/10.1016/j.fct.2022.113205. PMID: 37862867.
272. Chen YL, Chung SY, Chai HT, Chen CH, Liu CF, Chen YL, Huang TH, Zhen YY, Sung PH, Sun CK, Chua S, Lu HI, Lee FY, Sheu JJ, Yip HK. Early administration of carvedilol protected against doxorubicin-induced cardiomyopathy. J Pharmacol Exp Ther. 2015;355(3):516–27. https://doi.org/10.1124/jpet.115.225375. Epub 2015 Oct 28. Erratum in: J Pharmacol Exp Ther. 2017;360(2):312. https://doi.org/10.1124/jpet.115.225375err. PMID: 26511374.
273. Saleh Ahmed AS. Potential protective effect of catechin on doxorubicin-induced cardiotoxicity in adult male albino rats. Toxicol Mech Methods. 2021;32(2):97–105. https://doi.org/10.1080/15376516.2021.1972375.
274. Tian B, Hua S, Tian Y, Liu J. Chemical and physical chitosan hydrogels as prospective carriers for drug delivery: a review. J Mater Chem B. 2020;8(44):10050–64. https://doi.org/10.1039/d0tb01869d. PMID: 33073280.
275. Mantawy EM, El-Bakly WM, Esmat A, Badr AM, El-Demerdash E. Chrysin alleviates acute doxorubicin cardiotoxicity in rats via suppression of oxidative stress, inflammation and apoptosis. Eur J Pharmacol. 2014;728:107–18. https://doi.org/10.1016/j.ejphar.2014.01.065. Epub 2014 Feb 6. PMID: 24509133.
276. Lu J, Li J, Hu Y, Guo Z, Sun D, Wang P, Guo K, Duan DD, Gao S, Jiang J, Wang J, Liu P. Chrysophanol protects against doxorubicin-induced cardiotoxicity by suppressing cellular PARylation. Acta Pharm Sin B. 2019;9(4):782–93. https://doi.org/10.1016/j.apsb.2018.10.008. Epub 2018 Nov 1. PMID: 31384538; PMCID: PMC6663922.
277. Han D, Wang Y, Wang Y, Dai X, Zhou T, Chen J, Tao B, Zhang J, Cao F. The tumor-suppressive human circular RNA CircITCH sponges miR-330-5p to ameliorate doxorubicin-induced cardiotoxicity through upregulating SIRT6, survivin, and SERCA2a. Circ Res. 2020;127(4):e108–25. https://doi.org/10.1161/CIRCRESAHA.119.316061. Epub 2020 May 11. PMID: 32392088.
278. Liu X, Qiu Y, Huang N, Liu YH, Wang HH, Yu YN, Song YT, Wan GR, Wang SX, Li P, Yin YL. Citronellal alleviates doxorubicin-induced cardiotoxicity by suppressing oxidative stress and apoptosis via Na^+ /H^+ exchanger-1 inhibition. J Biochem Mol Toxicol. 2022;36(3):e22971. https://doi.org/10.1002/jbt.22971. Epub 2021 Nov 23. PMID: 34813134.
279. Wu R, Yao PA, Wang HL, Gao Y, Yu HL, Wang L, Cui XH, Xu X, Gao JP. Effect of fermented Cordyceps sinensis on doxorubicin-induced cardiotoxicity in rats. Mol Med Rep. 2018;18(3):3229–41. https://doi.org/10.3892/mmr.2018.9310. Epub 2018 Jul 24. PMID: 30066944; PMCID: PMC6102656.
280. Wu ZZ, Rao M, Xu S, Hu HY, Tang QZ. Coumestrol ameliorates doxorubicin-induced cardiotoxicity via activating AMPKα. Free Radic Res. 2020;54(8–9):629–39. https://doi.org/10.1080/10715762.2020.1822525.

281. He H, Luo Y, Qiao Y, Zhang Z, Yin D, Yao J, You J, He M. Curcumin attenuates doxorubicin-induced cardiotoxicity via suppressing oxidative stress and preventing mitochondrial dysfunction mediated by 14-3-3γ. Food Funct. 2018;9(8):4404–18. https://doi.org/10.1039/c8fo00466h. PMID: 30063064.
282. Beak J, Huang W, Parker JS, Hicks ST, Patterson C, Simpson PC, Ma A, Jin J, Jensen BC. An oral selective alpha-1A adrenergic receptor agonist prevents doxorubicin cardiotoxicity. JACC Basic Transl Sci. 2017;2(1):39–53. https://doi.org/10.1016/j.jacbts.2016.10.006. PMID: 28286875; PMCID: PMC5343290.
283. Hsieh PL, Chu PM, Cheng HC, Huang YT, Chou WC, Tsai KL, Chan SH. Dapagliflozin mitigates doxorubicin-caused myocardium damage by regulating AKT-mediated oxidative stress, cardiac remodeling, and inflammation. Int J Mol Sci. 2022;23(17):10146. https://doi.org/10.3390/ijms231710146. PMID: 36077544; PMCID: PMC9456438.
284. Sheibani M, Nezamoleslami S, Faghir-Ghanesefat H, Emami AH, Dehpour AR. Cardioprotective effects of dapsone against doxorubicin-induced cardiotoxicity in rats. Cancer Chemother Pharmacol. 2020;85(3):563–71. https://doi.org/10.1007/s00280-019-04019-6. Epub 2020 Jan 8. PMID: 31915967.
285. Sheibani M, Azizi Y, Shayan M, Nezamoleslami S, Eslami F, Farjoo MH, Dehpour AR. Doxorubicin-induced cardiotoxicity: an overview on pre-clinical therapeutic approaches. Cardiovasc Toxicol. 2022;22(4):292–310. https://doi.org/10.1007/s12012-022-09721-1. Epub 2022 Jan 21. PMID: 35061218.
286. Pecoraro M, Ciccarelli M, Fiordelisi A, Iaccarino G, Pinto A, Popolo A. Diazoxide improves mitochondrial connexin 43 expression in a mouse model of doxorubicin-induced cardiotoxicity. Int J Mol Sci. 2018;19(3):757. https://doi.org/10.3390/ijms19030757. PMID: 29518932; PMCID: PMC5877618.
287. Chen R, Sun G, Xu L, Zhang X, Zeng W, Sun X. Didymin attenuates doxorubicin-induced cardiotoxicity by inhibiting oxidative stress. Chin Herb Med. 2021;14(1):70–8. https://doi.org/10.1016/j.chmed.2021.07.002. PMID: 36120130; PMCID: PMC9476736.
288. Al-Malky HS, Osman AM, Damanhouri ZA, Alkreathy HM, Al Aama JY, Ramadan WS, Al Qahtani AA, Al Mahdi HB. Modulation of doxorubicin-induced expression of the multidrug resistance gene in breast cancer cells by diltiazem and protection against cardiotoxicity in experimental animals. Cancer Cell Int. 2019;19:191. https://doi.org/10.1186/s12935-019-0912-0. PMID: 31367189; PMCID: PMC6657176.
289. Chang D, Li H, Qian C, Wang Y. DiOHF protects against doxorubicin-induced cardiotoxicity through ERK1 signaling pathway. Front Pharmacol. 2019;10:1081. https://doi.org/10.3389/fphar.2019.01081. PMID: 31611788; PMCID: PMC6777440.
290. Sonawane VK, Mahajan UB, Shinde SD, Chatterjee S, Chaudhari SS, Bhangale HA, Ojha S, Goyal SN, Kundu CN, Patil CR. A chemosensitizer drug: disulfiram prevents doxorubicin-induced cardiac dysfunction and oxidative stress in rats. Cardiovasc Toxicol. 2018;18(5):459–70. https://doi.org/10.1007/s12012-018-9458-y. PMID: 29779119.
291. Ongnok B, Khuanjing T, Chunchai T, Pantiya P, Kerdphoo S, Arunsak B, Nawara W, Jaiwongkam T, Apaijai N, Chattipakorn N, Chattipakorn SC. Donepezil protects against doxorubicin-induced chemobrain in rats via attenuation of inflammation and oxidative stress without interfering with doxorubicin efficacy. Neurotherapeutics. 2021;18(3):2107–25. https://doi.org/10.1007/s13311-021-01092-9. Epub 2021 Jul 26. Erratum in: Neurotherapeutics. 2023;20(2):602–3. https://doi.org/10.1007/s13311-023-01347-7. PMID: 34312765; PMCID: PMC8608968.
292. Dantas D, Pereira AG, Fujimori ASS, Ribeiro APD, de Almeida Silva CCV, Monte MG, Corrêa CR, Fernandes AA, Bazan SGZ, Azevedo PS, Minicucci MF, de Paiva SAR, Zornoff LAM, Polegato BF. Doxycycline attenuates doxorubicin-induced cardiotoxicity by improving myocardial energy metabolism in rats. J Cardiovasc Dev Dis. 2022;9(8):254. https://doi.org/10.3390/jcdd9080254. PMID: 36005418; PMCID: PMC9410319.
293. Zhao L, Zhang B. Doxorubicin induces cardiotoxicity through upregulation of death receptors mediated apoptosis in cardiomyocytes. Sci Rep. 2017;7:44735. https://doi.org/10.1038/srep44735. PMID: 28300219; PMCID: PMC5353581.

294. Singla DK, Johnson TA, Tavakoli Dargani Z. Exosome treatment enhances anti-inflammatory M2 macrophages and reduces inflammation-induced pyroptosis in doxorubicin-induced cardiomyopathy. Cells. 2019;8(10):1224. https://doi.org/10.3390/cells8101224. PMID: 31600901; PMCID: PMC6830113.
295. Janbabai G, Nabati M, Faghihinia M, Azizi S, Borhani S, Yazdani J. Effect of enalapril on preventing anthracycline-induced cardiomyopathy. Cardiovasc Toxicol. 2017;17(2):130–9. https://doi.org/10.1007/s12012-016-9365-z. PMID: 27003392.
296. Shaker RA, Abboud SH, Assad HC, Hadi N. Enoxaparin attenuates doxorubicin induced cardiotoxicity in rats via interfering with oxidative stress, inflammation and apoptosis. BMC Pharmacol Toxicol. 2018;19(1):3. https://doi.org/10.1186/s40360-017-0184-z. PMID: 29321061; PMCID: PMC5763526.
297. Rjiba-Touati K, Ayed-Boussema I, Guedri Y, Achour A, Bacha H, Abid-Essefi S. Effect of recombinant human erythropoietin on mitomycin C-induced oxidative stress and genotoxicity in rat kidney and heart tissues. Hum Exp Toxicol. 2016;35(1):53–62. https://doi.org/10.1177/0960327115577521. Epub 2015 Mar 2. PMID: 25733728.
298. Bansal N, Artinian NT, Bakris G, Chang T, Cohen J, Flythe J, Lea J, Vongpatanasin W, Chertow GM, American Heart Association Council on the Kidney in Cardiovascular Disease; Council on Cardiovascular and Stroke Nursing; and Council on Epidemiology and Prevention. Hypertension in patients treated with in-center maintenance hemodialysis: current evidence and future opportunities: a scientific statement from the American Heart Association. Hypertension. 2023;80(6):e112–22. https://doi.org/10.1161/HYP.0000000000000230. Epub 2023 Apr 24. PMID: 37092336.
299. Zhang X, Hu C, Kong CY, Song P, Wu HM, Xu SC, Yuan YP, Deng W, Ma ZG, Tang QZ. FNDC5 alleviates oxidative stress and cardiomyocyte apoptosis in doxorubicin-induced cardiotoxicity via activating AKT. Cell Death Differ. 2020;27(2):540–55. https://doi.org/10.1038/s41418-019-0372-z. Epub 2019 Jun 17. PMID: 31209361; PMCID: PMC7206111.
300. Ji Y, Jin D, Qi J, Wang X, Zhang C, An P, Luo Y, Luo J. Fucoidan protects against doxorubicin-induced cardiotoxicity by reducing oxidative stress and preventing mitochondrial function injury. Int J Mol Sci. 2022;23(18):10685. https://doi.org/10.3390/ijms231810685. PMID: 36142635; PMCID: PMC9504360.
301. Zhang Z, Shen C, Wu N, Wang J, Du W, Chen X. Gab1 overexpression alleviates doxorubicin-induced cardiac oxidative stress, inflammation, and apoptosis through PI3K/Akt signaling pathway. J Cardiovasc Pharmacol. 2022;80(6):804–12. https://doi.org/10.1097/FJC.0000000000001333. PMID: 35856909.
302. Wang Y, Lei T, Yuan J, Wu Y, Shen X, Gao J, Feng W, Lu Z. GCN2 deficiency ameliorates doxorubicin-induced cardiotoxicity by decreasing cardiomyocyte apoptosis and myocardial oxidative stress. Redox Biol. 2018;17:25–34. https://doi.org/10.1016/j.redox.2018.04.009. Epub 2018 Apr 7. PMID: 29660505; PMCID: PMC6006681.
303. Chen M, Samuel VP, Wu Y, Dang M, Lin Y, Sriramaneni R, Sah SK, Chinnaboina GK, Zhang G. Nrf2/HO-1 mediated protective activity of genistein against doxorubicin-induced cardiac toxicity. J Environ Pathol Toxicol Oncol. 2019;38(2):143–52. https://doi.org/10.1615/JEnvironPatholToxicolOncol.2019029341. PMID: 31679277.
304. Elhadidy MG, Elmasry A, Rabei MR, Eladel AE. Effect of ghrelin on VEGF-B and connexin-43 in a rat model of doxorubicin-induced cardiomyopathy. J Basic Clin Physiol Pharmacol. 2019;31(1):20180212. https://doi.org/10.1515/jbcpp-2018-0212. PMID: 31730522.
305. Jasim ST, Al-Kuraishy HM, Al-Gareeb AI. Gingko Biloba protects cardiomyocytes against acute doxorubicin induced cardiotoxicity by suppressing oxidative stress. J Pak Med Assoc. 2019;69(Suppl 3)(8):S103–7. PMID: 31603888.
306. Wang H, Yu P, Gou H, Zhang J, Zhu M, Wang ZH, Tian JW, Jiang YT, Fu FH. Cardioprotective effects of 20(S)-ginsenoside Rh2 against doxorubicin-induced cardiotoxicity in vitro and in vivo. Evid Based Complement Alternat Med. 2012;2012:506214. https://doi.org/10.1155/2012/506214. Epub 2012 Oct 17. PMID: 23125868; PMCID: PMC3483725.

307. Upadhyay S, Mantha AK, Dhiman M. Glycyrrhiza glabra (Licorice) root extract attenuates doxorubicin-induced cardiotoxicity via alleviating oxidative stress and stabilising the cardiac health in H9c2 cardiomyocytes. J Ethnopharmacol. 2020;258:112690. https://doi.org/10.1016/j.jep.2020.112690. Epub 2020 Feb 24. PMID: 32105749.
308. Dos Santos JM, Alfredo TM, Antunes KÁ, da Cunha JDSM, Costa EMA, Lima ES, Silva DB, Carollo CA, Schmitz WO, Boleti APA, Dos Santos EL, de Picoli Souza K. *Guazuma ulmifolia* Lam. decreases oxidative stress in blood cells and prevents doxorubicin-induced cardiotoxicity. Oxidative Med Cell Longev. 2018;2018:2935051. https://doi.org/10.1155/2018/2935051. PMID: 30050650; PMCID: PMC6046128.
309. Hull TD, Boddu R, Guo L, Tisher CC, Traylor AM, Patel B, Joseph R, Prabhu SD, Suliman HB, Piantadosi CA, Agarwal A, George JF. Heme oxygenase-1 regulates mitochondrial quality control in the heart. JCI Insight. 2016;1(2):e85817. https://doi.org/10.1172/jci.insight.85817. PMID: 27110594; PMCID: PMC4838906.
310. Lien CY, Chuang TY, Hsu CH, Lin CL, Wang SE, Sheu SJ, Chien CT, Wu CH. Oral treatment with the herbal formula B307 alleviates cardiac toxicity in doxorubicin-treated mice via suppressing oxidative stress, inflammation, and apoptosis. Onco Targets Ther. 2015;8:1193–210. https://doi.org/10.2147/OTT.S82936. PMID: 26060405; PMCID: PMC4454207.
311. Chen YL, Zhuang XD, Xu ZW, Lu LH, Guo HL, Wu WK, Liao XX. Higenamine combined with [6]-gingerol suppresses doxorubicin-triggered oxidative stress and apoptosis in cardiomyocytes via upregulation of PI3K/Akt pathway. Evid Based Complement Alternat Med. 2013;2013:970490. https://doi.org/10.1155/2013/970490. Epub 2013 Jun 5. PMID: 23861719; PMCID: PMC3687593.
312. Pillai VB, Kanwal A, Fang YH, Sharp WW, Samant S, Arbiser J, Gupta MP. Honokiol, an activator of Sirtuin-3 (SIRT3) preserves mitochondria and protects the heart from doxorubicin-induced cardiomyopathy in mice. Oncotarget. 2017;8(21):34082–98. https://doi.org/10.18632/oncotarget.16133. PMID: 28423723; PMCID: PMC5470953.
313. Sirangelo I, Liccardo M, Iannuzzi C. Hydroxytyrosol prevents doxorubicin-induced oxidative stress and apoptosis in cardiomyocytes. Antioxidants (Basel). 2022;11(6):1087. https://doi.org/10.3390/antiox11061087. PMID: 35739984; PMCID: PMC9220035.
314. Kanno SI, Hara A. The mRNA expression of *Il6* and *Pdcd1* are predictive and protective factors for doxorubicin-induced cardiotoxicity. Mol Med Rep. 2021;23(2):113. https://doi.org/10.3892/mmr.2020.11752. Epub 2020 Dec 10. PMID: 33300057; PMCID: PMC7723161.
315. Ye D, Wang Z, Xu Y, Ye J, Wang M, Liu J, Zhang J, Zhao M, Chen J, Wan J. Interleukin-9 aggravates doxorubicin-induced cardiotoxicity by promoting inflammation and apoptosis in mice. Life Sci. 2020;255:117844. https://doi.org/10.1016/j.lfs.2020.117844. Epub 2020 May 25. PMID: 32464124.
316. Ye J, Wang Y, Xu Y, Wang Z, Liu L, Wang M, Ye D, Zhang J, Yang Z, Lin Y, Ji Q, Wan J. Interleukin-22 deficiency alleviates doxorubicin-induced oxidative stress and cardiac injury via the p38 MAPK/macrophage/Fizz3 axis in mice. Redox Biol. 2020;36:101636. https://doi.org/10.1016/j.redox.2020.101636. Epub 2020 Jul 7. Retraction in: Redox Biol. 2024;76:103357. https://doi.org/10.1016/j.redox.2024.103357. Erratum in: Redox Biol. 2021;38:101770. https://doi.org/10.1016/j.redox.2020.101770. PMID: 32863209; PMCID: PMC7371904.
317. Guo L, Zheng X, Wang E, Jia X, Wang G, Wen J. Irigenin treatment alleviates doxorubicin (DOX)-induced cardiotoxicity by suppressing apoptosis, inflammation and oxidative stress via the increase of miR-425. Biomed Pharmacother. 2020;125:109784. https://doi.org/10.1016/j.biopha.2019.109784. Epub 2020 Feb 25. PMID: 32092815.
318. Zhuo C, Xin J, Huang W, Zhang D, Yan X, Li R, Li H, Lan J, Lin L, Li L, Wang X, Liu L, Wang Y, Li X, Mao Y, Chen H, Wu S, Yang X, Jiang W. Irisin protects against doxorubicin-induced cardiotoxicity by improving AMPK-Nrf2 dependent mitochondrial fusion and strengthening endogenous anti-oxidant defense mechanisms. Toxicology. 2023;494:153597. https://doi.org/10.1016/j.tox.2023.153597. Epub 2023 Jul 25. PMID: 37499777.

319. Chen C, Jiang L, Zhang M, Pan X, Peng C, Huang W, Jiang Q. Isodunnianol alleviates doxorubicin-induced myocardial injury by activating protective autophagy. Food Funct. 2019;10(5):2651–7. https://doi.org/10.1039/c9fo00063a. PMID: 31025676.
320. Alzahrani AM, Rajendran P, Veeraraghavan VP, Hanieh H. Cardiac protective effect of kirenol against doxorubicin-induced cardiac hypertrophy in H9c2 cells through Nrf2 signaling via PI3K/AKT pathways. Int J Mol Sci. 2021;22(6):3269. https://doi.org/10.3390/ijms22063269. PMID: 33806909; PMCID: PMC8004766.
321. Zhou H, Pu S, Zhou H, Guo Y. Klotho as potential autophagy regulator and therapeutic target. Front Pharmacol. 2021;12:755366. https://doi.org/10.3389/fphar.2021.755366. PMID: 34737707; PMCID: PMC8560683.
322. Xia Y, Chen Z, Chen A, Fu M, Dong Z, Hu K, Yang X, Zou Y, Sun A, Qian J, Ge J. LCZ696 improves cardiac function via alleviating Drp1-mediated mitochondrial dysfunction in mice with doxorubicin-induced dilated cardiomyopathy. J Mol Cell Cardiol. 2017;108:138–48. https://doi.org/10.1016/j.yjmcc.2017.06.003. Epub 2017 Jun 15. PMID: 28623750.
323. Liang X, Wang S, Wang L, Ceylan AF, Ren J, Zhang Y. Mitophagy inhibitor liensinine suppresses doxorubicin-induced cardiotoxicity through inhibition of Drp1-mediated maladaptive mitochondrial fission. Pharmacol Res. 2020;157:104846. https://doi.org/10.1016/j.phrs.2020.104846. Epub 2020 Apr 25. PMID: 32339784.
324. Chen D, Ma X, Zhu J, Wang Y, Guo S, Qin J. Pectin based hydrogel with covalent coupled doxorubicin and limonin loading for lung tumor therapy. Colloids Surf B Biointerfaces. 2024;234:113670. https://doi.org/10.1016/j.colsurfb.2023.113670. Epub 2023 Nov 28. PMID: 38042108.
325. Khalilzadeh M, Abdollahi A, Abdolahi F, Abdolghaffari AH, Dehpour AR, Jazaeri F. Protective effects of magnesium sulfate against doxorubicin induced cardiotoxicity in rats. Life Sci. 2018;207:436–41. https://doi.org/10.1016/j.lfs.2018.06.022. Epub 2018 Jun 22. PMID: 29940240.
326. Hu C, Zhang X, Wei W, Zhang N, Wu H, Ma Z, Li L, Deng W, Tang Q. Matrine attenuates oxidative stress and cardiomyocyte apoptosis in doxorubicin-induced cardiotoxicity *via* maintaining AMPKα/UCP2 pathway. Acta Pharm Sin B. 2019;9(4):690–701. https://doi.org/10.1016/j.apsb.2019.03.003. Epub 2019 Mar 16. PMID: 31384530; PMCID: PMC6664099.
327. Gharanei M, Hussain A, Janneh O, Maddock H. Attenuation of doxorubicin-induced cardiotoxicity by mdivi-1: a mitochondrial division/mitophagy inhibitor. PLoS One. 2013;8(10):e77713. https://doi.org/10.1371/journal.pone.0077713. PMID: 24147064; PMCID: PMC3798380.
328. Kobashigawa LC, Xu YC, Padbury JF, Tseng YT, Yano N. Metformin protects cardiomyocyte from doxorubicin induced cytotoxicity through an AMP-activated protein kinase dependent signaling pathway: an in vitro study. PLoS One. 2014;9(8):e104888. https://doi.org/10.1371/journal.pone.0104888. PMID: 25127116; PMCID: PMC4134245.
329. Ahmed AZ, Satyam SM, Shetty P, D'Souza MR. Methyl gallate attenuates doxorubicin-induced cardiotoxicity in rats by suppressing oxidative stress. Scientifica (Cairo). 2021;2021:6694340. https://doi.org/10.1155/2021/6694340. PMID: 33510932; PMCID: PMC7822703.
330. Sánchez-Sánchez R, Reinal I, Peiró-Molina E, Buigues M, Tejedor S, Hernándiz A, Selva M, Hervás D, Cañada AJ, Dorronsoro A, Santaballa A, Salvador C, Caiment F, Kleinjans J, Martínez-Dolz L, Moscoso I, Lage R, González-Juanatey JR, Panadero J, Aparicio-Puerta E, Bernad A, Sepúlveda P. MicroRNA-4732-3p is dysregulated in breast cancer patients with cardiotoxicity, and its therapeutic delivery protects the heart from doxorubicin-induced oxidative stress in rats. Antioxidants (Basel). 2022;11(10):1955. https://doi.org/10.3390/antiox11101955. PMID: 36290678; PMCID: PMC9599023.
331. Liu Y, Li Y, Ni J, Shu Y, Wang H, Hu T. MiR-124 attenuates doxorubicin-induced cardiac injury via inhibiting p66Shc-mediated oxidative stress. Biochem Biophys Res Commun. 2020;521(2):420–6. https://doi.org/10.1016/j.bbrc.2019.10.157. Epub 2019 Oct 28. PMID: 31672275.

332. Yan M, Cao Y, Wang Q, Xu K, Dou L, Huang X, Chen B, Tang W, Lan M, Liu B, Zhu K, Yang Y, Sun S, Zhang X, Man Y, Hei M, Shen T, Li J. miR-488-3p protects cardiomyocytes against doxorubicin-induced cardiotoxicity by inhibiting cyclinG1. Oxidative Med Cell Longev. 2022;2022:5184135. https://doi.org/10.1155/2022/5184135. PMID: 35186188; PMCID: PMC8853758.
333. Meng J, Xu C. MicroRNA-495-3p diminishes doxorubicin-induced cardiotoxicity through activating AKT. J Cell Mol Med. 2022;26(7):2076–88. https://doi.org/10.1111/jcmm.17230. Epub 2022 Feb 13. PMID: 35152537; PMCID: PMC8980898.
334. Wang R, Xu Y, Niu X, Fang Y, Guo D, Chen J, Zhu H, Dong J, Zhao R, Wang Y, Qi B, Ren G, Li X, Liu L, Zhang M. MiR-22 inhibition alleviates cardiac dysfunction in doxorubicin-induced cardiomyopathy by targeting the sirt1/PGC-1α pathway. Front Physiol. 2021;12:646903. https://doi.org/10.3389/fphys.2021.646903. Retraction in: Front Physiol. 2022;13:1020568. https://doi.org/10.3389/fphys.2022.1020568. PMID: 33868015; PMCID: PMC8047466.
335. Wang JX, Zhang XJ, Feng C, Sun T, Wang K, Wang Y, Zhou LY, Li PF. MicroRNA-532-3p regulates mitochondrial fission through targeting apoptosis repressor with caspase recruitment domain in doxorubicin cardiotoxicity. Cell Death Dis. 2015;6(3):e1677. https://doi.org/10.1038/cddis.2015.41. PMID: 25766316; PMCID: PMC4385919.
336. Yin J, Guo J, Zhang Q, Cui L, Zhang L, Zhang T, Zhao J, Li J, Middleton A, Carmichael PL, Peng S. Doxorubicin-induced mitophagy and mitochondrial damage is associated with dysregulation of the PINK1/parkin pathway. Toxicol In Vitro. 2018;51:1–10. https://doi.org/10.1016/j.tiv.2018.05.001. Epub 2018 May 3. PMID: 29729358.
337. Das S, Janardhanan KK, Thampi BSH. Bioactive extract of morel mushroom, Morchella esculenta (Ascomycota) attenuates doxorubicin-induced oxidative stress leading to myocardial injury. Int J Med Mushrooms. 2022;24(8):31–44. https://doi.org/10.1615/IntJMedMushrooms.2022044516. PMID: 35997093.
338. Kuzu M, Kandemir FM, Yildirim S, Kucukler S, Caglayan C, Turk E. Morin attenuates doxorubicin-induced heart and brain damage by reducing oxidative stress, inflammation and apoptosis. Biomed Pharmacother. 2018;106:443–53. https://doi.org/10.1016/j.biopha.2018.06.161. Epub 2018 Jul 11. PMID: 29990832.
339. Zamanian G, Shayan M, Rahimi N, Bahremand T, Shafaroodi H, Ejtemaei-Mehr S, Aghaei I, Dehpour AR. Interaction of morphine tolerance with pentylenetetrazole-induced seizure threshold in mice: the role of NMDA-receptor/NO pathway. Epilepsy Behav. 2020;112:107343. https://doi.org/10.1016/j.yebeh.2020.107343. Epub 2020 Aug 2. PMID: 32755816.
340. Sun J, Sun G, Cui X, Meng X, Qin M, Sun X. Myricitrin protects against doxorubicin-induced cardiotoxicity by counteracting oxidative stress and inhibiting mitochondrial apoptosis via ERK/P53 pathway. Evid Based Complement Alternat Med. 2016;2016:6093783. https://doi.org/10.1155/2016/6093783. Epub 2016 Sep 14. PMID: 27703489; PMCID: PMC5039279.
341. Sangomla S, Saifi MA, Khurana A, Godugu C. Nanoceria ameliorates doxorubicin induced cardiotoxicity: possible mitigation via reduction of oxidative stress and inflammation. J Trace Elem Med Biol. 2018;47:53–62. https://doi.org/10.1016/j.jtemb.2018.01.016. Epub 2018 Feb 1. PMID: 29544808.
342. Yarmohammadi F, Rezaee R, Karimi G. Natural compounds against doxorubicin-induced cardiotoxicity: a review on the involvement of Nrf2/ARE signaling pathway. Phytother Res. 2021;35(3):1163–75. https://doi.org/10.1002/ptr.6882. Epub 2020 Sep 28. PMID: 32985744.
343. Sandamali JAN, Hewawasam RP, Jayatilaka KAPW, Mudduwa LKB. *Nauclea orientalis* (L.) bark extract protects rat cardiomyocytes from doxorubicin-induced oxidative stress, inflammation, apoptosis, and DNA fragmentation. Oxidative Med Cell Longev. 2022;2022:1714841. https://doi.org/10.1155/2022/1714841. PMID: 35198093; PMCID: PMC8860544.
344. Arunachalam S, Nagoor Meeran MF, Azimullah S, Sharma C, Goyal SN, Ojha S. Nerolidol attenuates oxidative stress, inflammation, and apoptosis by modulating Nrf2/MAPK signaling pathways in doxorubicin-induced acute cardiotoxicity in rats. Antioxidants (Basel). 2021;10(6):984. https://doi.org/10.3390/antiox10060984. PMID: 34205494; PMCID: PMC8235529.

345. Xu N, Lu Y, Yao X, Zhao R, Li Z, Li J, Zhang Y, Li B, Zhou Y, Shen H, Wang L, Chen K, Yang L, Lu S. NMCP-2 polysaccharide purified from *Morchella conica* effectively prevents doxorubicin-induced cardiotoxicity by decreasing cardiomyocyte apoptosis and myocardial oxidative stress. Food Sci Nutr. 2021;9(11):6262–73. https://doi.org/10.1002/fsn3.2586. PMID: 34760256; PMCID: PMC8565241.
346. Al-Salam S, Kandhan K, Sudhadevi M, Tariq S. Nootkatone ameliorates doxorubicin induced myocardial injury through modulation of NF-κB signals and oxidative stress. Cell Physiol Biochem. 2022;56(4):401–17. https://doi.org/10.33594/000000559. PMID: 36001774.
347. Dai GF, Wang Z, Zhang JY. Octreotide protects doxorubicin-induced cardiac toxicity via regulating oxidative stress. Eur Rev Med Pharmacol Sci. 2018;22(18):6139–48. https://doi.org/10.26355/eurrev_201809_15954. PMID: 30280802.
348. Karakoç MD, Sekkin S. Effects of oleuropein on epirubicin and cyclophosphamide combination treatment in rats. Turk J Pharm Sci. 2021;18(4):420–9. https://doi.org/10.4274/tjps.galenos.2020.69008. PMID: 34496482; PMCID: PMC8430407.
349. Li J, Wu Y, Wang D, Zou L, Fu C, Zhang J, Leung GP. Oridonin synergistically enhances the anti-tumor efficacy of doxorubicin against aggressive breast cancer via pro-apoptotic and anti-angiogenic effects. Pharmacol Res. 2019;146:104313. https://doi.org/10.1016/j.phrs.2019.104313. Epub 2019 Jun 13. PMID: 31202781.
350. Cheng X, Liu D, Xing R, Song H, Tian X, Yan C, Han Y. Orosomucoid 1 attenuates doxorubicin-induced oxidative stress and apoptosis in cardiomyocytes via Nrf2 signaling. Biomed Res Int. 2020;2020:5923572. https://doi.org/10.1155/2020/5923572. Erratum in: Biomed Res Int. 2022;2022:9828942. https://doi.org/10.1155/2022/9828942. PMID: 33134382; PMCID: PMC7591952.
351. Qin Y, Lv C, Zhang X, Ruan W, Xu X, Chen C, Ji X, Lu L, Guo X. Neuraminidase1 inhibitor protects against doxorubicin-induced cardiotoxicity *via* suppressing Drp1-dependent mitophagy. Front Cell Dev Biol. 2021;9:802502. https://doi.org/10.3389/fcell.2021.802502. PMID: 34977042; PMCID: PMC8719652.
352. Hu C, Zhang X, Zhang N, Wei WY, Li LL, Ma ZG, Tang QZ. Osteocrin attenuates inflammation, oxidative stress, apoptosis, and cardiac dysfunction in doxorubicin-induced cardiotoxicity. Clin Transl Med. 2020;10(3):e124. https://doi.org/10.1002/ctm2.124. Epub 2020 Jul 3. PMID: 32618439; PMCID: PMC7418805.
353. Carvalho PB, Gonçalves AF, Alegre PH, Azevedo PS, Roscani MG, Bergamasco CM, Modesto PN, Fernandes AA, Minicucci MF, Paiva SA, Antonio L, Zornoff M, Polegato BF. Pamidronate attenuates oxidative stress and energetic metabolism changes but worsens functional outcomes in acute doxorubicin-induced cardiotoxicity in rats. Cell Physiol Biochem. 2016;40(3–4):431–42. https://doi.org/10.1159/000452558. Epub 2016 Nov 25. PMID: 27889760.
354. Gyöngyösi M, Lukovic D, Zlabinger K, Spannbauer A, Gugerell A, Pavo N, Traxler D, Pils D, Maurer G, Jakab A, Riesenhuber M, Pircher A, Winkler J, Bergler-Klein J. Liposomal doxorubicin attenuates cardiotoxicity via induction of interferon-related DNA damage resistance. Cardiovasc Res. 2020;116(5):970–82. https://doi.org/10.1093/cvr/cvz192. PMID: 31346605.
355. Liu D, Ma Z, Xu L, Zhang X, Qiao S, Yuan J. PGC1α activation by pterostilbene ameliorates acute doxorubicin cardiotoxicity by reducing oxidative stress via enhancing AMPK and SIRT1 cascades. Aging (Albany NY). 2019;11(22):10061–73. https://doi.org/10.18632/aging.102418. Epub 2019 Nov 16. PMID: 31733141; PMCID: PMC6914429.
356. Russo M, Guida F, Paparo L, Trinchese G, Aitoro R, Avagliano C, Fiordelisi A, Napolitano F, Mercurio V, Sala V, Li M, Sorriento D, Ciccarelli M, Ghigo A, Hirsch E, Bianco R, Iaccarino G, Abete P, Bonaduce D, Calignano A, Berni Canani R, Tocchetti CG. The novel butyrate derivative phenylalanine-butyramide protects from doxorubicin-induced cardiotoxicity. Eur J Heart Fail. 2019;21(4):519–28. https://doi.org/10.1002/ejhf.1439. Epub 2019 Mar 6. PMID: 30843309.
357. Wang C, Hu L, Guo S, Yao Q, Liu X, Zhang B, Meng X, Yang X. Phosphocreatine attenuates doxorubicin-induced cardiotoxicity by inhibiting oxidative stress and activating TAK1 to

promote myocardial survival in vivo and in vitro. Toxicology. 2021;460:152881. https://doi.org/10.1016/j.tox.2021.152881. Epub 2021 Aug 3. PMID: 34358621.

358. El-Agamy DS, El-Harbi KM, Khoshhal S, Ahmed N, Elkablawy MA, Shaaban AA, Abo-Haded HM. Pristimerin protects against doxorubicin-induced cardiotoxicity and fibrosis through modulation of Nrf2 and MAPK/NF-kB signaling pathways. Cancer Manag Res. 2018;11:47–61. https://doi.org/10.2147/CMAR.S186696. PMID: 30588110; PMCID: PMC6304079.
359. Sadek KM, Mahmoud SFE, Zeweil MF, Abouzed TK. Proanthocyanidin alleviates doxorubicin-induced cardiac injury by inhibiting NF-kB pathway and modulating oxidative stress, cell cycle, and fibrogenesis. J Biochem Mol Toxicol. 2021;35(4):e22716. https://doi.org/10.1002/jbt.22716. Epub 2021 Jan 23. PMID: 33484087.
360. Du XY, Xiang DC, Gao P, Peng H, Liu YL. Inhibition of (pro)renin receptor-mediated oxidative stress alleviates doxorubicin-induced heart failure. Front Oncol. 2022;12:874852. https://doi.org/10.3389/fonc.2022.874852. PMID: 35574363; PMCID: PMC9106363.
361. Chularojmontri L, Gerdprasert O, Wattanapitayakul SK. Pummelo protects Doxorubicin-induced cardiac cell death by reducing oxidative stress, modifying glutathione transferase expression, and preventing cellular senescence. Evid Based Complement Alternat Med. 2013;2013:254835. https://doi.org/10.1155/2013/254835. Epub 2013 Jan 21. PMID: 23401708; PMCID: PMC3564430.
362. Chen X, Peng X, Luo Y, You J, Yin D, Xu Q, He H, He M. Quercetin protects cardiomyocytes against doxorubicin-induced toxicity by suppressing oxidative stress and improving mitochondrial function via 14-3-3γ. Toxicol Mech Methods. 2019;29(5):344–54. https://doi.org/10.1080/15376516.2018.1564948. Epub 2019 Feb 18. PMID: 30636491.
363. Zhang J, Wang M, Ding W, Zhao M, Ye J, Xu Y, Wang Z, Ye D, Li D, Liu J, Wan J. Resolvin E1 protects against doxorubicin-induced cardiotoxicity by inhibiting oxidative stress, autophagy and apoptosis by targeting AKT/mTOR signaling. Biochem Pharmacol. 2020;180:114188. https://doi.org/10.1016/j.bcp.2020.114188. Epub 2020 Aug 1. PMID: 32750329.
364. Liu X, Zhao H, Jin Q, You W, Cheng H, Liu Y, Song E, Liu G, Tan X, Zhang X, Wan F. Resveratrol induces apoptosis and inhibits adipogenesis by stimulating the SIRT1-AMPKα-FOXO1 signalling pathway in bovine intramuscular adipocytes. Mol Cell Biochem. 2018;439(1–2):213–23. https://doi.org/10.1007/s11010-017-3149-z. Epub 2017 Aug 17. PMID: 28819881.
365. Liao ZQ, Jiang YN, Su ZL, Bi HL, Li JT, Li CL, Yang XL, Zhang Y, Xie X. Rutaecarpine inhibits doxorubicin-induced oxidative stress and apoptosis by activating AKT signaling pathway. Front Cardiovasc Med. 2022;8:809689. https://doi.org/10.3389/fcvm.2021.809689. PMID: 35071368; PMCID: PMC8766983.
366. Wang F, Han L. Upregulation of serum and glucocorticoid-regulated kinase 1 (SGK1) ameliorates doxorubicin-induced cardiotoxic injury, apoptosis, inflammation and oxidative stress by suppressing glucose regulated protein 78 (GRP78)-mediated endoplasmic reticulum stress. Bioengineered. 2022;13(1):844–55. https://doi.org/10.1080/21655979.2021.2013109. PMID: 34898378; PMCID: PMC8805919.
367. Yao L, Gui M, Li J, Lu B, Wang J, Zhou X, Fu D. Shengxian decoction decreases doxorubicin-induced cardiac apoptosis by regulating the TREM1/NF-κB signaling pathway. Mol Med Rep. 2021;23(3):219. https://doi.org/10.3892/mmr.2021.11858. Epub 2021 Jan 26. Retraction in: Mol Med Rep. 2024;29(3):35. https://doi.org/10.3892/mmr.2024.13159. PMID: 33495812; PMCID: PMC7845587.
368. Wu YP, Zhang S, Xin YF, Gu LQ, Xu XZ, Zhang CD, You ZQ. Evidences for the mechanism of Shenmai injection antagonizing doxorubicin-induced cardiotoxicity. Phytomedicine. 2021;88:153597. https://doi.org/10.1016/j.phymed.2021.153597. Epub 2021 May 21. PMID: 34111614.
369. Bin Jardan YA, Ansari MA, Raish M, Alkharfy KM, Ahad A, Al-Jenoobi FI, Haq N, Khan MR, Ahmad A. Sinapic acid ameliorates oxidative stress, inflammation, and apoptosis in acute doxorubicin-induced cardiotoxicity via the NF-κB-mediated pathway. Biomed Res

Int. 2020;2020:3921796. https://doi.org/10.1155/2020/3921796. PMID: 32258120; PMCID: PMC7085847.

370. Ruan Y, Dong C, Patel J, Duan C, Wang X, Wu X, Cao Y, Pu L, Lu D, Shen T, Li J. SIRT1 suppresses doxorubicin-induced cardiotoxicity by regulating the oxidative stress and p38MAPK pathways. Cell Physiol Biochem. 2015;35(3):1116–24. https://doi.org/10.1159/000373937. Epub 2015 Feb 6. PMID: 25766524.
371. Guo R, Wu K, Chen J, Mo L, Hua X, Zheng D, Chen P, Chen G, Xu W, Feng J. Exogenous hydrogen sulfide protects against doxorubicin-induced inflammation and cytotoxicity by inhibiting p38MAPK/NFκB pathway in H9c2 cardiac cells. Cell Physiol Biochem. 2013;32(6):1668–80. https://doi.org/10.1159/000356602. Epub 2013 Dec 13. PMID: 24356372.
372. Yoon CS, Kim HK, Mishchenko NP, Vasileva EA, Fedoreyev SA, Stonik VA, Han J. Spinochrome D attenuates doxorubicin-induced cardiomyocyte death via improving glutathione metabolism and attenuating oxidative stress. Mar Drugs. 2018;17(1):2. https://doi.org/10.3390/md17010002. PMID: 30577438; PMCID: PMC6356724.
373. Liu G, Liu Y, Wang R, Hou T, Chen C, Zheng S, Dong Z. Spironolactone attenuates doxorubicin-induced cardiotoxicity in rats. Cardiovasc Ther. 2016;34(4):216–24. https://doi.org/10.1111/1755-5922.12189. PMID: 27097055.
374. Wu Z, Zhao X, Miyamoto A, Zhao S, Liu C, Zheng W, Wang H. Effects of steroidal saponins extract from Ophiopogon japonicus root ameliorates doxorubicin-induced chronic heart failure by inhibiting oxidative stress and inflammatory response. Pharm Biol. 2019;57(1):176–83. https://doi.org/10.1080/13880209.2019.1577467. PMID: 30860934; PMCID: PMC6419681.
375. Sheibani M, Faghir-Ghanesefat H, Dehpour S, Keshavarz-Bahaghighat H, Sepand MR, Ghahremani MH, Azizi Y, Rahimi N, Dehpour AR. Sumatriptan protects against myocardial ischaemia-reperfusion injury by inhibition of inflammation in rat model. Inflammopharmacology. 2019;27(5):1071–80. https://doi.org/10.1007/s10787-019-00586-5. Epub 2019 Mar 30. PMID: 30929154.
376. Abu-Khudir R, Ibrahim WM, Shams ME, Salama AF. Trehalose alleviates doxorubicin-induced cardiotoxicity in female Swiss albino mice by suppression of oxidative stress and autophagy. J Biochem Mol Toxicol. 2021;35(9):e22859. https://doi.org/10.1002/jbt.22859. Epub 2021 Jul 30. PMID: 34328254.
377. Eid BG, El-Shitany NAE, Neamatallah T. Trimetazidine improved adriamycin-induced cardiomyopathy by downregulating TNF-α, BAX, and VEGF immunoexpression via an antioxidant mechanism. Environ Toxicol. 2021;36(6):1217–25. https://doi.org/10.1002/tox.23120. Epub 2021 Mar 11. PMID: 33704910.
378. Wang Z, Wang M, Liu J, Ye J, Jiang H, Xu Y, Ye D, Wan J. Inhibition of TRPA1 attenuates doxorubicin-induced acute cardiotoxicity by suppressing oxidative stress, the inflammatory response, and endoplasmic reticulum stress. Oxidative Med Cell Longev. 2018;2018:5179468. https://doi.org/10.1155/2018/5179468. PMID: 29682158; PMCID: PMC5850896.
379. Sirangelo I, Sapio L, Ragone A, Naviglio S, Iannuzzi C, Barone D, Giordano A, Borriello M. Vanillin prevents doxorubicin-induced apoptosis and oxidative stress in rat H9c2 cardiomyocytes. Nutrients. 2020;12(8):2317. https://doi.org/10.3390/nu12082317. PMID: 32752227; PMCID: PMC7468857.
380. Alhowail A. Varenicline enhances the survival of doxorubicin-treated mice. Eur Rev Med Pharmacol Sci. 2020;24(21):11350–5. https://doi.org/10.26355/eurrev_202011_23626. PMID: 33215455.
381. Liu Y, Asnani A, Zou L, Bentley VL, Yu M, Wang Y, Dellaire G, Sarkar KS, Dai M, Chen HH, Sosnovik DE, Shin JT, Haber DA, Berman JN, Chao W, Peterson RT. Visnagin protects against doxorubicin-induced cardiomyopathy through modulation of mitochondrial malate dehydrogenase. Sci Transl Med. 2014;6(266):266ra170. https://doi.org/10.1126/scitranslmed.3010189. PMID: 25504881; PMCID: PMC4360984.

382. Latic N, Erben RG. Vitamin D and cardiovascular disease, with emphasis on hypertension, atherosclerosis, and heart failure. Int J Mol Sci. 2020;21(18):6483. https://doi.org/10.3390/ijms21186483. PMID: 32899880; PMCID: PMC7555466.
383. Sun Z, Yan B, Yu WY, Yao X, Ma X, Sheng G, Ma Q. Vitexin attenuates acute doxorubicin cardiotoxicity in rats via the suppression of oxidative stress, inflammation and apoptosis and the activation of FOXO3a. Exp Ther Med. 2016;12(3):1879–84. https://doi.org/10.3892/etm.2016.3518. Epub 2016 Jul 13. PMID: 27588105; PMCID: PMC4997971.
384. Jiang Y, Liu Y, Xiao W, Zhang D, Liu X, Xiao H, You S, Yuan L. Xinmailong attenuates doxorubicin-induced lysosomal dysfunction and oxidative stress in H9c2 cells via HO-1. Oxidative Med Cell Longev. 2021;2021:5896931. https://doi.org/10.1155/2021/5896931. PMID: 33854694; PMCID: PMC8019640.
385. Lin H, Zhang J, Ni T, Lin N, Meng L, Gao F, Luo H, Liu X, Chi J, Guo H. Yellow wine polyphenolic compounds prevents doxorubicin-induced cardiotoxicity through activation of the Nrf2 signalling pathway. J Cell Mol Med. 2019;23(9):6034–47. https://doi.org/10.1111/jcmm.14466. Epub 2019 Jun 21. PMID: 31225944; PMCID: PMC6714138.
386. Gu Y, Ju A, Jiang B, Zhang J, Man S, Liu C, Gao W. Yiqi Fumai lyophilized injection attenuates doxorubicin-induced cardiotoxicity, hepatotoxicity and nephrotoxicity in rats by inhibition of oxidative stress, inflammation and apoptosis. RSC Adv. 2018;8(71):40894–911. https://doi.org/10.1039/c8ra07163b. Retraction in: RSC Adv. 2025;15(2):947. https://doi.org/10.1039/d5ra90003d. PMID: 35557896; PMCID: PMC9091596.
387. Yu X, Cui L, Zhang Z, Zhao Q, Li S. α-Linolenic acid attenuates doxorubicin-induced cardiotoxicity in rats through suppression of oxidative stress and apoptosis. Acta Biochim Biophys Sin Shanghai. 2013;45(10):817–26. https://doi.org/10.1093/abbs/gmt082. Epub 2013 Jul 28. PMID: 23896563.
388. Al-Taee H, Azimullah S, Meeran MFN, Alaraj Almheiri MK, Al Jasmi RA, Tariq S, Ab Khan M, Adeghate E, Ojha S. β-caryophyllene, a dietary phytocannabinoid attenuates oxidative stress, inflammation, apoptosis and prevents structural alterations of the myocardium against doxorubicin-induced acute cardiotoxicity in rats: An in vitro and in vivo study. Eur J Pharmacol. 2019;858:172467. https://doi.org/10.1016/j.ejphar.2019.172467. Epub 2019 Jun 16. PMID: 31216443.
389. Attia AA, Sorour JM, Mohamed NA, Mansour TT, Al-Eisa RA, El-Shenawy NS. Biochemical, histological, and ultrastructural studies of the protective role of vitamin E on cyclophosphamide-induced cardiotoxicity in male rats. Biomedicines. 2023;11:390.
390. Zare MFR, Rakhshan K, Aboutaleb N, Nikbakht F, Naderi N, Bakhshesh M, Azizi Y. Apigenin attenuates doxorubicin induced cardiotoxicity via reducing oxidative stress and apoptosis in male rats. Life Sci. 2019;232:116623. https://doi.org/10.1016/j.lfs.2019.116623. Epub 2019 Jul 4. PMID: 31279781.
391. Jiao Y, Li Y, Zhang J, Zhang S, Zha Y, Wang J. RRM2 Alleviates doxorubicin-induced cardiotoxicity through the AKT/mTOR signaling pathway. Biomolecules. 2022;12(2):299. https://doi.org/10.3390/biom12020299. PMID: 35204799; PMCID: PMC8869767.
392. Karakikes I, Ameen M, Termglinchan V, Wu JC. Human induced pluripotent stem cell-derived cardiomyocytes: insights into molecular, cellular, and functional phenotypes. Circ Res. 2015;117(1):80–8. https://doi.org/10.1161/CIRCRESAHA.117.305365. PMID: 26089365; PMCID: PMC4546707.

Optimizing Cardiovascular Safety: Cardiotoxicity Prevention in Tyrosine Kinase Inhibitor, Monoclonal Antibody, Proteasome Inhibitor, and 5-Fluorouracil Anticancer Therapies

Jonathan Wright, Evans Osuji, Tanvi Brar, Shane S. Scott, and Sakima A. Smith

Abbreviations

5-FU	5-Fluorouracil
ACE	Angiotensin-converting enzyme inhibitors
ALK	Anaplastic lymphoma kinase
AMPK	AMP-activated protein kinase
ARB	Angiotensin II receptor blockers
ATP	Adenosine triphosphate
BB	Beta-blockers
BCRABL	Break point cluster-Abelson
BH4	Tetrahydrobiopterin
BTK	Bruton tyrosine kinase
CaMKII	Calcium/calmodulin-dependent protein kinase II
CAT	Carnitine acetyltransferase
CCB	Calcium channel blocker
c-KIT	Stem cell factor receptor
CoQ10	Co-enzyme Q10

J. Wright
Department of Internal Medicine, The Ohio State University Wexner Medical Center, Columbus, OH, USA
e-mail: jonathan.wright@osumc.edu

E. Osuji
Department of Internal Medicine, Johns Hopkins Bayview Medical Center, Baltimore, MD, USA
e-mail: eosuji2@jh.edu

T. Brar
Dorothy M. Davis Heart and Lung Research Institute, The Ohio State University College of Medicine, Columbus, OH, USA
e-mail: tanvi.brar@osumc.edu

S. A. Smith, S. S. Scott (eds.), *Myocardial Oxidative Stress in Anti-Cancer Therapies*, Oxidative Stress in Applied Basic Research and Clinical Practice,
https://doi.org/10.1007/978-3-031-97517-2_7

COX-2	Cyclooxygenase-2
DICT	Drug-induced cardiotoxicity
EGFR	Epidermal growth factor receptor
eNOS	Endothelial nitric oxide synthase
ERK	Extracellular signal-regulated kinase
ET-1	Endothelin-1
ETC	Electron transport chain
GLP-1	Glucagon-like peptide 1
GSH	Glutathione
GSH-Px	Glutathione peroxidase
HER2	Human epidermal growth factor receptor 2
hERG	Human ether-a-go-go related gene
JNK	Jun N-terminal kinase
LVEF	Left ventricular ejection fraction
LVD	Left ventricular dysfunction
MAbs	Monoclonal antibodies
MAPK	Mitogen-activated protein kinases
MET	Mesenchymal epithelial transition factor receptor
Mito-TEMPO	Mitochondrial-specific antioxidant
MPG	N-2-mercaptopropyonyl glycine
NAC	N-acetyl cysteine
NADPH	Nicotinamide adenine dinucleotide phosphate
NF-kB	Nuclear factor-kappa B
NLRP3	NOD-like receptor protein 3
NO	Nitric oxide
NRF2	Nuclear factor erythroid 2-related factor 2
PARP-1	Poly-ADP ribose polymerase 1
PDGF	Platelet-derived growth factor
PDGFR	Platelet-derived growth factor receptor
PI	Proteasome inhibitors
PI3K	Phosphoinositide 3 kinase
PPAR	Peroxisome proliferator-activated receptor
QTc	Corrected QT interval
RAAS	Renin-angiotensin system

S. S. Scott (✉)
Dorothy M. Davis Heart and Lung Research Institute, The Ohio State University College of Medicine, Columbus, OH, USA

The Ohio State University Medical Scientist Training Program (MSTP), Columbus, OH, USA
e-mail: shane.scott@osumc.edu

S. A. Smith
Dorothy M. Davis Heart and Lung Research Institute, The Ohio State University College of Medicine, Columbus, OH, USA
Sakima.smith@osumc.edu

RCT	Randomized control trial
RNS	Reactive nitrogen species
ROS	Reactive oxygen species
SGK1	Serum/glucocorticoid-regulated kinase 1
SGLT2	Sodium-glucose cotransporter-2
SOD	Superoxide dismutase
TKI	Tyrosine kinase inhibitors
TLR4	Toll-like receptor 4
TNFα	Tumor necrosis factor alpha
TRPV1	Transient receptor potential vanilloid 1
TXNIP	Thioredoxin-interacting protein
VEGF	Vascular endothelial growth factor

Introduction

Drug-induced cardiotoxicity (DICT) is of growing concern, particularly with anti-cancer chemotherapies. Mechanisms underlying DICT in chemotherapies are multifaceted, where oxidative stress plays a key role in the pathogenesis of cardiac damage and disease. Oxidative stress results from an imbalance in the production of reactive oxygen species (ROS) and the cellular antioxidant defense mechanisms. These highly reactive molecules target essential cellular components such as DNA, structural proteins, and lipid membranes [1]. Further ROS can impact myocardial calcium homeostasis, leading to arrhythmias, while also triggering pro-inflammatory signals that result in cardiac remodeling and hypertrophy [2]. The heart is particularly sensitive to oxidative stress due to its high oxygen demand; thus, even slight imbalances in the cardiac redox balance can result in cellular damage [2].

Key components in maintaining cardiomyocyte oxidative stress balance include mitochondria, NADPH oxidase, and uncoupled endothelial nitric oxide synthase (eNOS). Mitochondria are florid ROS generators given their role in oxidative phosphorylation, where the electron transport chain (ETC) produces ROS-like superoxides (O_2^-) and hydroxyl radicals ($OH^\bullet$) during cellular metabolism [3]. NADPH oxidase generates ROS during cardiomyocyte stretch, whereby the resultant ROS sensitize ryanodine receptors to be increasingly prone to aberrant firing and resultant arrhythmias [4]. Finally, eNOS produces nitric oxide (NO) as a potent vasodilator, often in conjunction with the reducing agent tetrahydrobiopterin (BH_4). However, when the cellular redox is unbalanced, BH_4 can be consumed, and resultant uncoupled eNOS is prone to superoxide formation [5]. Overall, the heart is highly susceptible to the buildup of ROS and subsequent damage, especially considering its limited ability to regenerate.

A critical signaling pathway involved in a cell's defense against oxidative stress and reactive oxygen species is mitogen-activated protein kinases (MAPKs), a serine and threonine kinase responsive to a broad range of extracellular stimuli. MAPK signaling pathways, including ERK, JNK, and p38, provide an elaborate and interconnected regulation of antioxidant defenses and redox balance, ultimately influencing overall cellular homeostasis. Many anticancer chemotherapies cause aberration of such signaling in cardiomyocytes, resulting in myocardial inflammation, cell death, and scarring leading to impaired cardiac function and conductance [6].

Chemotherapy-induced myocardial oxidative stress can also result in mitochondrial dysregulation, proteasome inhibition, and endothelial dysfunction.

Numerous pharmacotherapies have been investigated to combat the cardiotoxic oxidative stress from chemotherapy. While traditional chemotherapies such as anthracyclines directly generate ROS, tyrosine kinase inhibitors (TKIs) and monoclonal antibodies (MAbs) generate ROS via mitochondrial dysregulation and the resulting redox imbalance. Further, 5-fluorouracil (5-FU) has a unique mechanism where DICT results from endothelial dysregulation, which subsequently produces ROS. Finally, proteasome inhibitor's (PI) DICT manifests secondary to off-targeting, causing proteasome dysfunction and proteotoxic agents in cardiomyocytes, resulting in oxidative stress.

Therapeutic agents target oxidative stress by acting either directly or indirectly on downstream effects. Direct agents reduce stress by either acting as an antioxidant or ROS scavenger, or by targeting other cellular mediators involved in oxidative damage. In contrast, indirect agents mitigate oxidative stress by modulating apoptotic signaling, restoring calcium homeostasis, attenuating ferroptosis, or increasing autophagy.

While the incidence of DICT from anticancer chemotherapy is relatively low, in light of the ever-increasing prognosis for cancer patients, it is of growing prevalence and concern for patients and clinicians. Oxidative stress and its downstream signaling offer prime pharmacologic targets to mitigate DICT. This chapter serves as a broad resource to consolidate and compare multiple mechanisms behind the oxidative stress caused by TKIs, MAbs, PIs, and 5-FU. Further, it will outline the current literature on the pharmacologic treatment strategies for oxidative stress induced by these novel anticancer chemotherapies. By identifying treatment options, we can potentially mitigate oxidative stress in patients and improve the safety of chemotherapies and patient morbidity and prognosis.

Tyrosine Kinase Inhibitors

Tyrosine kinase inhibitors (TKIs) are a class of anticancer molecules which have revolutionized the treatment of multiple neoplasms, including solid tumors, leukemia, breast cancer, adenocarcinoma, sarcoma, and other stromal tumors cancers [7–12]. Many TKIs have been shown to cause oxidative stress in the heart via mitochondrial dysregulation and increased intracellular ROS [13–17]. While sunitinib and sorafenib have been studied extensively, several other TKIs, including imatinib, gefitinib, and ibrutinib, also have established evidence for oxidative stress-mediated

DICT. Direct agents alleviate stress by acting as antioxidants, scavenging ROS, or targeting other forms of direct oxidative damage.

Additionally, oxidative stress has been implicated in DICT associated with nilotinib, crizotinib, lenvatinib, regorafenib, lapatinib, erlotinib, and trametinib; however, treatment options targeting this mechanism have yet to be explored.

Sunitinib

Sunitinib was approved by the FDA in 2006 and is the first drug in its class for vascular endothelial growth factor (VEGF) TKIs. It has since been approved for a wide range of cancers including renal cell carcinoma, gastrointestinal stromal tumors, and pancreatic neuroendocrine tumors, where its deployment as chemotherapy continues to grow. Sunitinib's specific mechanism of inducing oxidative stress includes mitochondrial energy dysregulation leading to superoxide accumulation, causing AMP-activated protein kinase (AMPK) inhibition, lipid oxidation, and human ether-a-go-go related gene (hERG) block culminating in cardiomyocyte death and disrupted myocyte contractility (Fig. 1) [13–15, 18]. In human cardiomyocytes, sunitinib has demonstrated a dose-dependent negative inotropic effect associated with increased intracellular ROS [19]. AMPK has been shown to be a specific mitochondrial metabolic and ROS regulator, with inhibition leading to increased mitochondrial ROS [20].

Targets for sunitinib-mediated oxidative stress treatment have been studied in a variety of preclinical models (Table 1). Cellular models have focused on H9c2 cells, a rat heart tissue derivative. One potential target is calcium/calmodulin-dependent protein kinase II (CaMKII), an upstream regulator of AMPK signaling pathways in myocardium and serves as an AMPK binding partner [21]. CaMKII plays a key role in initiating cardiac remodeling and inflammation through the nuclear factor-kappa B (NF-kB) inflammation cascade and ROS generation [22]. Furthermore, a downstream target of CaMKII is the NLRP3, an inflammasome gene triggered by ROS which also contributes to cardiac remodeling and inflammation (Fig. 1) [22]. Polydatin (trans-resveratrol-3-O-glucoside) is a ROS scavenger and signaling pathway inhibitor [23]. Polydatin was shown to reduce intracellular ROS and biomarkers of oxidative stress [24]. Additionally, KN-93, a targeted CaMKII inhibitor, was shown to reverse TKI-induced mitochondrial damage in cardiac fibroblasts [15].

Further studies targeting oxidative stress for treatment of sunitinib-mediated cardiotoxicity have mostly remained in animal models, targeting AMPK pathway, which is downregulated by sunitinib. Trimetazidine is a fatty acid oxidation inhibitor initially developed as an anti-anginal drug [25]. In one study, trimetazidine was shown to activate the AMPK pathway in H9c2 cardiomyocytes and reversed sunitinib-induced cardiotoxicity in 129S1/SvImJ mice [26]. In a separate study, empagliflozin, a sodium-glucose cotransporter-2 (SGLT2) inhibitor with known cardioprotective properties, improved sunitinib-induced hypertension in C57BL/6J mice via restoration of dysregulated AMPK signaling [27–29]. Further, supplementation with carnitine, a metabolic partner in fatty acid oxidation, provided relief for

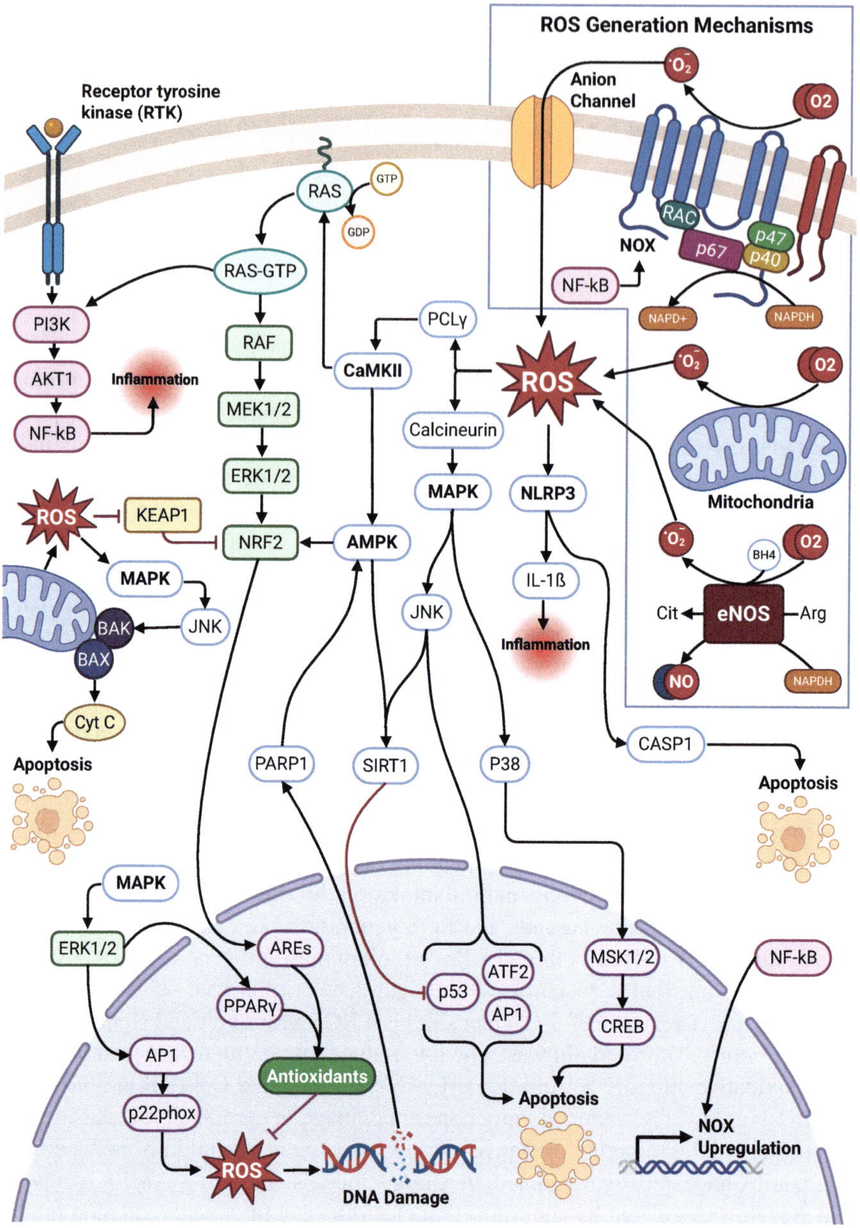

Fig. 1 Confluence of important intracellular signaling in the regulation of oxidative stress mediated by cardiotoxic anti-cancer therapies. *Abbreviations:* AKT1 Protein kinase B, AMPK AMP activated protein kinase, AP1 Activator protein 1, AREs Antioxidant response elements, ATF2 Activating transcription factor-2, BAK Bcl-2 antagonist killer 1, BAX Bcl-2-associated X protein, BH4 Tetrahydrobiopterin, CaMKII Calcium/calmodulin-dependent protein kinase II, CREB cAMP response element-binding protein, ERK Extracellular signal-regulated kinase, IL-1B Interleukin 1 beta, JNK Jun N-terminal kinase, KEAP1 Kelch-like ECH-associated protein 1, MAPK Mitogen-activated protein kinases, MEK1/2 Mitogen-activated protein kinase kinase 1/2, MITO

sunitinib-induced DICT in Wistar rats through a similar restoration of the AMPK signaling pathway [30]. Serum/glucocorticoid regulated kinase 1 (SGK1) is a downstream target in AMPK-mediated activation of Orai1/STIM1, an endoplasmic reticulum calcium channel regulatory complex [31]. Berberine, a phytochemical and an AMPK activator provides protection against sunitinib-associated DICT via SGK1 activation, which normalizes sunitinib-induced calcium dysregulation [32–34]. Other phytochemicals, including isoquercetin, an antioxidant flavanol, also possess AMPK stimulating activity [35]. In a phase I clinical trial ($n = 12$), isoquercetin demonstrated improved patient-reported fatigue scores and patient-reported adverse events scores [36]. While AMPK is a wide signaling partner and is involved many cellular processes, many proposed therapies converge on protection through AMPK upregulation, which promotes mitochondrial protection from ROS.

Beyond AMPK signaling, one study examined the role of rivaroxaban, a direct factor Xa inhibitor, in myocardial protection. Prior studies provided evidence of rivaroxaban having anti-inflammatory properties, through blockade of ROS generation. In Wistar rats, rivaroxaban was found to mitigate sunitinib-induced cardiac injury via restoration of antioxidant enzyme levels and blocking the proinflammatory signaling cascades [37–39]. Further research, conducted in H9c2 cells with MitoTEMPO, a mitochondrial-specific ROS scavenger, composed of a combination of piperidine nitroxide and triphenylphosphonium [40]. When co-exposed with sunitinib, MitoTEMPO prevented the depletion of ATP and glutathione (GSH), as well as decreased the rise in hydroperoxide and caspase 3/7, compared to sunitinib-only treated cells [14].

Sorafenib

Sorafenib, another TKI targeting VEGF, has revolutionized the treatment landscape for various cancers, particularly advanced renal cell carcinoma and hepatocellular carcinoma [41, 42]. Despite its remarkable anticancer efficacy, sorafenib is associated with significant DICT. The precise mechanisms are complex and multifactorial, with emerging evidence suggesting a key role for oxidative stress. Several studies have implicated sorafenib in promoting mitochondrial ROS toxicity, potentially through inhibition of the mitochondrial electron transport chain (ETC) (Table 1) [16, 17]. Sorafenib also triggered CaMKII overactivation, disrupting

Fig. 1 (continued) Mitochondria, MSK1/2 Mitogen- and stress-activated kinases-1 and -2, NADP+ Nicotinamide adenine dinucleotide phosphate, NADPH Nicotinamide adenine dinucleotide phosphate, NF-kB Nuclear factor-kappa B, NLRP3 NOD-like receptor protein 3, NO Nitric oxide, Fig. 1 (continued) NOX NADPH oxidase, NRF2 Nuclear factor erythroid 2-related factor 2, p22phox Human neutrophil cytochrome b light chain, P38 Cytokinin Specific Binding Protein, PARP1 Poly-ADP ribose polymerase 1, PCLy Phosphoinositide-specific phospholipase C gamma, PI3K Phosphoinositide 3 kinase, PPARy Peroxisome proliferator-activated receptor, RAF Rapidly accelerated fibrosarcoma kinase, RAS Rat sarcoma virus GTPase, ROS Reactive oxygen species, RTK Receptor tyrosine kinase, SIRT1 Sirtuin (silent mating type information regulation 2 homolog) 1

Table 1 List of published studies which target oxidative stress for treating TKI-induced DICT

Tyrosine kinase inhibitors					
Drug	Treatment	Mechanism	Level of evidence	Author	Year
Sunitinib	KN-93	Inhibits CaMKII signaling	Cardiac fibroblasts	McMullen	2021
	Polydatin	Reduces ROS production	AC16/H9c2 cells	Quagliariello	2021
	Trimetazidine	Activates AMPK-mTOR-autophagy signaling	129S1/SvImJ mice	Yang	2019
	Empagliflozin	Activates AMPK-mTOR-autophagy signaling	C57BL/6 J mice	Ren	2021
	Carnitine	Activates AMPK-mTOR-autophagy signaling	Wistar rats	Sayed-Ahmed	2019
	Isoquercetin	Activates AMPK-mTOR-autophagy signaling	Phase I RCT	Buonerba	2018
	Berberine	SGK1 activation to regulate calcium	Mice	Li	2023
	Rivaroxaban	Increased antioxidant enzymes and blocked proinflammatory signaling	Wistar rats	Imam	2020
	Mito-TEMPO	Mitochondrial ROS scavenger	H9c2 cells	Bouitbir	2019
Sorafenib	N-2-mercaptopropyonyl glycine	Primary ROS scavenger	H9c2 cells	Liang	2022
	Hesperetin	Inhibits TLR4-NLRP3 signaling	Swiss mice	Zaafar	2022
	Losartan	Primary antioxidant	Rat	Abdelgalil	2020
Imatinib	KN-93	Inhibits CaMKII signaling	Cardiac fibroblasts	McMullen	2021
	Berberine	Inhibits Nrf2-dependent ferroptosis	H9c2 cells	Song	2023
	Carnitine	Activates PDGF-PPARγ-MAPK signaling	Wistar rats	Mansour	2021
	BGP-15	Activates AMPK-mTOR-autophagy signaling	Wistar rats	Sarszegi	2012
Gefitinib	Liraglutide	Increases SOD and decreases NF-κB	Wistar rats	Alasmari	2020
	ARBs	Activates JNK-p38-MAPK signaling	Wistar rats	Alanazi	2022
Ibrutinib	Apocynin	Inhibits NADPH oxidase	C57BL/6 mice	Yang	2020

calcium homeostasis [17]. In human cardiomyocytes, sorafenib was also shown to cause ROS generation through downregulation of stanniocalcin-1 [43]. Finally, ferroptosis, a mechanism of cell death secondary to accumulation of intracellular iron followed by subsequent lipid peroxidation, was shown to be a key mechanism in sorafenib-mediated myocardial stress and damage [44, 45].

Studies seeking to reduce sorafenib-induced oxidative stress are limited to preclinical models. Liang and colleagues evaluated the use of N-2-mercaptopropionyl glycine (MPG), a ROS-scavenging antioxidant, in reversing sorafenib-induced cardiomyocyte apoptosis. The authors showed that MPG reversed sorafenib-induced damage by reducing ROS generation and calcium accumulation, preventing apoptosis [46]. While MPG is not utilized as a pharmacotherapy, its potential therapeutic effect via ROS scavenging requires further exploration. Hesperetin is a flavonoid found in citrus fruits and has demonstrated ROS scavenging properties [16, 47]. However, there is some evidence that hesperetin may also promote mitochondrial ROS generation [48, 49]. Zaafar et al. reported that hesperetin also acts as an anti-inflammatory agent by reducing NOD-like receptor protein 3 (NLRP3)/toll-like receptor 4 (TLR4) signaling and inhibiting apoptosis [50]. In the absence of hesperetin, sorafenib increased NLRP3, TNF-α, IL-1β, and IL-6 in male Swiss mice [50]. Cardiac tissue from mice treated with hesperetin plus exhibited reduced TNF-α, IL-1β, and IL-6 levels and lower NLRP3/TLR4 expressions, in addition to decreased collagen deposition and caspase-3, and increased Bcl-2 expression [50]. However, prior to use of hesperetin as an adjuvant to sorafenib, further clinical studies are required.

Losartan has also been explored as a potential antioxidant, showing an ability to decrease myocardial oxidative stress following myocardial infarction in Sprague-Dawley rats through blocking of superoxide production [51, 52]. In Wistar rats, losartan co-treatment prevented sorafenib-mediated reductions in the left ventricular pressure, myocardial contraction and relaxation, and myocardial metabolic changes, but its direct effect on oxidative stress was not explored [53]. While these studies do not directly evaluate the treatment of oxidative and mitochondrial stress induced by sorafenib, the aforementioned drugs portend promising therapeutic potential and require further preclinical and clinical validation.

Imatinib

Imatinib is among the first-generation TKIs with a wide range of uses. Traditionally sought after for its BCR-ABL inhibition, it was approval in 2001 for the treatment of chronic myeloid leukemia. Imatinib also targets stem cell factor receptor (c-KIT) and platelet-derived growth factor receptor (PDGFR) [54, 55]. Imatinib-mediated cardiotoxicity remains understudied due to the rare prevalence of disease within the treatment population [56–58]. Similar to sorafenib, imatinib promotes DICT via impairment of the ETC, leading to mitochondrial ROS accumulation [17, 59]. Additionally, imatinib, along with sunitinib, results in mitochondrial superoxide formation in a concentration-dependent manner in vivo [15]. Recently, ferroptosis was also associated with imatinib's DICT [60].

Therapies implicated in the prevention of imatinib-induced myocardial oxidative stress and toxicity include KN-93, berberine hydrochloride, and carnitine (Table 1) [15, 61, 62]. KN-93 was described in cardiac fibroblasts to provide protection through targeted CaMKII inhibition, reversing imatinib-induced mitochondrial damage [15]. Nuclear factor erythroid 2-related factor 2 (Nrf2) is a transcription factor which regulates cellular antioxidant response and is a regulator of lipid peroxidation and ferroptosis [63]. Berberine hydrochloride attenuated imatinib-induced cardiotoxicity through inhibition of Nrf2-mediated ferroptosis [61]. Carnitine provided oxidative stress protection in imatinib-treated Wistar rats via activation of the PDGF/PPARy/MAPK pathway [62]. Upregulation of PPARy enhanced expression of antioxidants like superoxide dismutase (SOD), catalase, and glutathione peroxidase (GSH-Px), thus promoting the clearance of ROS and maintaining redox balance [64]. Finally, in Wistar rats, BGP-15, a poly(ADP-ribose) polymerase-1 (PARP-1) inhibitor, regulated imatinib-induced cardiac dysfunction by targeting AMPK signaling [65].

Others

Gefitinib, an epidermal growth factor receptor (EGFR) TKI first approved in 2003, is primarily deployed for the treatment of non-small cell lung cancer [66]. In the literature, gefitinib has been shown to cause cardiac hypertrophy, cardiomyocyte apoptosis, and oxidative stress in a rat model [67]. Liraglutide, a glucagon-like peptide 1 (GLP-1) receptor agonist, was shown in Wistar rats to increase the antioxidant enzyme SOD and decrease NF-kB levels induced by gefitinib (Table 1) [68]. Angiotensin receptor blockade in Wistar rats treated with gefitinib was also shown to reduce cardiac hypertrophy via abatement of angiotensin II-mediated oxidative stress via the JNK/p38/MAPK pathway [69].

Nilotinib, a second-generation BCR-ABL TKI, is also associated with DICT. Similar to sunitinib, nilotinib increases ROS generation and hERG channel blockade, culminating in arrhythmias [18]. One study found nilotinib-induced ER stress and ROS generation, which led to increased H9c2 cell apoptosis [70]. A meta-analysis of cardiovascular adverse events found nilotinib to have a higher incidence of DICT than imatinib [71]. Crizotinib, an ALK/MET inhibitor, was also found to cause increased ROS generation, in addition to cholesterol accumulation and arrhythmias via ion channel blockade [18]. Crizotinib treatment also promotes QTc prolongation and DICT in patients [72, 73]. To date, while no studies have explored putative cardioprotective drugs against nilotinib- and crizotinib-induced myocardial oxidative stress, the use of liposomal formulations of both drugs has been shown to reduce DICT in rabbits [74].

Finally, ibrutinib, a Bruton tyrosine kinase (BTK) inhibitor, is used primarily to treat chronic lymphocytic leukemia, lymphomas, and Waldenström macroglobulinemia [75–78]. Ibrutinib is associated with higher rates of atrial fibrillation and increased incidence and severity of hypertension [79–81]. One study using human atrial cardiomyocytes attributed the development of ibrutinib-mediated atrial fibrillation to increased oxidative stress [82]. This same study demonstrated that treatment with apocynin, an NADPH oxidase inhibitor, both prevented and reversed atrial fibrillation while increasing expression of antioxidant signaling and enzymes (Table 1) [82, 83]. While apocynin is not commercially available, this study provides a potential therapy mechanism to combat ibrutinib-mediated atrial fibrillation.

Several other TKI medications have also been shown to trigger DICT via increased myocardial oxidative stress, including lenvatinib, regorafenib, lapatinib, erlotinib, and trametinib. However, no studies have directly investigated potential pharmacotherapy options for mitigating their induced oxidative stress [84–93].

Monoclonal Antibodies

Introduction

The development of monoclonal antibodies (MAbs) has transformed and revolutionized the treatment of many diseases, namely autoimmune diseases and cancer. Despite their strong anticancer therapeutic effects, MAbs also have a range of adverse effects, including DICT secondary to oxidative stress. This section of the chapter will focus on two MAbs, bevacizumab and trastuzumab, due to robust evidence suggesting that both promote DICT via ROS generation and mitochondrial toxicity. Bevacizumab targets VEGF and is used in the treatment of breast cancer, lung cancer, hepatocellular carcinoma, and sarcomas [10, 94–96]. Trastuzumab targets human epidermal growth factor receptor 2 (HER2) and is primarily used in patients with HER2 positive breast cancer, as well as colorectal, endometrial, and gastric cancer with HER2 positivity [97–100]. These drugs offer great treatment options for patients, but it is important to consider their DICT profile and possibilities for prevention and management.

Bevacizumab

First approved in 2004, bevacizumab is among the first anticancer mAbs. Bevacizumab has demonstrated adverse reactions in patients, including hypertension, thromboembolisms, left ventricular dysfunction (LVD), heart failure, and myocardial infarction [101]. While the mechanism underlying bevacizumab cardiotoxicity is multifaceted, oxidative stress and impaired angiogenesis are considered primary culprits [102–104]. In preclinical models including H9C2 cells and Wistar rats, bevacizumab primarily incites oxidative stress via generation of mitochondrial

ROS [105, 106], above the threshold for normal mitochondrial function and cellular signaling [107]. Mitochondrial ROS overload results in mitochondrial DNA and organelle damage, impairing ATP production. The subsequent energy depletion and oxidative stress contribute to cell death [108, 109]. Preclinical studies seeking to investigate therapeutics that block bevacizumab cardiotoxicity seek to attenuate these mitochondrial alterations (Table 2).

Among the putative therapeutic drugs for combating bevacizumab DICT, selenium and capsazepine have shown great potential. Selenium is an essential mineral and key component in many antioxidant enzymes, such as GSH-Px, and its deficiency has been implicated in cardiomyopathy [110–112]. Capsazepine is a competitive inhibitor of transient receptor potential vanilloid 1 (TRPV1), a nonselective cation channel [113]. In mouse models, TRPV1 inhibition protects against pressure overload, while TRPV1 activation promotes resistance against oxidative stress via activation of autophagy signaling [114, 115]. Both selenium and capsazepine have been shown to independently reduce cardiomyocyte apoptosis, ROS generation, and mitochondrial cell death signaling in human cardiomyocytes treated with bevacizumab, with a compound effect when co-administered [116].

Curcumin is a naturally occurring polyphenol in turmeric with ROS scavenging and cardioprotective properties [117, 118]. In Wistar rat cardiomyocytes exposed to bevacizumab, curcumin protected against lipid peroxidation and mitochondrial ROS formation, thereby preventing cardiomyocyte destruction [106]. Ellagic acid is another phytochemical and ROS scavenger which promotes cardioprotection in rat cardiomyocytes treated with bevacizumab via restoration of ETC complex II function [119].

Calcium channel blockers (CCBs) have also been investigated. For example, benidipine is a dihydropyridine with the capacity to block L-, T-, and N-type calcium channels and was shown to inhibit oxidative stress and decrease blood pressure in hypertensive patients [120]. In the Wistar rat model of isoproterenol-induced myocardial infarction, benidipine prevented oxidative stress, inflammatory changes, and apoptosis-related myofibril damage [121]. Specifically, benidipine prevented consumption of antioxidant enzymes such as GSH, CAT, SOD, and lipid peroxidases [121]. Similarly, in Wistar rats exposed to bevacizumab, benidipine administered in combination with ATP resulted in decreased lipid peroxidation and GSH consumption and prevented increases in cardiac enzymes [122]. Supporting this, CCBs have been shown to possess antioxidant effects on cardiac tissue by inhibiting calcium influx from L-type calcium channels and the resultant oxidative enzyme activation [123]. While benidipine is not FDA approved, inhibition of L-type calcium channels is a class-wide property of CCBs and may provide a novel mechanism of protection against bevacizumab-related DICT.

Table 2 List of published studies that target oxidative stress for treating MAb-induced DICT in preclinical models

Monoclonal antibodies mechanisms					
Drug	Treatment	Mechanism	Level of evidence	Author	Year
Bevacizumab	Curcumin	Primary ROS scavenger	Wistar rats	Sabet	2020
	Ellagic acid	Primary ROS scavenger	Wistar rats	Khanlou	2022
	Benidipine	Prevents depletion of antioxidant enzymes	Wistar rats	Yildirim	2022
	Selenium	Component of glutathione peroxidase	AC16 human cardiomyocytes	Oncel	2019
	Capsazepine	Inhibits TRPV1	AC16 human cardiomyocytes	Oncel	2019
Trastuzumab	Metoprolol	Increases ROS via sirtuin-3 inhibition	Male C57Bl6 wild-type mice	Nicol	2021
	Lisinopril	Prevents depletion of SOD, CAT, GST, GPx, and GSH levels	Wistar rats	Olorundare	2020
	Valsartan	Prevents depletion of SOD, CAT, GST, GPx, and GSH levels	Wistar rats	Olorundare	2020
	Candesartan	Prevents depletion of SOD, CAT, GST, GPx, and GSH levels	Sprague-Dawley rats	Cho	2020
	Perindopril	Attenuates NF-κB	C57Bl/6 female mice	Eekhoudt	2022
	Rosuvastatin	Prevents depletion of CAT and GPx	Male Balb/c mice	Kabel	2017
	Rosuvastatin	Reduces oxidative stress marker GSH	Sprague-Dawley rats	Cho	2020
	Ranolazine	Inhibits ROS formation	C57Bl/6 mice	Riccio	2018
	Empagliflozin	Attenuates ferroptosis	Adult male C57BL/6 J	Min	2023
	Amlodipine	Prevents depletion of SOD, CAT, GST, GPx, and GSH levels	Wistar rats	Olorundare	2020
	Selenium	Component of antioxidant enzymes	White New Zealand rabbits	Laird-Fick	2020
	Metformin	Attenuates mitochondrial ROS	Wistar rats	Arinno	2023
	N-acetyl cysteine amide	Reduces lipid peroxidation	C57BI/6 female mice	Goyal	2016
	Melatonin	Attenuates mitochondrial ROS	Wistar rats	Arinno	2023
	Curcumin	Inhibits ROS formation	Wistar rats	Shirmard	2022
	Chrysin	Inhibits ROS formation	Wistar rats	Shirmard	2022
	Thymoquinone	Inhibits ROS formation	Wistar rats	Shirmard	2022
	CoQ10	Inhibits ROS formation and lipid peroxidation	Human fetal Cardiomyocytes (HFC cells)	Quagliariello	2020
	Ubiquinone	Prevents depletion of CAT and GPx	Male Balb/c mice	Kabel	2017

Trastuzumab

In post-market surveillance studies, trastuzumab treatment has been associated with decreased left-ventricular ejection fraction (LVEF), cardiomyopathy, arrhythmia, heart failure, and hypertension [124–127]. The HER2 receptor is a member of the EGFR family and plays a critical role in cardiomyocyte pro-survival signaling, acting as an upstream receptor of MAPK and phosphoinositide 3 kinase (PI3K) [128]. As mentioned previously, MAPK is a critical signaling intermediate for cellular defense against oxidative stress, where PI3K also regulates cellular redox balance (Fig. 1) [129]. Several preclinical and clinical studies show that trastuzumab promotes oxidative stress, especially when co-administered with anthracyclines (Tables 2 and 3). For example, trastuzumab inhibits HER2 receptor pro-survival cellular signaling, leading to mitochondrial ROS generation and oxidative stress, inducing ferroptosis [130, 131]. Clinical studies have also demonstrated trastuzumab-induced oxidative stress [132].

β-blockers (BB), such as carvedilol, metoprolol, and bisoprolol, have become cornerstones of heart failure treatment with vast cardioprotective properties. BBs also provide mortality benefit in patients with heart failure by preventing cardiac remodeling [133]. For example, BBs have the ability to lower myocardial oxygen demand, reduce ROS generation from negative chronotropy and afterload reduction, and reduce cardiac lipid peroxidation [134–138]. While BBs provide strong myocardial oxidative stress protection, studies investigating BBs for prevention of trastuzumab cardiotoxicity are mixed. In one study, metoprolol failed to protect against oxidative stress and mitochondrial dysfunction in C57BL/6 wild-type mice treated with doxorubicin and trastuzumab [139]. In a randomized controlled trial (RCT) of 31 patients, bisoprolol provided significant protection against LVEF decline from trastuzumab but did not prevent cardiac remodeling [140]. Finally, in a case-control RCT with 60 patients, HER2+ breast cancer patients treated with trastuzumab and carvedilol, no significant protective effect from trastuzumab DICT was reported [141].

Renin-angiotensin-aldosterone system (RAAS) inhibitors, including angiotensin-converting enzyme (ACE) inhibitors and angiotensin II receptor blockers (ARBs), are critical for the treatment of heart failure. ACE and ARB treatment has shown significant mortality benefit in heart failure treatment due to blockage of RAAS-induced myocardial remodeling [133]. ACE inhibitors provide oxidative stress protection through several mechanisms, including (1) reduced vascular inflammation via decreased angiotensin II and increased bradykinin, (2) reduced NADPH activity, and (3) preservation of SOD activity [142]. ARBs also protect against oxidative stress through reduced angiotensin II action and NADPH activity [143]. In murine models, treatment with ACE inhibitors (e.g., lisinopril and perindopril) and ARBs (e.g., valsartan and candesartan) protects against trastuzumab DICT and improves cellular response to trastuzumab-induced oxidative stress [144–146]. A retrospective study of 14 patients found that patients who were concurrently prescribed RAAS inhibitors had preserved LVEF, but not to a significant level [147]. In a RCT

Table 3 List of published studies which target oxidative stress for treating MAb-induced DICT in human studies

Monoclonal antibodies human studies					
Drug	Treatment	Finding	Level of evidence	Author	Year
Trastuzumab	Carvedilol	No significant protect effect against DICT	RCT	Esfandbod	2021
	Bisoprolol	Protects against LVEF decline	RCT	Pituskin	2017
	RAASi	No significant protect effect	Retrospective cohort	Moey	2019
	Perindopril	Protects against LVEF decline	RCT	Pituskin	2017
	Candesartan	Protects against LVEF decline	RCT	Gulati	2016
	Candesartan	No significant protect effect	RCT	Boekhout	2016
	BB or RAASi	Treatment associated with recovery of LV function	Retrospective cohort	Ohtani	2019
	BB or RAASi	Treatment associated with recovery of LV function	Retrospective cohort	Oliva	2012
	BB or RAASi	Treatment associated with recovery of LV function	Prospective cohort	Geuna	2020
	BB or RAASi	Prevented against DICT and improved survival	Retrospective cohort	Wittayanukorn	2017
	BB or RAASi	Treatment attenuates mild DICT	Phase I clinical trial	Leong	2019
	Lisinopril + carvedilol	No difference in survival in patients co-administered anthracyclines	RCT	Guglin	2019
	Statins	Lower risk of heart failure	Retrospective cohort	Abdel-Qadir	2021
	Statins	Lower risk of heart failure	Retrospective cohort	Seicean	2012
	Statins	Lower risk of DICT	Retrospective case-control	Calvillo-Arguelles	2019
	Statins	Lower risk of DICT	Meta-analysis/sys review	Obasi	2021

of 33 patients, perindopril was mildly protective against LVEF decline from trastuzumab but did not prevent cardiac remodeling [150]. In a separate RCT with 120 women, candesartan, but not metoprolol, protected against early LVEF decline in patients prescribed doxorubicin with and without trastuzumab [148]. Similarly, in a RCT trial of 206 women who received doxorubicin and trastuzumab, candesartan decreased cardiac events and protected against changes in LVEF [149].

In most studies, anti-cardiotoxicity effects of BBs and RAAS inhibitors are frequently evaluated together. In a retrospective study of 160 patients receiving trastuzumab, BBs and RAAS inhibitors independently predicted reversal of LVD in patients experiencing DICT [150]. Further, a retrospective study of 6542 breast cancer patients who received anthracyclines or trastuzumab reported that ACE inhibitor and BB exposure was associated with a 3.4% lower risk of developing DICT and a 3.9% lower 5-year all-cause mortality [151]. A prospective study of breast cancer patients with mild trastuzumab DICT showed improved asymptomatic LVEF recovery when treated with carvedilol or enalapril [152]. However, these studies did not stratify the patient cohorts by RAAS inhibitors and BBs. Further, a multicenter retrospective study of 499 patients treated with adjuvant trastuzumab found that patients with early LVEF decline also received a combination RAAS inhibitor and BB, but this cohort also had subsequent LVEF recovery [153]. Carvedilol ($n = 149$) and lisinopril ($n = 149$) also protected against cardiotoxicity in patients treated with trastuzumab and anthracyclines but did not reduce DICT in patients treated with trastuzumab in non-anthracycline regimens [154]. In a phase I clinical trial of 20 patients, both RAAS inhibitors and BBs were also associated with improved trastuzumab induced mild LVEF decline [155]. As discussed herein, the evidence for RAAS inhibitors- and BB-mediated cardiotoxicity prevention of trastuzumab is mixed and therefore warrants further investigation.

Statins are another flagship treatment for cardiovascular disease and are commonly prescribed to combat atherosclerotic disease. In addition to lipid-lowering ability, statins also mitigate oxidative stress by increasing NO availability, inhibiting NADPH oxidase, and upregulating catalase and paraoxonase activity [156, 157]. In murine models exposed to trastuzumab, rosuvastatin administration increased antioxidant enzymes (e.g., catalase and GSH-Px) and decreased oxidative stress markers such as GSH and malondialdehyde [146, 158]. In two separate retrospective cohort studies, women prescribed doxorubicin had a significantly lower incidence of heart failure when concurrently prescribed statins. However, a similar correlation was not seen with DICT secondary to trastuzumab therapy [159, 160]. In contrast, a separate study found that statin coadministration lowered the risk of DICT in patients prescribed trastuzumab, even after adjusting for concurrent anthracycline use [161]. Supporting this, a prospective controlled trial of 89 patients treated with anthracyclines, with or without adjuvant trastuzumab, demonstrated significantly higher LVEF in patients receiving rosuvastatin compared to placebo. However, it should be noted that the physician evaluating changes in LVEF was not blinded, limiting the strength of the findings [162]. Taken together, since these studies typically evaluated cardioprotection of statins in patients cotreated with trastuzumab and anthracyclines, it is difficult to determine whether statin-mediated protection provided a benefit specific to trastuzumab DICT. Future blinded, prospective studies are needed to isolate the specific impact of statins on trastuzumab-induced cardiac dysfunction.

Beyond BBs, RAAS inhibitors, and statins, recent preclinical studies have demonstrated a role for other established medications for mitigating oxidative stress. Among these medications are the two diabetic drugs empagliflozin and metformin. Both empagliflozin and metformin were found to attenuate trastuzumab-associated ROS formation, ferroptosis, and mitochondrial stress [163, 164]. Regarding ion channel blockade, ranolazine, a sodium channel blocker, blunted trastuzumab-associated ROS formation, and amlodipine attenuated antioxidant consumption [144, 165]. Melatonin has also been investigated due to its direct effect as an ROS and RNS scavenger, as well as its ability to promote redox enzyme balance [166]. Similar to selenium, melatonin utilizes its native antioxidant action to attenuate trastuzumab-induced mitochondrial oxidative stress [164, 167]. N-acetyl cysteine (NAC), which serves as a precursor to reduced GSH, is utilized in acetaminophen overdose due to its direct ability to counteract drug-induced oxidative stress [168]. Goyal and colleagues showed that NAC provided protection against phospholipid oxidation and ROS formation in C57BL/6 mice treated with doxorubicin and trastuzumab [169]. Notwithstanding, more clinical studies are required to evaluate the cardioprotective potential of metformin, empagliflozin, and NAC in trastuzumab-induced cardiotoxicity.

Some preclinical studies have also investigated phytochemicals with demonstrated antioxidant properties in combating trastuzumab-mediated cardiotoxicity. Chrysin, a three-ringed flavonoid found in honey, and thymoquinone, a phytochemical found in black cumin, have demonstrated antioxidant properties and ROS scavenging abilities [170, 171]. Chrysin and thymoquinone, along with curcumin, were all reported to have independent cardioprotection against trastuzumab-mediated cardiotoxicity by increasing antioxidant signaling and attenuating drug-induced mitochondrial stress [172]. In addition, co-enzyme Q10 (CoQ10), which is oxidized to ubiquinone and reduced to ubiquinol, functions as an electron acceptor in the ETC. CoQ10 is a native antioxidant due to ubiquinol's ability to donate electrons and neutralize ROS [173]. In human fetal cardiomyocytes, CoQ10 reduced lipid peroxidation and ROS formation caused by combination treatment with trastuzumab and doxorubicin, with an even greater effect when loaded as a nano-emulsification [174]. In male Blab/c mice, oral ubiquinone alone or in combination with rosuvastatin attenuated trastuzumab-induced oxidative stress and cardiotoxicity [158]. These data support a role for phytochemicals and organic supplements as a potential avenue for antioxidant defense in the prevention of trastuzumab-mediated cardiotoxicity.

In summary, oxidative stress plays a significant role in the DICT of two widely used mAbs, bevacizumab and trastuzumab. While the mechanisms underlying bevacizumab- and trastuzumab-induced cardiotoxicity are multifactorial, the literature provides clear evidence that oxidative stress and subsequent mitochondrial instability are significant contributors to cardiac toxicity. As such, continued investigation of potential therapeutic targets is critical for developing therapy options for patients.

Fluorouracil

Antimetabolites are a class of anticancer chemotherapies that target metabolic pathways essential for cellular division. Chief among these therapeutics are the pyrimidine analogs: 5-fluorouracil (5-FU), 5-FU's oral formulation capecitabine, and their relative gemcitabine. These therapies have been used extensively for a range of cancers since 5-FU's approval in 1962, including breast, gastrointestinal, genitourinary, and squamous cell cancers [175–179]. 5-FU acts to halt the cell cycle by acting as a pyrimidine analogue, inserting into RNA and DNA strands and halting polymerization [180]. While remarkably effective for cancer therapy, a key manifestation of 5-FU cardiotoxicity is angina pectoris from coronary artery vasospasms, which may result in myocardial ischemia [181–183]. The mechanism behind this endothelial dysfunction and resulting vasospasm is multifaceted, whereby 5-FU metabolites directly promote endothelial injury, leading to endothelial inflammation and thrombosis [184]. A proposed mechanism of interest centers on the ability of 5-FU to induce production of NO and ROS, leading to endothelial dysfunction, vasospasm, and thrombosis [185–187]. Furthermore, 5-FU upregulates thioredoxin-interacting protein (TXNIP), a mediator of the thioredoxin redox system, which increases ROS accumulation and oxidative stress [188]. In this section, the potential therapeutic options that mitigate 5-FU's drug-induced oxidative stress on cardiac tissue and resultant coronary artery vasospasm are explored (Table 4).

Treating 5-FU-induced oxidative stress requires targeted interventions to mitigate ROS and preserve redox balance. Both verapamil and diltiazem have been shown to downregulate TXNIP [189]. Owing to 5'FU-mediated upregulation of TXNIP expression, a role for the non-dihydropyridine CCBs verapamil and diltiazem has been investigated. However, treatment with verapamil and diltiazem had no effect on 5'FU-induced coronary vasospasm in an in vitro rabbit model despite observed protection from 5-FU-related DICT in a single case study [190, 191].

Endothelin-1 (ET-1), a powerful vasoconstrictor that also elicits ROS generation and superoxide formation, is upregulated in preclinical work modeling 5-FU DICT and represents another proposed mechanism behind 5-FU cardiotoxicity [192–194]. Bosentan, a competitive ET-1 receptor inhibitor, protected against 5-FU DICT via downregulation of endothelin and the NF-kB in Wistar rats [195]. In a different study using Wistar rats, coadministration of 5-FU with simvastatin similarly blunted the ET-1 and NF-kB upregulation [192].

Beyond these two specific mechanisms, the remaining treatment strategies discussed are centered on direct treatment of oxidative stress and inflammatory reduction. Two separate studies using mouse models have also shown that cellular and mitochondrial ROS scavengers, such as MitoTEMPO and the indole alkaloid derivative B, respectively, reduced 5-FU-associated oxidative and mitochondrial stress, preventing cardiotoxicity [196, 197]. In Wistar rats, fenofibrate offered cardioprotection from 5-FU through attenuation of antioxidant consumption, reduced inflammation, and inhibited cell death signaling [198]. Cyclooxygenase-2 (COX-2), a pro-inflammatory rate-limiting enzyme involved in prostaglandin signaling, is also

Table 4 List of published studies which target oxidative stress for treating 5-FU-induced DICT

Antimetabolites					
Drug	Treatment	Mechanism	Level of evidence	Author	Year
5-fluorouracil	Bosentan	Modulation of TLR4-MyD88-NFκB signaling, eNOS, and endothelin signaling	Wistar rats	Refaie	2022
	Simvastatin	Modulation of ROCK-NF-κB, Akt-eNOS, and ET-1-ERK1/2 signaling	Wistar rats	Muhammad	2020
	Diltiazem	Prevent coronary artery vasospasm	Case report	Boldig	2022
	Mito-TEMPO	Mitochondrial ROS scavenger	Male BALB/C mice	Tambe	2023
	Indole alkaloid	Primary ROS scavenger	Male ICR mice	Bi	2018
	Fenofibrate	Upregulation of PPARα and inhibition of IL-6/STAT signaling	Wistar rats	Refaie	2022
	Colchicine	Reduced oxidative stress and COX2 expression	Wistar rats	Safarpour	2022
	Propolis	Reduced oxidative stress and COX2 expression	Wistar rats	Barary	2022
	Silymarin	Reduced oxidative stress and COX2 expression	Wistar rats	Safarpour	2022
	Kaempferol	Reduced oxidative stress and COX2 expression	Wistar rats	Safarpour	2022
	Myricetin	Primary antioxidant and inflammatory signaling downregulation	Wistar rats	Arafah	2022
	Naringin	Primary antioxidant and caspase-3 downregulation	Wistar rats	Gui	2023
	Quercetin	Primary antioxidant and caspase-3 downregulation	Wistar rats	Lokman	2023
	Nitroglycerin	Prevent coronary artery vasospasm	Case report	Allison	2020
	Nitroglycerin + morphine	Prevent coronary artery vasospasm	Case report	Vargo	2015
Capecitabine	Baicalein	Modulation of TLR4/MyD88/NF-κB signaling	Wistar rats	Ibrahim	2023

upregulated by ROS [199, 200]. In rat models, colchicine (an anti-inflammatory medication), propolis (a natural product derived from beeswax), silymarin (natural product derived from milk thistle), and kaempferol (a polyphenol derived from multiple brassica species) have demonstrated protection against 5-FU DICT via down-regulation of COX-2, pro-inflammatory signaling, and oxidative stress while restoring antioxidants [201–204]. Other therapeutic targets include antioxidant flavonoids, including myricetin, naringin, quercetin, and baicalein, which all attenuate 5-FU-associated DICT in murine models via their antioxidant action and downregulation of pro-apoptotic signaling [205–208].

In summary, studies investigating oxidative stress resulting from 5-FU have been limited to preclinical studies. Furthermore, TXNIP and ET-1 signaling pathways seem to underlie oxidative stress, ROS generation, and subsequent vasospasm observed in the presence of 5-FU. Fenofibrate, colchicine, and phytochemicals have shown some promise in directly combating oxidative stress induced by 5-FU. Three case reports explored the prophylactic use of diltiazem or nitroglycerin to reduce coronary vasospasm by reducing myocardial oxygen demand in patients treated with 5-FU [209, 210]. Given that 5-FU-associated angina pectoris can lead to medication discontinuation of this life-saving cancer treatment, more research is warranted to investigate its underlying mechanism and therapeutic targets.

Proteasome Inhibitors

Proteasome inhibitors have emerged as a promising class of anticancer agents, exerting their therapeutic effects by disrupting proteasomal degradation of proteins critical for cell survival and proliferation. Among these inhibitors, carfilzomib has gained significant attention for its efficacy in the treatment of multiple myeloma [211]. However, the clinical utility of carfilzomib is often tempered due to the emergence of DICT. Carfilzomib-associated cardiotoxicity is due to off-targeted proteasomal inhibition in non-neoplastic cells, causing protein aggregation and inflammatory cascades resulting in pro-apoptotic signals in cardiomyocytes and vascular smooth muscle cells [212]. Subsequent cellular dysfunction may result in hypertension, left ventricular dysfunction, and myocardial ischemia in 7–27% of patients treated with carfilzomib [213]. Furthermore, loss of proteasome function creates proteotoxic byproducts that elicit oxidative stress in cardiomyocytes, which can be particularly damaging in patients with pre-existing cardiovascular disease [213].

Current studies investigating therapies that prevent carfilzomib-related DICT remain limited to preclinical models (Table 5). One promising medication, metformin, was studied by Efentakis et al., given metformin's ability to activate AMPKα, a critical signaling mediator of proteasome function [214–216]. The authors demonstrated that in C57BL/6 mice treated with carfilzomib, metformin prevented the decline in fractional shortening, reduced autophagy-related proteins, and AMPKα signaling in myocardial and smooth muscle cells, promoting protection against DICT associated with carfilzomib treatment [214–216]. Beyond metformin, dexrazoxane was investigated owing to its iron-chelating effects and use in the treatment of doxorubicin cardiotoxicity. Dexrazoxane resulted in significant reversal in carfilzomib-induced oxidative stress and antioxidant consumption, probably through its established function as a ROS scavenger [217, 218]. Apremilast, a phosphodiesterase 4 inhibitor used in the treatment of Behçet disease, elevates

Table 5 List of present studies which target oxidative stress for treating PI-induced DICT

Proteasome inhibitors					
Drug	Treatment	Mechanism	Level of evidence	Author	Year
Carfilzomib	Metformin	Increased AMPKα and LC3B-dependent autophagy	C57BL/6 mice	Efentakis	2019
	Metformin	Increased AMPKα and LC3B-dependent autophagy	C57BL/6 mice	Efentakis	2020
	Metformin	Increased AMPKα and LC3B-dependent autophagy	C57BL/6 mice	Efentakis	2021
	Dexrazoxane	Reduced antioxidant consumption and lipid peroxidation	Wistar rats	Al-Harbi	2016
	Apremilast	Reduced antioxidant consumption and lipid peroxidation	Wistar rats	Imam	2016
	Rutin	Reduced antioxidant consumption and lipid peroxidation	Wistar rats	Imam	2015
	Rutin	Reduced antioxidant consumption and lipid peroxidation	Wistar rats	Al-Harbi	2019
	Dexamethasone	Reduced ER stress and attenuated HSP90 activity	H9c2 cardiomyocytes	Jannuzzi	2023
	Zingerone	Reduced oxidative stress and proinflammatory signaling	Wistar rats	Alam	2022

intracellular cAMP, offering anti-inflammatory and reduced ROS burden [219, 220]. In Wistar rats treated with carfilzomib, apremilast reversed carfilzomib-mediated increase in oxidative stress markers and restored GSH levels [221]. Dexamethasone, a synthetic glucocorticoid, was evaluated in combination treatment alongside carfilzomib and ixazomib. Coadministration of dexamethasone with these proteasome inhibitors reduced oxidative stress by attenuating ER stress and reducing oxidative phosphorylation of proteins [222]. Phytochemicals were also investigated as putative therapeutics in combating proteasome inhibitor DICT. Rutin, a bioflavonoid found in citrus plants with antioxidant capacity, attenuated carfilzomib-associated increase in oxidative stress biomarkers and restored cellular expression of GSH and catalase [223–225]. Finally, zingerone, a vanillylacetone found in ginger with antioxidant and anti-inflammatory properties, provided a reduction in carfilzomib oxidative stress similar to rutin, while also reversing the carfilzomib-induced rise in tumor necrosis factor-α (TNFα) and caspase 3 [226].

In conclusion, carfilzomib DICT poses a challenge when considering the clinical utility of carfilzomib. Interestingly, drugs that target oxidative stress pathways represent a viable therapeutic intervention for attenuating DICT caused by carfilzomib. Notwithstanding, more preclinical and clinical investigation is required for the development of targeted treatments for preventing carfilzomib-mediated cardiotoxicity.

Conclusion

There is an intricate and evolving relationship between anticancer chemotherapies and oxidative stress-induced cardiotoxicity. A significant body of evidence implicates oxidative stress as a central mechanism in DICT associated with anthracyclines, tyrosine kinase inhibitors, monoclonal antibodies, proteasome inhibitors, and fluorouracil. Through an exploration of a variety of medications, ranging from established cardiovascular agents such as BBs, RAAS inhibitors, and statins, to emerging therapies including metformin and phytochemical derivatives, we have gained valuable insights into the mechanisms underlying oxidative stress-induced myocardial injury. It is now evident that oxidative stress contributes to cellular dysfunction, inflammatory signaling, and ultimately, adverse cardiovascular patient outcomes in cancer patients. A comprehensive summary of the key signaling pathways involved in cardiotoxic oxidative stress is described in Fig. 1. By understanding these mechanisms, identification of potential therapeutic targets can be developed to minimize the cardiotoxic effects of these lifesaving anticancer therapies and enhance patient outcomes. Continued research in this field is crucial to further unravel the complexities of drug-induced oxidative stress and pave the way for more effective strategies for preventing and managing cardiotoxicity in clinical practice.

Disclosures The authors have no relevant affiliations or financial involvement to disclose.

References

1. Moris D, Spartalis M, Spartalis E, Karachaliou GS, Karaolanis GI, Tsourouflis G, et al. The role of reactive oxygen species in the pathophysiology of cardiovascular diseases and the clinical significance of myocardial redox. Ann Transl Med. 2017;5(16):326.
2. Münzel T, Camici GG, Maack C, Bonetti NR, Fuster V, Kovacic JC. Impact of oxidative stress on the heart and vasculature part 2 of a 3-part series. J Am Coll Cardiol. 2017;70(2):212–29.
3. Ide T, Tsutsui H, Kinugawa S, Suematsu N, Hayashidani S, Ichikawa K, et al. Direct evidence for increased hydroxyl radicals originating from superoxide in the failing myocardium. Circ Res. 2000;86(2):152–7.
4. Prosser BL, Ward CW, Lederer WJ. X-ROS signaling: rapid mechano-chemo transduction in heart. Science. 2011;333(6048):1440–5.
5. Förstermann U, Münzel T. Endothelial nitric oxide synthase in vascular disease: from marvel to menace. Circulation. 2006;113(13):1708–14.
6. Papaconstantinou J. The role of signaling pathways of inflammation and oxidative stress in development of senescence and aging phenotypes in cardiovascular disease. Cells. 2019;8(11):1383.
7. Sternberg CN, Davis ID, Mardiak J, Szczylik C, Lee E, Wagstaff J, et al. Pazopanib in locally advanced or metastatic renal cell carcinoma: results of a randomized phase III trial. J Clin Oncol. 2010;28(6):1061–8.
8. van der Graaf WTA, Blay JY, Chawla SP, Kim DW, Bui-Nguyen B, Casali PG, et al. Pazopanib for metastatic soft-tissue sarcoma (PALETTE): a randomised, double-blind, placebo-controlled phase 3 trial. Lancet Lond Engl. 2012;379(9829):1879–86.
9. Demetri GD, van Oosterom AT, Garrett CR, Blackstein ME, Shah MH, Verweij J, et al. Efficacy and safety of sunitinib in patients with advanced gastrointestinal stromal tumour after failure of imatinib: a randomised controlled trial. Lancet Lond Engl. 2006;368(9544):1329–38.
10. Miller K, Wang M, Gralow J, Dickler M, Cobleigh M, Perez EA, et al. Paclitaxel plus bevacizumab versus paclitaxel alone for metastatic breast cancer. N Engl J Med. 2007;357(26):2666–76.
11. Pfeiffer P, Yilmaz M, Möller S, Zitnjak D, Krogh M, Petersen LN, et al. TAS-102 with or without bevacizumab in patients with chemorefractory metastatic colorectal cancer: an investigator-initiated, open-label, randomised, phase 2 trial. Lancet Oncol. 2020;21(3):412–20.
12. Cortes JE, Kim DW, Pinilla-Ibarz J, le Coutre P, Paquette R, Chuah C, et al. A phase 2 trial of ponatinib in Philadelphia chromosome-positive leukemias. N Engl J Med. 2013;369(19):1783–96.
13. Zv V, P F, L L, P P. Drug-induced mitochondrial dysfunction and cardiotoxicity. Am J Physiol Heart Circ Physiol [Internet]. 2015 Nov [cited 2023 Mar 22];309(9). Available from: https://pubmed.ncbi.nlm.nih.gov/26386112/
14. Bouitbir J, Alshaikhali A, Panajatovic MV, Abegg VF, Paech F, Krähenbühl S. Mitochondrial oxidative stress plays a critical role in the cardiotoxicity of sunitinib: running title: Sunitinib and oxidative stress in hearts. Toxicology. 2019;426:152281.
15. McMullen CJ, Chalmers S, Wood R, Cunningham MR, Currie S. Sunitinib and Imatinib display differential cardiotoxicity in adult rat cardiac fibroblasts that involves a role for calcium/Calmodulin dependent protein Kinase II. Front Cardiovasc Med. 2020;7:630480.
16. Ma W, Liu M, Liang F, Zhao L, Gao C, Jiang X, et al. Cardiotoxicity of sorafenib is mediated through elevation of ROS level and CaMKII activity and dysregulation of calcium homoeostasis. Basic Clin Pharmacol Toxicol. 2020;126(2):166–80.
17. Bouitbir J, Panajatovic MV, Krähenbühl S. Mitochondrial toxicity associated with Imatinib and Sorafenib in isolated rat heart fibers and the Cardiomyoblast H9c2 cell line. Int J Mol Sci. 2022;23(4):2282.
18. Doherty KR, Wappel RL, Talbert DR, Trusk PB, Moran DM, Kramer JW, et al. Multi-parameter in vitro toxicity testing of crizotinib, sunitinib, erlotinib, and nilotinib in human cardiomyocytes. Toxicol Appl Pharmacol. 2013;272(1):245–55.

19. Rainer PP, Doleschal B, Kirk JA, Sivakumaran V, Saad Z, Groschner K, et al. Sunitinib causes dose-dependent negative functional effects on myocardium and cardiomyocytes. BJU Int. 2012;110(10):1455–62.
20. Rabinovitch RC, Samborska B, Faubert B, Ma EH, Gravel SP, Andrzejewski S, et al. AMPK maintains cellular metabolic homeostasis through regulation of mitochondrial reactive oxygen species. Cell Rep. 2017;21(1):1–9.
21. Meng Y, Ding P, Wang H, Yang X, Wang Z, Nie D, et al. Ca2+/calmodulin-dependent protein kinase II inhibition reduces myocardial fatty acid uptake and oxidation after myocardial infarction. Biochim Biophys Acta Mol Cell Biol Lipids. 2022;1867(6):159120.
22. Suetomi T, Willeford A, Brand CS, Cho Y, Ross RS, Miyamoto S, et al. Inflammation and NLRP3 Inflammasome activation initiated in response to pressure overload by Ca2+/Calmodulin-dependent protein Kinase II δ signaling in Cardiomyocytes are essential for adverse cardiac remodeling. Circulation. 2018;138(22):2530–44.
23. Zou J, Yang Y, Yang Y, Liu X. Polydatin suppresses proliferation and metastasis of non-small cell lung cancer cells by inhibiting NLRP3 inflammasome activation via NF-κB pathway. Biomed Pharmacother. 2018;108:130–6.
24. Quagliariello V, Berretta M, Buccolo S, Iovine M, Paccone A, Cavalcanti E, et al. Polydatin reduces cardiotoxicity and enhances the anticancer effects of Sunitinib by decreasing pro-oxidative stress, pro-inflammatory cytokines, and NLRP3 Inflammasome expression. Front Oncol. 2021;11:680758.
25. Glezer MG. On behalf of the choice-study participants OBOTCSP. [Treatment of patients with stable Ischemic heart disease in real clinical practice in Russia. The CHOICE-2 Program]. Kardiologiia. 2016;56(5):5–11.
26. Yang Y, Li N, Chen T, Zhang C, Liu L, Qi Y, et al. Trimetazidine ameliorates sunitinib-induced cardiotoxicity in mice via the AMPK/mTOR/autophagy pathway. Pharm Biol. 2019;57(1):625–31.
27. Ren C, Sun K, Zhang Y, Hu Y, Hu B, Zhao J, et al. Sodium-glucose CoTransporter-2 inhibitor Empagliflozin Ameliorates Sunitinib-induced cardiac dysfunction via regulation of AMPK-mTOR signaling pathway-mediated autophagy. Front Pharmacol. 2021;12:664181.
28. Anker SD, Butler J, Filippatos G, Ferreira JP, Bocchi E, Böhm M, et al. Empagliflozin in heart failure with a preserved ejection fraction. N Engl J Med. 2021;385(16):1451–61.
29. Voors AA, Angermann CE, Teerlink JR, Collins SP, Kosiborod M, Biegus J, et al. The SGLT2 inhibitor empagliflozin in patients hospitalized for acute heart failure: a multinational randomized trial. Nat Med. 2022;28(3):568–74.
30. Sayed-Ahmed MM, Alrufaiq BI, Alrikabi A, Abdullah ML, Hafez MM, Al-Shabanah OA. Carnitine supplementation attenuates Sunitinib-induced inhibition of AMP-activated protein Kinase downstream signals in cardiac tissues. Cardiovasc Toxicol. 2019;19(4):344–56.
31. Lang F, Eylenstein A, Shumilina E. Regulation of Orai1/STIM1 by the kinases SGK1 and AMPK. Cell Calcium. 2012;52(5):347–54.
32. Jin Y, Liu S, Ma Q, Xiao D, Chen L. Berberine enhances the AMPK activation and autophagy and mitigates high glucose-induced apoptosis of mouse podocytes. Eur J Pharmacol. 2017;794:106–14.
33. Chen K, Li G, Geng F, Zhang Z, Li J, Yang M, et al. Berberine reduces ischemia/reperfusion-induced myocardial apoptosis via activating AMPK and PI3K-Akt signaling in diabetic rats. Apoptosis Int J Program Cell Death. 2014;19(6):946–57.
34. Li C, Wu W, Xing J, Yan W, Zhang J, Sun J, et al. Berberine attenuates sunitinib-induced cardiac dysfunction by normalizing calcium regulation disorder via SGK1 activation. Food Chem Toxicol Int J Publ Br Ind Biol Res Assoc. 2023;175:113743.
35. Jayachandran M, Wu Z, Ganesan K, Khalid S, Chung SM, Xu B. Isoquercetin upregulates antioxidant genes, suppresses inflammatory cytokines and regulates AMPK pathway in streptozotocin-induced diabetic rats. Chem Biol Interact. 2019;303:62–9.

36. Buonerba C, De Placido P, Bruzzese D, Pagliuca M, Ungaro P, Bosso D, et al. Isoquercetin as an adjunct therapy in patients with kidney cancer receiving first-line Sunitinib (QUASAR): results of a phase I trial. Front Pharmacol. 2018;9:189.
37. Hara T, Fukuda D, Tanaka K, Higashikuni Y, Hirata Y, Nishimoto S, et al. Rivaroxaban, a novel oral anticoagulant, attenuates atherosclerotic plaque progression and destabilization in ApoE-deficient mice. Atherosclerosis. 2015;242(2):639–46.
38. Ishibashi Y, Matsui T, Ueda S, Fukami K, Yamagishi S, ichi. Advanced glycation end products potentiate citrated plasma-evoked oxidative and inflammatory reactions in endothelial cells by up-regulating protease-activated receptor-1 expression. Cardiovasc Diabetol. 2014;13:60.
39. Imam F, Al-Harbi NO, Khan MR, Qamar W, Alharbi M, Alshamrani AA, et al. Protective effect of RIVA against Sunitinib-induced cardiotoxicity by inhibiting oxidative stress-mediated inflammation: probable role of TGF-β and Smad signaling. Cardiovasc Toxicol. 2020;20(3):281–90.
40. Trnka J, Blaikie FH, Smith RAJ, Murphy MP. A mitochondria-targeted nitroxide is reduced to its hydroxylamine by ubiquinol in mitochondria. Free Radic Biol Med. 2008;44(7):1406–19.
41. Rini BI, Escudier B, Tomczak P, Kaprin A, Szczylik C, Hutson TE, et al. Comparative effectiveness of axitinib versus sorafenib in advanced renal cell carcinoma (AXIS): a randomised phase 3 trial. Lancet Lond Engl. 2011;378(9807):1931–9.
42. Llovet JM, Ricci S, Mazzaferro V, Hilgard P, Gane E, Blanc JF, et al. Sorafenib in advanced hepatocellular carcinoma. N Engl J Med. 2008;359(4):378–90.
43. Kawabata M, Umemoto N, Shimada Y, Nishimura Y, Zhang B, Kuroyanagi J, et al. Downregulation of stanniocalcin 1 is responsible for sorafenib-induced cardiotoxicity. Toxicol Sci. 2015;143(2):374–84.
44. Jiang H, Wang C, Zhang A, Li Y, Li J, Li Z, et al. ATF4 protects against sorafenib-induced cardiotoxicity by suppressing ferroptosis. Biomed Pharmacother. 2022;153:113280.
45. Li Y, Yan J, Zhao Q, Zhang Y, Zhang Y. ATF3 promotes ferroptosis in sorafenib-induced cardiotoxicity by suppressing Slc7a11 expression. Front Pharmacol. 2022;13:904314.
46. Liang F, Zhang K, Ma W, Zhan H, Sun Q, Xie L, et al. Impaired autophagy and mitochondrial dynamics are involved in Sorafenib-induced cardiomyocyte apoptosis. Toxicology. 2022;481:153348.
47. Yang B, Xu B, Zhao H, Wang YB, Zhang J, Li CW, et al. Dioscin protects against coronary heart disease by reducing oxidative stress and inflammation via Sirt1/Nrf2 and p38 MAPK pathways. Mol Med Rep. 2018;18(1):973–80.
48. Wu D, Zhang J, Wang J, Li J, Liao F, Dong W. Hesperetin induces apoptosis of esophageal cancer cells via mitochondrial pathway mediated by the increased intracellular reactive oxygen species. Tumour Biol J Int Soc Oncodevelopmental Biol Med. 2016;37(3):3451–9.
49. Zhang J, Wu D, Null V, Song J, Wang J, Yi J, et al. Hesperetin induces the apoptosis of gastric cancer cells via activating mitochondrial pathway by increasing reactive oxygen species. Dig Dis Sci. 2015;60(10):2985–95.
50. Zaafar D, Khalil HMA, Rasheed RA, Eltelbany RFA, Zaitone SA. Hesperetin mitigates sorafenib-induced cardiotoxicity in mice through inhibition of the TLR4/NLRP3 signaling pathway. PLoS One. 2022;17(8):e0271631.
51. Khaper N, Singal PK. Modulation of oxidative stress by a selective inhibition of angiotensin II type 1 receptors in MI rats. J Am Coll Cardiol. 2001;37(5):1461–6.
52. Fortuño A, Bidegain J, Robador PA, Hermida J, López-Sagaseta J, Beloqui O, et al. Losartan metabolite EXP3179 blocks NADPH oxidase-mediated superoxide production by inhibiting protein kinase C: potential clinical implications in hypertension. Hypertens Dallas Tex 1979. 2009;54(4):744–50.
53. Abdelgalil AA, Mohamed OY, Ahamad SR, Al-Jenoobi FI. The protective effect of losartan against sorafenib induced cardiotoxicity: ex-vivo isolated heart and metabolites profiling studies in rat. Eur J Pharmacol. 2020;882:173229.

54. Nagar B, Hantschel O, Young MA, Scheffzek K, Veach D, Bornmann W, et al. Structural basis for the autoinhibition of c-Abl tyrosine kinase. Cell. 2003;112(6):859–71.
55. Waller CF. Imatinib mesylate. Recent Results Cancer Res Fortschritte Krebsforsch Progres Dans Rech Sur Cancer. 2014;201:1–25.
56. Orphanos GS, Ioannidis GN, Ardavanis AG. Cardiotoxicity induced by tyrosine kinase inhibitors. Acta Oncol Stockh Swed. 2009;48(7):964–70.
57. Kerkelä R, Grazette L, Yacobi R, Iliescu C, Patten R, Beahm C, et al. Cardiotoxicity of the cancer therapeutic agent imatinib mesylate. Nat Med. 2006;12(8):908–16.
58. Hatfield A, Owen S, Pilot PR. In reply to "Cardiotoxicity of the cancer therapeutic agent imatinib mesylate". Nat Med. 2007;13(1):13;author reply 15-16.
59. Chambers TP, Santiesteban L, Gomez D, Chambers JW. Sab mediates mitochondrial dysfunction involved in imatinib mesylate-induced cardiotoxicity. Toxicology. 2017;382:24–35.
60. Song C, Li D, Zhang J, Zhao X. Role of ferroptosis in promoting cardiotoxicity induced by Imatinib Mesylate via down-regulating Nrf2 pathways in vitro and in vivo. Toxicol Appl Pharmacol. 2022;435:115852.
61. Song C, Li D, Zhang J, Zhao X. Berberine hydrochloride alleviates imatinib mesylate - induced cardiotoxicity through the inhibition of Nrf2-dependent ferroptosis. Food Funct. 2023;14(2):1087–98.
62. Mansour HH, El Kiki SM, Ibrahim AB, Omran MM. Effect of l-carnitine on cardiotoxicity and apoptosis induced by imatinib through PDGF/ PPARγ /MAPK pathways. Arch Biochem Biophys. 2021;704:108866.
63. Dodson M, Castro-Portuguez R, Zhang DD. NRF2 plays a critical role in mitigating lipid peroxidation and ferroptosis. Redox Biol. 2019;23:101107.
64. Kim T, Yang Q. Peroxisome-proliferator-activated receptors regulate redox signaling in the cardiovascular system. World J Cardiol. 2013;5(6):164–74.
65. Sarszegi Z, Bognar E, Gaszner B, Kónyi A, Gallyas F, Sumegi B, et al. BGP-15, a PARP-inhibitor, prevents imatinib-induced cardiotoxicity by activating Akt and suppressing JNK and p38 MAP kinases. Mol Cell Biochem. 2012;365(1–2):129–37.
66. Douillard JY, Ostoros G, Cobo M, Ciuleanu T, McCormack R, Webster A, et al. First-line gefitinib in Caucasian EGFR mutation-positive NSCLC patients: a phase-IV, open-label, single-arm study. Br J Cancer. 2014;110(1):55–62.
67. Korashy HM, Attafi IM, Ansari MA, Assiri MA, Belali OM, Ahmad SF, et al. Molecular mechanisms of cardiotoxicity of gefitinib in vivo and in vitro rat cardiomyocyte: role of apoptosis and oxidative stress. Toxicol Lett. 2016;252:50–61.
68. AlAsmari AF, Ali N, AlAsmari F, AlAnazi WA, AlShammari MA, Al-Harbi NO, et al. Liraglutide attenuates gefitinib-induced cardiotoxicity and promotes cardioprotection through the regulation of MAPK/NF-κB signaling pathways. Saudi Pharm J. 2020;28(4):509–18.
69. Alanazi WA, Alhamami HN, Alharbi M, Alhazzani K, Alanazi AS, Alsanea S, et al. Angiotensin II type 1 receptor blockade attenuates gefitinib-induced cardiac hypertrophy via adjusting angiotensin II-mediated oxidative stress and JNK/P38 MAPK pathway in a rat model. Saudi Pharm J. 2022;30(8):1159–69.
70. Lekes D, Szadvari I, Krizanova O, Lopusna K, Rezuchova I, Novakova M, et al. Nilotinib induces ER stress and cell death in H9c2 cells. Physiol Res. 2016;65(Suppl 4):S505–14.
71. Li S, He J, Zhang X, Cai Y, Liu J, Nie X, et al. Cardiovascular adverse events in chronic myeloid leukemia patients treated with nilotinib or imatinib: a systematic review, meta-analysis and integrative bioinformatics analysis. Front Cardiovasc Med. 2022;9:966182.
72. Kobat H, Elkonaissi I, Foreman E, O'Brien M, Dorak MT, Nabhani-Gebara S. Investigating the efficacy of osimertinib and crizotinib in phase 3 clinical trials on anticancer treatment-induced cardiotoxicity: are real-world studies the way forward? J Oncol Pharm Pract. 2023;29(3):646–62.
73. Narayan HK, Sheline K, Wong V, Kuo D, Choo S, Yoon J, et al. Cardiovascular toxicities with pediatric tyrosine kinase inhibitor therapy: An analysis of adverse events reported to the Food and Drug Administration. Pediatr Blood Cancer. 2023;70(2):e30059.

74. Shopp GM, Helson L, Bouchard A, Salvail D, Majeed M. Liposomes ameliorate Crizotinib- and Nilotinib-induced inhibition of the cardiac IKr channel and QTc prolongation. Anticancer Res. 2014;34(9):4733–40.
75. Byrd JC, Brown JR, O'Brien S, Barrientos JC, Kay NE, Reddy NM, et al. Ibrutinib versus ofatumumab in previously treated chronic lymphoid leukemia. N Engl J Med. 2014;371(3):213–23.
76. Wang ML, Rule S, Martin P, Goy A, Auer R, Kahl BS, et al. Targeting BTK with ibrutinib in relapsed or refractory mantle-cell lymphoma. N Engl J Med. 2013;369(6):507–16.
77. Noy A, de Vos S, Thieblemont C, Martin P, Flowers CR, Morschhauser F, et al. Targeting Bruton tyrosine kinase with ibrutinib in relapsed/refractory marginal zone lymphoma. Blood. 2017;129(16):2224–32.
78. Dimopoulos MA, Tedeschi A, Trotman J, García-Sanz R, Macdonald D, Leblond V, et al. Phase 3 trial of Ibrutinib plus Rituximab in Waldenström's Macroglobulinemia. N Engl J Med. 2018;378(25):2399–410.
79. Abdel-Qadir H, Sabrie N, Leong D, Pang A, Austin PC, Prica A, et al. Cardiovascular risk associated with Ibrutinib use in chronic lymphocytic leukemia: a population-based Cohort Study. J Clin Oncol. 2021;39(31):3453–62.
80. Dickerson T, Wiczer T, Waller A, Philippon J, Porter K, Haddad D, et al. Hypertension and incident cardiovascular events following ibrutinib initiation. Blood. 2019;134(22):1919–28.
81. Chen ST, Azali L, Rosen L, Zhao Q, Wiczer T, Palettas M, et al. Hypertension and incident cardiovascular events after next-generation BTKi therapy initiation. J Hematol Oncol. 2022;15(1):92.
82. Yang X, An N, Zhong C, Guan M, Jiang Y, Li X, et al. Enhanced cardiomyocyte reactive oxygen species signaling promotes ibrutinib-induced atrial fibrillation. Redox Biol. 2020;30:101432.
83. Heumüller S, Wind S, Barbosa-Sicard E, Schmidt HHHW, Busse R, Schröder K, et al. Apocynin is not an inhibitor of vascular NADPH oxidases but an antioxidant. Hypertension. 2008;51(2):211–7.
84. Gunaydin Akyildiz A, Boran T, Jannuzzi AT, Alpertunga B. Mitochondrial dynamics imbalance and mitochondrial dysfunction contribute to the molecular cardiotoxic effects of lenvatinib. Toxicol Appl Pharmacol. 2021;423:115577.
85. Liu J, Huang L, Wan M, Chen G, Su M, Han F, et al. Lenvatinib induces cardiac developmental toxicity in zebrafish embryos through regulation of Notch mediated-oxidative stress generation. Environ Toxicol. 2022;37(6):1310–20.
86. Liang C, Zhu D, Xia W, Hong Z, Wang QS, Sun Y, et al. Inhibition of YAP by lenvatinib in endothelial cells increases blood pressure through ferroptosis. Biochim Biophys Acta Mol basis Dis. 2023;1869(1):166586.
87. Sueta D, Suyama K, Sueta A, Tabata N, Yamashita T, Tomiguchi M, et al. Lenvatinib, an oral multi-kinases inhibitor, -associated hypertension: potential role of vascular endothelial dysfunction. Atherosclerosis. 2017;260:116–20.
88. Boran T, Akyildiz AG, Jannuzzi AT, Alpertunga B. Extended regorafenib treatment can be linked with mitochondrial damage leading to cardiotoxicity. Toxicol Lett. 2021;336:39–49.
89. Zhang S, Xu X, Li Z, Yi T, Ma J, Zhang Y, et al. Analysis and validation of differentially expressed Ferroptosis-related genes in Regorafenib-induced cardiotoxicity. Oxidative Med Cell Longev. 2022;2022:2513263.
90. Sun L, Wang H, Xu D, Yu S, Zhang L, Li X. Lapatinib induces mitochondrial dysfunction to enhance oxidative stress and ferroptosis in doxorubicin-induced cardiomyocytes via inhibition of PI3K/AKT signaling pathway. Bioengineered. 2022;13(1):48–60.
91. Hsu WT, Huang CY, Yen CYT, Cheng AL, Hsieh PCH. The HER2 inhibitor lapatinib potentiates doxorubicin-induced cardiotoxicity through iNOS signaling. Theranostics. 2018;8(12):3176–88.

92. Mak IT, Kramer JH, Chmielinska JJ, Spurney CF, Weglicki WB. EGFR-TKI, erlotinib, causes hypomagnesemia, oxidative stress, and cardiac dysfunction: attenuation by NK-1 receptor blockade. J Cardiovasc Pharmacol. 2015;65(1):54–61.
93. Beck TC, Arhontoulis DC, Morningstar JE, Hyams N, Stoddard A, Springs K, et al. Cellular and molecular mechanisms of MEK1 inhibitor-induced cardiotoxicity. JACC Cardio Oncol. 2022;4(4):535–48.
94. Gordan JD, Kennedy EB, Abou-Alfa GK, Beg MS, Brower ST, Gade TP, et al. Systemic therapy for advanced hepatocellular carcinoma: ASCO guideline. J Clin Oncol. 2020;38(36):4317–45.
95. Agulnik M, Yarber JL, Okuno SH, von Mehren M, Jovanovic BD, Brockstein BE, et al. An open-label, multicenter, phase II study of bevacizumab for the treatment of angiosarcoma and epithelioid hemangioendotheliomas. Ann Oncol. 2013;24(1):257–63.
96. Socinski MA, Jotte RM, Cappuzzo F, Orlandi F, Stroyakovskiy D, Nogami N, et al. Atezolizumab for first-line treatment of metastatic nonsquamous NSCLC. N Engl J Med. 2018;378(24):2288–301.
97. Gianni L, Pienkowski T, Im YH, Roman L, Tseng LM, Liu MC, et al. Efficacy and safety of neoadjuvant pertuzumab and trastuzumab in women with locally advanced, inflammatory, or early HER2-positive breast cancer (NeoSphere): a randomised multicentre, open-label, phase 2 trial. Lancet Oncol. 2012;13(1):25–32.
98. Sartore-Bianchi A, Trusolino L, Martino C, Bencardino K, Lonardi S, Bergamo F, et al. Dual-targeted therapy with trastuzumab and lapatinib in treatment-refractory, KRAS codon 12/13 wild-type, HER2-positive metastatic colorectal cancer (HERACLES): a proof-of-concept, multicentre, open-label, phase 2 trial. Lancet Oncol. 2016;17(6):738–46.
99. Fader AN, Roque DM, Siegel E, Buza N, Hui P, Abdelghany O, et al. Randomized phase II trial of Carboplatin-Paclitaxel versus Carboplatin-Paclitaxel-Trastuzumab in uterine serous carcinomas that overexpress human epidermal growth factor receptor 2/neu. J Clin Oncol. 2018;36(20):2044–51.
100. Janjigian YY, Kawazoe A, Yañez P, Li N, Lonardi S, Kolesnik O, et al. The KEYNOTE-811 trial of dual PD-1 and HER2 blockade in HER2-positive gastric cancer. Nature. 2021;600(7890):727–30.
101. Van Leeuwen MT, Luu S, Gurney H, Brown MR, Pearson SA, Webber K, et al. Cardiovascular toxicity of targeted therapies for cancer: An overview of systematic reviews. JNCI Cancer Spectr. 2020;4(6):pkaa076.
102. Shiojima I, Sato K, Izumiya Y, Schiekofer S, Ito M, Liao R, et al. Disruption of coordinated cardiac hypertrophy and angiogenesis contributes to the transition to heart failure. J Clin Invest. 2005;115(8):2108–18.
103. Chen MH, Kerkelä R, Force T. Mechanisms of cardiac dysfunction associated with tyrosine kinase inhibitor cancer therapeutics. Circulation. 2008;118(1):84–95.
104. Izzedine H, Ederhy S, Goldwasser F, Soria JC, Milano G, Cohen A, et al. Management of hypertension in angiogenesis inhibitor-treated patients. Ann Oncol. 2009;20(5):807–15.
105. Li Y, Tian W, Yue D, Chen C, Li C, Zhang Z, et al. Bevacizumab-induced mitochondrial dysfunction, endoplasmic reticulum stress, and ERK inactivation contribute to cardiotoxicity. Oxidative Med Cell Longev. 2021;2021:5548130.
106. Sabet NS, Atashbar S, Khanlou EM, Kahrizi F, Salimi A. Curcumin attenuates bevacizumab-induced toxicity via suppressing oxidative stress and preventing mitochondrial dysfunction in heart mitochondria. Naunyn Schmiedeberg's Arch Pharmacol. 2020;393(8):1447–57.
107. Thannickal VJ, Fanburg BL. Reactive oxygen species in cell signaling. Am J Physiol Lung Cell Mol Physiol. 2000;279(6):L1005–28.
108. Zorov DB, Juhaszova M, Sollott SJ. Mitochondrial reactive oxygen species (ROS) and ROS-induced ROS release. Physiol Rev. 2014;94(3):909–50.
109. Redza-Dutordoir M, Averill-Bates DA. Activation of apoptosis signalling pathways by reactive oxygen species. Biochim Biophys Acta. 2016;1863(12):2977–92.

110. Kiełczykowska M, Kocot J, Paździor M, Musik I. Selenium - a fascinating antioxidant of protective properties. Adv Clin Exp Med. 2018;27(2):245–55.
111. Massoure PL, Camus O, Fourcade L, Simon F. Bilateral leg oedema after bariatric surgery: a selenium-deficient cardiomyopathy. Obes Res Clin Pract. 2017;11(5):622–6.
112. Rayman MP. The importance of selenium to human health. Lancet Lond Engl. 2000;356(9225):233–41.
113. Finn D. Capsazepine. In: Enna SJ, Bylund DB, editors. xPharm: the comprehensive pharmacology reference [Internet]. New York: Elsevier; 2007. p. 1–5. [cited 2023 May 25]. Available from: https://www.sciencedirect.com/science/article/pii/B9780080552323613765.
114. Wei J, Lin J, Zhang J, Tang D, Xiang F, Cui L, et al. TRPV1 activation mitigates hypoxic injury in mouse cardiomyocytes by inducing autophagy through the AMPK signaling pathway. Am J Physiol Cell Physiol. 2020;318(5):C1018–29.
115. Buckley CL, Stokes AJ. Mice lacking functional TRPV1 are protected from pressure overload cardiac hypertrophy. Channels Austin Tex. 2011;5(4):367–74.
116. Oncel CR, Ovey IS. The role of selenium in bevacizumab induced cardiotoxicity. Bratisl Lek Listy. 2019;120(2):131–8.
117. Jabczyk M, Nowak J, Hudzik B, Zubelewicz-Szkodzińska B. Curcumin in metabolic health and disease. Nutrients. 2021;13(12):4440.
118. Singh L, Sharma S, Xu S, Tewari D, Fang J. Curcumin as a natural remedy for atherosclerosis: a pharmacological review. Mol Basel Switz. 2021;26(13):4036.
119. Mohammad Khanlou E, Atashbar S, Kahrizi F, Shokouhi Sabet N, Salimi A. Bevacizumab as a monoclonal antibody inhibits mitochondrial complex II in isolated rat heart mitochondria: ameliorative effect of ellagic acid. Drug Chem Toxicol. 2022;45(1):456–463. https://doi.org/10.1080/01480545.2020.1715423
120. Yasunari K, Maeda K, Nakamura M, Watanabe T, Yoshikawa J. Benidipine, a long-acting calcium channel blocker, inhibits oxidative stress in polymorphonuclear cells in patients with essential hypertension. Hypertens Res. 2005;28(2):107–12.
121. Hassan MQ, Akhtar MS, Akhtar M, Ansari SH, Ali J, Haque SE, et al. Benidipine prevents oxidative stress, inflammatory changes and apoptosis related myofibril damage in isoproterenol-induced myocardial infarction in rats. Toxicol Mech Methods. 2015;25(1):26–33.
122. Yıldırım E, Yıldırım N, Cengiz M, Yazıcı GN, Coskun R, Suleyman B, et al. Protective effect of adenosine triphosphate and benidipine separately or together against cardiotoxicity caused by bevacizumab. Biotech Histochem. 2023;98(3):193–200.
123. Godfraind T. Antioxidant effects and the therapeutic mode of action of calcium channel blockers in hypertension and atherosclerosis. Philos Trans R Soc B Biol Sci. 2005;360(1464):2259–72.
124. Alhussein MM, Mokbel A, Cosman T, Aghel N, Yang EH, Mukherjee SD, et al. Pertuzumab cardiotoxicity in patients with HER2-positive cancer: a systematic review and meta-analysis. CJC Open. 2021;3(11):1372–82.
125. Koulaouzidis G, Yung AE, Yung DE, Skonieczna-Żydecka K, Marlicz W, Koulaouzidis A, et al. Conventional cardiac risk factors associated with trastuzumab-induced cardiotoxicity in breast cancer: systematic review and meta-analysis. Curr Probl Cancer. 2021;45(5):100723.
126. Wu YT, Xu Z, Zhang K, Wu JS, Li X, Arshad B, et al. Efficacy and cardiac safety of the concurrent use of trastuzumab and anthracycline-based neoadjuvant chemotherapy for HER2-positive breast cancer: a systematic review and meta-analysis. Ther Clin Risk Manag. 2018;14:1789–97.
127. Eiger D, Franzoi MA, Pondé N, Brandão M, de Angelis C, Schmitt Nogueira M, et al. Cardiotoxicity of trastuzumab given for 12 months compared to shorter treatment periods: a systematic review and meta-analysis of six clinical trials. ESMO Open. 2020;5(1):e000659.
128. Negro A, Brar BK, Lee KF. Essential roles of Her2/erbB2 in cardiac development and function. Recent Prog Horm Res. 2004;59:1–12.
129. Koundouros N, Poulogiannis G. Phosphoinositide 3-Kinase/Akt signaling and Redox metabolism in cancer. Front Oncol. 2018;8:160.

130. Sun L, Wang H, Yu S, Zhang L, Jiang J, Zhou Q. Herceptin induces ferroptosis and mitochondrial dysfunction in H9c2 cells. Int J Mol Med. 2022;49(2):17.
131. Pecoraro M, Marzocco S, Franceschelli S, Popolo A. Trastuzumab and Doxorubicin sequential administration increases oxidative stress and phosphorylation of Connexin 43 on Ser368. Int J Mol Sci. 2022;23(12):6375.
132. Dirican A, Levent F, Alacacioglu A, Kucukzeybek Y, Varol U, Kocabas U, et al. Acute cardiotoxic effects of adjuvant trastuzumab treatment and its relation to oxidative stress. Angiology. 2014;65(10):944–9.
133. Heidenreich PA, Bozkurt B, Aguilar D, Allen LA, Byun JJ, Colvin MM, et al. 2022 AHA/ACC/HFSA guideline for the management of heart failure: a report of the American College of Cardiology/American Heart Association Joint Committee on clinical practice guidelines. Circulation. 2022;145(18):e895–1032.
134. Chin BSP, Langford NJ, Nuttall SL, Gibbs CR, Blann AD, Lip GYH. Anti-oxidative properties of beta-blockers and angiotensin-converting enzyme inhibitors in congestive heart failure. Eur J Heart Fail. 2003;5(2):171–4.
135. Kukin ML, Kalman J, Charney RH, Levy DK, Buchholz-Varley C, Ocampo ON, et al. Prospective, randomized comparison of effect of long-term treatment with metoprolol or carvedilol on symptoms, exercise, ejection fraction, and oxidative stress in heart failure. Circulation. 1999;99(20):2645–51.
136. Nakamura K, Kusano K, Nakamura Y, Kakishita M, Ohta K, Nagase S, et al. Carvedilol decreases elevated oxidative stress in human failing myocardium. Circulation. 2002;105(24):2867–71.
137. Ide T, Tsutsui H, Kinugawa S, Utsumi H, Kang D, Hattori N, et al. Mitochondrial electron transport complex I is a potential source of oxygen free radicals in the failing myocardium. Circ Res. 1999;85(4):357–63.
138. Pimentel DR, Amin JK, Xiao L, Miller T, Viereck J, Oliver-Krasinski J, et al. Reactive oxygen species mediate amplitude-dependent hypertrophic and apoptotic responses to mechanical stretch in cardiac myocytes. Circ Res. 2001;89(5):453–60.
139. Nicol M, Sadoune M, Polidano E, Launay JM, Samuel JL, Azibani F, et al. Doxorubicin-induced and trastuzumab-induced cardiotoxicity in mice is not prevented by metoprolol. ESC Heart Fail. 2021;8(2):928–37.
140. Pituskin E, Mackey JR, Koshman S, Jassal D, Pitz M, Haykowsky MJ, et al. Multidisciplinary approach to novel therapies in cardio-oncology research (MANTICORE 101-breast): a randomized trial for the prevention of Trastuzumab-associated cardiotoxicity. J Clin Oncol. 2017;35(8):870–7.
141. Esfandbod M, Naderi M, Sadatnaseri A, Ahmadi A, Noroozi M, Sadeghi JS. Evaluation of the preventive effects of carvedilol on Trastuzumab-induced cardiotoxicity in early-stage and locally advanced HER2-positive breast cancer patients. Int J Hematol-Oncol Stem Cell Res. 2021;15(4):206–12.
142. Münzel T, Keaney JF. Are ACE inhibitors a "magic bullet" against oxidative stress? Circulation. 2001;104(13):1571–4.
143. Nickenig G, Harrison DG. The AT1-type angiotensin receptor in oxidative stress and atherogenesis. Circulation. 2002;105(3):393–6.
144. Olorundare OE, Adeneye AA, Akinsola AO, Ajayi AM, Agede OA, Soyemi SS, et al. Therapeutic potentials of selected antihypertensive agents and their fixed-dose combinations against Trastuzumab-mediated cardiotoxicity. Front Pharmacol. 2020;11:610331.
145. Eekhoudt CR, Bortoluzzi T, Varghese SS, Cheung DYC, Christie S, Eastman S, et al. Comparing flaxseed and perindopril in the prevention of Doxorubicin and Trastuzumab-induced cardiotoxicity in C57Bl/6 mice. Curr Oncol Tor Ont. 2022;29(5):2941–53.
146. Cho DH, Lim IR, Kim JH, Kim MN, Kim YH, Park KH, et al. Protective effects of statin and angiotensin receptor blocker in a Rat Model of Doxorubicin- and Trastuzumab-induced cardiomyopathy. J Am Soc Echocardiogr. 2020;33(10):1253–63.

147. Moey MYY, Liles DK, Carabello BA. Concomitant use of renin-angiotensin-aldosterone system inhibitors prevent trastuzumab-induced cardiotoxicity in HER2+ breast cancer patients: an institutional retrospective study. Cardio-Oncol Lond Engl. 2019;5:9.
148. Gulati G, Heck SL, Ree AH, Hoffmann P, Schulz-Menger J, Fagerland MW, et al. Prevention of cardiac dysfunction during adjuvant breast cancer therapy (PRADA): a 2 × 2 factorial, randomized, placebo-controlled, double-blind clinical trial of candesartan and metoprolol. Eur Heart J. 2016;37(21):1671–80.
149. Boekhout AH, Gietema JA, Milojkovic Kerklaan B, van Werkhoven ED, Altena R, Honkoop A, et al. Angiotensin II-receptor inhibition with Candesartan to prevent Trastuzumab-related Cardiotoxic effects in patients with early breast cancer: a randomized clinical trial. JAMA Oncol. 2016;2(8):1030–7.
150. Ohtani K, Ide T, Hiasa KI, Sakamoto I, Yamashita N, Kubo M, et al. Cardioprotective effect of renin-angiotensin inhibitors and β-blockers in trastuzumab-related cardiotoxicity. Clin Res Cardiol. 2019;108(10):1128–39.
151. Wittayanukorn S, Qian J, Westrick SC, Billor N, Johnson B, Hansen RA. Prevention of Trastuzumab and Anthracycline-induced cardiotoxicity using angiotensin-converting enzyme inhibitors or β-blockers in older adults with breast cancer. Am J Clin Oncol. 2018;41(9):909–18.
152. Geuna E, Lombardi P, Martinello R, Garino D, Bonzano A, Galizia D, et al. Treatment with Beta-blockers and ACE-inhibitors in breast cancer patients receiving adjuvant Trastuzumab-based therapy and developing mild cardiac toxicity: a prospective study. Cancers. 2020;12(2):327.
153. Oliva S, Cioffi G, Frattini S, Simoncini EL, Faggiano P, Boccardi L, et al. Administration of angiotensin-converting enzyme inhibitors and β-blockers during adjuvant trastuzumab chemotherapy for nonmetastatic breast cancer: marker of risk or cardioprotection in the real world? Oncologist. 2012;17(7):917–24.
154. Guglin M, Krischer J, Tamura R, Fink A, Bello-Matricaria L, McCaskill-Stevens W, et al. Randomized trial of Lisinopril versus carvedilol to prevent Trastuzumab cardiotoxicity in patients with breast cancer. J Am Coll Cardiol. 2019;73(22):2859–68.
155. Leong DP, Cosman T, Alhussein MM, Kumar Tyagi N, Karampatos S, Barron CC, et al. Safety of continuing Trastuzumab despite mild cardiotoxicity: a phase I trial. JACC Cardio Oncol. 2019;1(1):1–10.
156. Davignon J, Jacob RF, Mason RP. The antioxidant effects of statins. Coron Artery Dis. 2004;15(5):251–8.
157. Shishehbor MH, Brennan ML, Aviles RJ, Fu X, Penn MS, Sprecher DL, et al. Statins Promote Potent systemic antioxidant effects through specific inflammatory pathways. Circulation. 2003;108(4):426–31.
158. Kabel AM, Elkhoely AA. Targeting proinflammatory cytokines, oxidative stress, TGF-β1 and STAT-3 by rosuvastatin and ubiquinone to ameliorate trastuzumab cardiotoxicity. Biomed Pharmacother. 2017;93:17–26.
159. Abdel-Qadir H, Bobrowski D, Zhou L, Austin PC, Calvillo-Argüelles O, Amir E, et al. Statin exposure and risk of heart failure after Anthracycline- or Trastuzumab-based chemotherapy for early breast cancer: a propensity score–matched Cohort Study. J Am Heart Assoc. 2021;10(2):e018393.
160. Seicean S, Seicean A, Plana JC, Budd GT, Marwick TH. Effect of statin therapy on the risk for incident heart failure in patients with breast cancer receiving anthracycline chemotherapy: an observational clinical cohort study. J Am Coll Cardiol. 2012;60(23):2384–90.
161. Calvillo-Argüelles O, Abdel-Qadir H, Michalowska M, Billia F, Suntheralingam S, Amir E, et al. Cardioprotective effect of statins in patients with HER2-positive breast cancer receiving Trastuzumab therapy. Can J Cardiol. 2019;35(2):153–9.
162. Nabati M, Janbabai G, Esmailian J, Yazdani J. Effect of Rosuvastatin in preventing chemotherapy-induced cardiotoxicity in women with breast cancer: a randomized, single-blind, placebo-controlled trial. J Cardiovasc Pharmacol Ther. 2019;24(3):233–41.

163. Min J, Wu L, Liu Y, Song G, Deng Q, Jin W, et al. Empagliflozin attenuates trastuzumab-induced cardiotoxicity through suppression of DNA damage and ferroptosis. Life Sci. 2023;312:121207.
164. Arinno A, Maneechote C, Khuanjing T, Prathumsap N, Chunchai T, Arunsak B, et al. Melatonin and metformin ameliorated trastuzumab-induced cardiotoxicity through the modulation of mitochondrial function and dynamics without reducing its anticancer efficacy. Biochim Biophys Acta Mol basis Dis. 2023;1869(2):166618.
165. Riccio G, Antonucci S, Coppola C, D'Avino C, Piscopo G, Fiore D, et al. Ranolazine attenuates Trastuzumab-induced heart dysfunction by modulating ROS production. Front Physiol. 2018;9:38.
166. Reiter RJ, Mayo JC, Tan DX, Sainz RM, Alatorre-Jimenez M, Qin L. Melatonin as an antioxidant: under promises but over delivers. J Pineal Res. 2016;61(3):253–78.
167. Laird-Fick HS, Tokala H, Kandola S, Kehdi M, Pelosi A, Wang L, et al. Early morphological changes in cardiac mitochondria after subcutaneous administration of trastuzumab in rabbits: possible prevention with oral selenium supplementation. Cardiovasc Pathol. 2020;44:107159.
168. Du K, Ramachandran A, Jaeschke H. Oxidative stress during acetaminophen hepatotoxicity: sources, pathophysiological role and therapeutic potential. Redox Biol. 2016;10:148–56.
169. Goyal V, Bews H, Cheung D, Premecz S, Mandal S, Shaikh B, et al. The Cardioprotective role of N-acetyl cysteine amide in the prevention of Doxorubicin and Trastuzumab-mediated cardiac dysfunction. Can J Cardiol. 2016;32(12):1513–9.
170. Farkhondeh T, Samarghandian S, Bafandeh F. The cardiovascular protective effects of Chrysin: a narrative review on experimental researches. Cardiovasc Hematol Agents Med Chem. 2019;17(1):17–27.
171. Liu H, Liu HY, Jiang YN, Li N. Protective effect of thymoquinone improves cardiovascular function, and attenuates oxidative stress, inflammation and apoptosis by mediating the PI3K/Akt pathway in diabetic rats. Mol Med Rep. 2016;13(3):2836–42.
172. Shirmard LR, Shabani M, Moghadam AA, Zamani N, Ghanbari H, Salimi A. Protective effect of curcumin, Chrysin and Thymoquinone injection on Trastuzumab-induced cardiotoxicity via mitochondrial protection. Cardiovasc Toxicol. 2022;22(7):663–75.
173. Ernster L, Forsmark-Andrée P. Ubiquinol: an endogenous antioxidant in aerobic organisms. Clin Investig. 1993;71(8 Suppl):S60–5.
174. Quagliariello V, Vecchione R, De Capua A, Lagreca E, Iaffaioli RV, Botti G, et al. Nano-encapsulation of coenzyme Q10 in secondary and tertiary nano-emulsions for enhanced cardioprotection and hepatoprotection in human cardiomyocytes and hepatocytes during exposure to Anthracyclines and Trastuzumab. Int J Nanomedicine. 2020;15:4859–76.
175. Ajani JA, Winter KA, Gunderson LL, Pedersen J, Benson AB, Thomas CR, et al. Fluorouracil, mitomycin, and radiotherapy vs fluorouracil, cisplatin, and radiotherapy for carcinoma of the anal canal: a randomized controlled trial. JAMA. 2008;299(16):1914–21.
176. Levine MN, Bramwell VH, Pritchard KI, Norris BD, Shepherd LE, Abu-Zahra H, et al. Randomized trial of intensive cyclophosphamide, epirubicin, and fluorouracil chemotherapy compared with cyclophosphamide, methotrexate, and fluorouracil in premenopausal women with node-positive breast cancer. National Cancer Institute of Canada Clinical Trials Group. J Clin Oncol. 1998;16(8):2651–8.
177. Eifel PJ, Winter K, Morris M, Levenback C, Grigsby PW, Cooper J, et al. Pelvic irradiation with concurrent chemotherapy versus pelvic and para-aortic irradiation for high-risk cervical cancer: an update of radiation therapy oncology group trial (RTOG) 90-01. J Clin Oncol. 2004;22(5):872–80.
178. Cassidy J, Clarke S, Díaz-Rubio E, Scheithauer W, Figer A, Wong R, et al. Randomized phase III study of capecitabine plus oxaliplatin compared with fluorouracil/folinic acid plus oxaliplatin as first-line therapy for metastatic colorectal cancer. J Clin Oncol. 2008;26(12):2006–12.
179. Bourhis J, Sire C, Graff P, Grégoire V, Maingon P, Calais G, et al. Concomitant chemoradiotherapy versus acceleration of radiotherapy with or without concomitant chemotherapy in locally advanced head and neck carcinoma (GORTEC 99-02): an open-label phase 3 randomised trial. Lancet Oncol. 2012;13(2):145–53.

180. Longley DB, Harkin DP, Johnston PG. 5-fluorouracil: mechanisms of action and clinical strategies. Nat Rev Cancer. 2003;3(5):330–8.
181. Wacker A, Lersch C, Scherpinski U, Reindl L, Seyfarth M. High incidence of angina pectoris in patients treated with 5-fluorouracil. A planned surveillance study with 102 patients. Oncology. 2003;65(2):108–12.
182. Tsibiribi P, Descotes J, Lombard-Bohas C, Barel C, Bui-Xuan B, Belkhiria M, et al. Cardiotoxicity of 5-fluorouracil in 1350 patients with no prior history of heart disease. Bull Cancer (Paris). 2006;93(3):E27–30.
183. de Forni M, Malet-Martino MC, Jaillais P, Shubinski RE, Bachaud JM, Lemaire L, et al. Cardiotoxicity of high-dose continuous infusion fluorouracil: a prospective clinical study. J Clin Oncol. 1992;10(11):1795–801.
184. Cwikiel M, Eskilsson J, Wieslander JB, Stjernquist U, Albertsson M. The appearance of endothelium in small arteries after treatment with 5-fluorouracil. An electron microscopic study of late effects in rabbits. Scanning Microsc. 1996;10(3):805–18; discussion 819.
185. Focaccetti C, Bruno A, Magnani E, Bartolini D, Principi E, Dallaglio K, et al. Effects of 5-fluorouracil on morphology, cell cycle, proliferation, apoptosis, autophagy and ROS production in endothelial cells and cardiomyocytes. PLoS One. 2015;10(2):e0115686.
186. Lubos E, Handy DE, Loscalzo J. Role of oxidative stress and nitric oxide in atherothrombosis. Front Biosci J Virtual Libr. 2008;13:5323–44.
187. Senkus E, Jassem J. Cardiovascular effects of systemic cancer treatment. Cancer Treat Rev. 2011;37(4):300–11.
188. Yamaguchi F, Kamitori K, Sanada K, Horii M, Dong Y, Sui L, et al. Rare sugar D-allose enhances anti-tumor effect of 5-fluorouracil on the human hepatocellular carcinoma cell line HuH-7. J Biosci Bioeng. 2008;106(3):248–52.
189. Chong CR, Chan WPA, Nguyen TH, Liu S, Procter NEK, Ngo DT, et al. Thioredoxin-interacting protein: pathophysiology and emerging pharmacotherapeutics in cardiovascular disease and diabetes. Cardiovasc Drugs Ther. 2014;28(4):347–60.
190. Mosseri M, Fingert HJ, Varticovski L, Chokshi S, Isner JM. In vitro evidence that myocardial ischemia resulting from 5-fluorouracil chemotherapy is due to protein kinase C-mediated vasoconstriction of vascular smooth muscle. Cancer Res. 1993;53(13):3028–33.
191. Boldig K, Ganguly A, Kadakia M, Rohatgi A. Managing life-threatening 5-fluorouracil cardiotoxicity. BMJ Case Rep. 2022;15(10):e251016.
192. Muhammad RN, Sallam N, El-Abhar HS. Activated ROCK/Akt/eNOS and ET-1/ERK pathways in 5-fluorouracil-induced cardiotoxicity: modulation by simvastatin. Sci Rep. 2020;10(1):14693.
193. Sedeek MH, Llinas MT, Drummond H, Fortepiani L, Abram SR, Alexander BT, et al. Role of reactive oxygen species in endothelin-induced hypertension. Hypertension. 2003;42(4):806–10.
194. Wilcox CS, Wang C, Wang D. Endothelin-1-induced microvascular ROS and contractility in Angiotensin-II-infused mice depend on COX and TP receptors. Antioxidants. 2019;8(6):193.
195. Refaie MMM, Abdel-Gaber SA, Rahman SAAE, Hafez SMNA, Khalaf HM. Cardioprotective effects of bosentan in 5-fluorouracil-induced cardiotoxicity. Toxicology. 2022;465:153042.
196. Bi W, Bi Y, Li P, Hou S, Yan X, Hensley C, et al. Indole alkaloid derivative B, a novel bifunctional agent that mitigates 5-fluorouracil-induced cardiotoxicity. ACS Omega. 2018;3(11):15850–64.
197. Tambe PK, Mathew AJ, Bharati S. Cardioprotective potential of mitochondria-targeted antioxidant, mito-TEMPO, in 5-fluorouracil-induced cardiotoxicity. Cancer Chemother Pharmacol. 2023;91:389.
198. Refaie MMM, Shehata S, Bayoumi AMA, El-Tahawy NFG, Abdelzaher WY. The IL-6/STAT signaling pathway and PPARα are involved in mediating the dose-dependent cardioprotective effects of Fenofibrate in 5-fluorouracil-induced cardiotoxicity. Cardiovasc Drugs Ther. 2022;36(5):817–27.

199. Onodera Y, Teramura T, Takehara T, Shigi K, Fukuda K. Reactive oxygen species induce Cox-2 expression via TAK1 activation in synovial fibroblast cells. FEBS Open Bio. 2015;5:492–501.
200. Barbieri SS, Eligini S, Brambilla M, Tremoli E, Colli S. Reactive oxygen species mediate cyclooxygenase-2 induction during monocyte to macrophage differentiation: critical role of NADPH oxidase. Cardiovasc Res. 2003;60(1):187–97.
201. Safarpour S, Safarpour S, Pirzadeh M, Moghadamnia AA, Ebrahimpour A, Shirafkan F, et al. Colchicine Ameliorates 5-Fluorouracil-induced cardiotoxicity in rats. Oxidative Med Cell Longev. 2022;2022:6194532.
202. Safarpour S, Pirzadeh M, Ebrahimpour A, Shirafkan F, Madani F, Hosseini M, et al. Protective effect of Kaempferol and its nanoparticles on 5-fluorouracil-induced cardiotoxicity in rats. Biomed Res Int. 2022;2022:2273000.
203. Safarpour S, Safarpour S, Moghadamnia AA, Kazemi S, Ebrahimpour A, Shirafkan F, et al. Cardioprotective effect of silymarin nanoemulsion on 5-fluorouracil-induced cardiotoxicity in rats. Arch Pharm (Weinheim). 2022;355(7):e2200060.
204. Barary M, Hosseinzadeh R, Kazemi S, Liang JJ, Mansoori R, Sio TT, et al. The effect of propolis on 5-fluorouracil-induced cardiac toxicity in rats. Sci Rep. 2022;12(1):8661.
205. Lokman MS, Althagafi HA, Alharthi F, Habotta OA, Hassan AA, Elhefny MA, et al. Protective effect of quercetin against 5-fluorouracil-induced cardiac impairments through activating Nrf2 and inhibiting NF-κB and caspase-3 activities. Environ Sci Pollut Res Int. 2023;30(7):17657–69.
206. Gui Y, Famurewa AC, Olatunji OJ. Naringin ameliorates 5-fluorouracil induced cardiotoxicity: an insight into its modulatory impact on oxidative stress, inflammatory and apoptotic parameters. Tissue Cell. 2023;81:102035.
207. Arafah A, Rehman MU, Ahmad A, AlKharfy KM, Alqahtani S, Jan BL, et al. Myricetin (3,3′,4′,5,5′,7-Hexahydroxyflavone) prevents 5-fluorouracil-induced cardiotoxicity. ACS Omega. 2022;7(5):4514–24.
208. Ibrahim HA, Abd El-Alim AEAF, El-Hafeez MA, Metwally MMM, Khamis T, Galal AAA. Baicalein prevents capecitabine-induced heart damage in female Wistar rats and enhances its anticancer potential in MCF-7 breast cancer cells. Life Sci. 2023;319:121523.
209. Vargo CA, Blazer M, Reardon J, Gulati M, Bekaii-Saab T. Successful completion of adjuvant chemotherapy in a patient with colon cancer experiencing 5-Fluorouracil-induced cardiac vasospasm. Clin Colorectal Cancer. 2016;15(2):e61–3.
210. Allison JD, Tanavin T, Yang Y, Birnbaum G, Khalid U. Various manifestations of 5-Fluorouracil cardiotoxicity: a multicenter case series and review of literature. Cardiovasc Toxicol. 2020;20(4):437–42.
211. Bringhen S, Petrucci MT, Larocca A, Conticello C, Rossi D, Magarotto V, et al. Carfilzomib, cyclophosphamide, and dexamethasone in patients with newly diagnosed multiple myeloma: a multicenter, phase 2 study. Blood. 2014;124(1):63–9.
212. Wu P, Oren O, Gertz MA, Yang EH. Proteasome inhibitor-related cardiotoxicity: mechanisms, diagnosis, and management. Curr Oncol Rep. 2020;22(7):66.
213. Georgiopoulos G, Makris N, Laina A, Theodorakakou F, Briasoulis A, Trougakos IP, et al. Cardiovascular toxicity of proteasome inhibitors: underlying mechanisms and management strategies: JACC: CardioOncology state-of-the-art review. JACC Cardio Oncol. 2023;5(1):1–21.
214. Efentakis P, Doerschmann H, Witzler C, Siemer S, Nikolaou PE, Kastritis E, et al. Investigating the vascular toxicity outcomes of the irreversible proteasome inhibitor Carfilzomib. Int J Mol Sci. 2020;21(15):5185.
215. Efentakis P, Kremastiotis G, Varela A, Nikolaou PE, Papanagnou ED, Davos CH, et al. Molecular mechanisms of carfilzomib-induced cardiotoxicity in mice and the emerging cardioprotective role of metformin. Blood. 2019;133(7):710–23.

216. Efentakis P, Psarakou G, Varela A, Papanagnou ED, Chatzistefanou M, Nikolaou PE, et al. Elucidating Carfilzomib's induced cardiotoxicity in an in vivo model of aging: prophylactic potential of metformin. Int J Mol Sci. 2021;22(20):10956.
217. Al-Harbi NO. Carfilzomib-induced cardiotoxicity mitigated by dexrazoxane through inhibition of hypertrophic gene expression and oxidative stress in rats. Toxicol Mech Methods. 2016;26(3):189–95.
218. Junjing Z, Yan Z, Baolu Z. Scavenging effects of Dexrazoxane on free radicals. J Clin Biochem Nutr. 2010;47(3):238–45.
219. Schafer PH, Parton A, Capone L, Cedzik D, Brady H, Evans JF, et al. Apremilast is a selective PDE4 inhibitor with regulatory effects on innate immunity. Cell Signal. 2014;26(9):2016–29.
220. Tsai YF, Chen CY, Yang SC, Syu YT, Hwang TL. Apremilast ameliorates acute respiratory distress syndrome by inhibiting neutrophil-induced oxidative stress. Biom J. 2022;S2319-4170(22):00135–4.
221. Imam F, Al-Harbi NO, Al-Harbi MM, Ansari MA, Almutairi MM, Alshammari M, et al. Apremilast reversed carfilzomib-induced cardiotoxicity through inhibition of oxidative stress, NF-κB and MAPK signaling in rats. Toxicol Mech Methods. 2016;26(9):700–8.
222. Jannuzzi AT, Korkmaz NS, Gunaydin Akyildiz A, Arslan Eseryel S, Karademir Yilmaz B, Alpertunga B. Molecular cardiotoxic effects of proteasome inhibitors Carfilzomib and Ixazomib and their combination with Dexamethasone involve mitochondrial dysregulation. Cardiovasc Toxicol. 2023;23(3–4):121–31.
223. Al-Harbi NO, Imam F, Al-Harbi MM, Al-Shabanah OA, Alotaibi MR, As Sobeai HM, et al. Rutin inhibits carfilzomib-induced oxidative stress and inflammation via the NOS-mediated NF-κB signaling pathway. Inflammopharmacology. 2019;27(4):817–27.
224. Imam F, Al-Harbi NO, Al-Harbia MM, Korashy HM, Ansari MA, Sayed-Ahmed MM, et al. Rutin attenuates Carfilzomib-induced cardiotoxicity through inhibition of NF-κB, hypertrophic gene expression and oxidative stress. Cardiovasc Toxicol. 2017;17(1):58–66.
225. Enogieru AB, Haylett W, Hiss DC, Bardien S, Ekpo OE. Rutin as a potent antioxidant: implications for neurodegenerative disorders. Oxidative Med Cell Longev. 2018;2018:6241017.
226. Alam MF, Hijri SI, Alshahrani S, Alqahtani SS, Jali AM, Ahmed RA, et al. Zingerone attenuates Carfilzomib-induced cardiotoxicity in rats through oxidative stress and inflammatory cytokine network. Int J Mol Sci. 2022;23(24):15617.

The Future of Cardio-Oncology

Shane S. Scott, Cora Albers, Ashley N. Greenlee, Thai H. Ho, Daniel Addison, and Sakima A. Smith

Abbreviations

ICI	Immune checkpoint inhibitors
CAR-T	Chimeric antigen receptor T cell
AIC	Anthracycline-induced cardiotoxicity
CRS	Cytokine release syndrome
CCT	Cardiac computed tomography
SGLT2	Sodium–glucose cotransporter 2 inhibitors
ACE	Angiotensin-converting enzyme
ARB	Angiotensin receptor blocker
RCT	Randomized controlled trial
CMRi	Cardiac magnetic resonance imaging
LVEF	Left ventricular ejection fraction
FDA	Food and drug administration
PET	Position emission tomography

S. S. Scott
Dorothy M. Davis Heart and Lung Research Institute, The Ohio State University College of Medicine, Columbus, OH, USA

The Ohio State Medical Scientist Training Program (MSTP), Columbus, OH, USA
e-mail: shane.scott@osumc.edu

C. Albers · A. N. Greenlee · S. A. Smith (✉)
Dorothy M. Davis Heart and Lung Research Institute, The Ohio State University College of Medicine, Columbus, OH, USA
e-mail: cora.albers@osumc.edu; ashley.greenlee@osumc.edu; Sakima.smith@osumc.edu

T. H. Ho
Division of Hematology and Medical Oncology, Hollings Cancer Center Medical University of South Carolina College of Medicine, Charleston, SC, USA
e-mail: hothai@musc.edu

D. Addison
Audre and Bernard Rapoport Chair in Cardiovascular Research, Department of Internal Medicine, Division of Cardiology, UT Southwestern Medical Center, Dallas, TX, USA
e-mail: Daniel.Addison@UTSouthwestern.edu

S. A. Smith, S. S. Scott (eds.), *Myocardial Oxidative Stress in Anti-Cancer Therapies*, Oxidative Stress in Applied Basic Research and Clinical Practice,
https://doi.org/10.1007/978-3-031-97517-2_8

SES	Socioeconomic status
hiPSC-CM	Human inducible pluripotent stem cell cardiomyocyte
AI	Artificial intelligence
ECG	Electrocardiogram

Introduction

Cardio-oncology emerged as a distinct subspecialty to address growing concerns of anticancer therapy-induced cardiotoxicity. The report of heart failure following treatment with the nonspecific anthracycline doxorubicin by Lefrak and colleagues first identified anticancer cardiotoxicity in breast cancer patients [1, 2]. Later, evidence of radiation therapy-induced cardiomyopathy was also described. The increasing burden of cardiotoxicity with the use of anthracyclines and radiation therapy among cancer patients and survivors demanded the need for a multidisciplinary approach to clinical management. Cardio-oncologists were initially optimistic that the development of targeted anticancer agents with limited off-target effects would reduce cardiotoxicity. Yet, trastuzumab, a targeted therapy for breast cancer, was found to be associated with cardiomyopathy and heart failure in a subset of patients [2]. Since then, other targeted therapies, including tyrosine kinase inhibitors and monoclonal antibodies, have also been linked to both acute cardiovascular complications, such as hypertension and arrhythmias, and chronic complications, including heart failure and cardiomyopathy. Recently, inflammatory syndromes (e.g., cytokine release syndrome, CRS) and inflammatory cardiomyopathies (e.g., myocarditis and pericarditis) attributed to the use of immune checkpoint inhibitors (ICIs) and chimeric antigen receptor T cell (CAR-T) therapies have also posed new challenges.

To curtail the impact of cardiotoxicity, imaging-based cardiac monitoring has been a cornerstone. Echocardiography remains the standard first-line method for detecting changes in left ventricular function secondary to cardiotoxicity at advanced stages [3, 4]. Recently, cardiac magnetic resonance imaging (CMRi) and nuclear imaging have been employed to detect subclinical function and tissue alterations that precede myocardial injury that is detectable by echocardiography [3]. In addition, cardiac computed tomography (CCT) is now being used as second-line imaging to assess cardiac health [3]. Cardiac biomarkers have also been investigated for their use in baseline risk assessment and diagnosis of cardiotoxicity [4–6]. Moreover, serum biomarker analysis is inexpensive and capable of detecting more subtle evidence of cardiomyocyte injury [4]. Current cardio-oncological literature endorses surveillance with the use of biomarker assessments at baseline, during, and post-chemotherapy for optimal benefit [5].

Management guidelines are limited due to the diversity of cancers and anticancer treatments, as well as poor understanding of cardiotoxicity mechanisms, highlighting prospective areas of research for basic and translational cardiovascular scientists [4]. While anecdotal use of β-blockers, angiotensin-converting enzyme (ACE) inhibitors and angiotensin receptor blocker (ARB) inhibitors has prevailed in clinical practice, recent basic science and translational preclinical data has sparked excitement for the use of sodium–glucose cotransporter 2 (SGLT2) inhibitors [7]. These data are further strengthened by the development of several randomized controlled clinical trials (RCT).

Despite significant advancements in identifying, detecting, and managing cardiotoxicity, challenges persist in developing therapeutic treatments and prevention guidelines. In this chapter, the significant strides made, existing gaps, promising ongoing work, and future directions in cardio-oncology are highlighted.

Significant Strides in Cardio-Oncology

Since the inception of cardio-oncology as a distinct field in the early 2000s, several milestones have bolstered the field. From the emergence of societies like the International Cardio-oncology Society (ICOS), high-impact peer-reviewed journals dedicated to the field, development of multidisciplinary care models, to advances in imaging, and recognition of novel anticancer therapy cardiovascular syndromes, the field is poised to impact and reduce cardiotoxicity burden.

Multidisciplinary Care Teams

The establishment of cardio-oncology clinics, subspecialty programs, and societies has fostered multidisciplinary collaboration and increased interfacing between oncologists and cardiologists for optimization of patient care. These clinics emphasize early detection of cardiac complications through baseline assessments and tailored monitoring protocols [8, 9]. Cardiologists subspecializing in multimodal imaging have demonstrated the use of CMR imaging, strain echocardiography, nuclear imaging, and CCT to improve early detection of subclinical and clinical cardiotoxicity associated with various therapies including anthracyclines, ICIs, and radiation therapy [10, 11]. These modalities are crucial for monitoring cardiomyopathy, vascular toxicity, and myocarditis [8, 10, 11]. Furthermore, imaging tools are being paired or optimized with artificial intelligence (AI) applications to improve diagnostic accuracy and cardiotoxicity risk prediction in patients with preexisting cardiovascular disease [9, 10].

Pharmacovigilance

Pharmacovigilance analysis of emerging therapies such as ICIs and CAR-T therapies has introduced unique challenges such as cytokine release syndrome (CRS) and inflammatory cardiomyopathies (e.g., myocarditis, pericarditis), necessitating innovative diagnostic and therapeutic approaches [12]. One study reported 21.4% incidence of major adverse cardiac events after CAR-T cell infusion [13]. Furthermore, CRS has also been linked to cardiac adverse events and cardiac injury [14]. Alvi et al. reported that 54% of patients receiving CAR-T had elevated troponins and reduced left ventricular ejection fraction (LVEF)—all of which occurred in the presence of Grade ≥2 CRS [15]. With the growing popularity of these drugs, it is vital to understand the mechanisms via which ICIs, CAR-Ts, and the accompanying inflammatory environment mediate myocardial injury and damage.

Imaging and Monitoring

Recent advances in cardiovascular imaging have allowed for earlier detection of cardiac dysfunction throughout all stages of cancer therapy, especially in patients who are at higher risk for developing cardiotoxicity. These imaging and monitoring techniques encompass a variety of technologies, including echocardiography, cardiac MRI, FDG PET/CT, and analysis of biomarker levels, including troponin and natriuretic peptides [16, 17]. As research surrounding the specific cardiotoxicities associated with certain cancer therapies progresses, modern clinical guidelines have begun to employ a role for continued cardiac imaging and monitoring throughout treatment [16, 18–20]. The increase in imaging frequency aims to detect subclinical changes in cardiac function, such as myocardial dysfunction and strain, decreases in LVEF, and arterial hypertension, prior to clinical presentation [19, 21, 22]. Additionally, these techniques are able to provide further detail on functional changes in concert with abnormal biomarker levels or ECG recordings [23, 24].

Milestones in Cardioprotection

Cardioprotective strategies have long been an important area of research within the cardio-oncology discipline. Dexrazoxane, a potent iron chelator, was the first FDA-approved drug for clinical use toward the prevention of doxorubicin-mediated cardiomyopathy, left ventricular dysfunction, and heart failure [25]. Conventional cardiogenic drugs such as β-blockers, ACE inhibitors, ARBs, and statins have been previously favored for cardioprotective therapy for prevention of asymptomatic to symptomatic disease [26, 27]. However, modern research has brought into question the efficacy of these therapies with respect to cardioprotection. Regarding the use of

β-blockers, neither preclinical (e.g., carvedilol) nor clinical studies (e.g., metoprolol) have shown cardioprotective effects of these drugs [28–30]. In a retrospective study of childhood survivors treated with doxorubicin, treatment with enalapril resulted in short-term improvement in LV structure and function [31]. Moreover, Georgakopoulos and colleagues evaluated cardioprotection of enalapril and metoprolol in 47 patients receiving doxorubicin and demonstrated that while heart failure was less frequent in patients treated with metoprolol and enalapril than in those with no treatment, no echocardiographic changes were observed [30]. However, this study was limited by the sample size. In a separate placebo-controlled trial (NCT01009918) of 468 patients, Guglin et al. showed that neither lisinopril nor carvedilol reduced cardiovascular events in patients with breast cancer [32]. However, both drugs reduced cardiotoxicity in HER2 breast cancer patients exposed to anthracyclines prior to receiving a trastuzumab treatment [32]. Owing to these data, several clinical trials are underway to further evaluate β-blockers including NCT03418961 and NCT02818517.

Studies investigating the ARB, candesartan, have shown the drug to prevent an early decline in global LVEF in two separate RCT trials with breast cancer patients receiving trastuzumab [29, 33]. There are preclinical and clinical data to suggest a role for rosuvastatin in the prevention of anticancer therapy-induced cardiotoxicity and cardiac left ventricular dysfunction [28, 34]. In a randomized controlled trial (RCT) of 89 breast cancer patients, prophylactic use of rosuvastatin prevented reduction in LVEF [34]. In a recent landmark trial titled *STOP-CA* (Statins To Prevent the Cardiotoxicity From Anthracyclines, NCT02943590), it was shown that treatment of lymphoma patients (N = 300) receiving anthracyclines with atorvastatin (40 mg/day, N = 150) significantly reduces the incidence of cardiac dysfunction compared to placebo (N = 150) [35]. As of this publication, several other ongoing trials, including NCT05559164, NCT05792293, and NCT06684509, are exploring the role of statins in reducing cardiotoxicity from anthracyclines and HER2-targeted therapy in a variety of cancers such as breast cancer, leukemia, and Hodgkin's disease.

New preclinical and observational studies suggest an important role for SGLT2 inhibitors in preventing cardiac dysfunction secondary to anticancer therapies [7, 36, 37]. Specifically, the SGLT2 inhibitors empagliflozin and dapagliflozin have gained notoriety. Currently, a randomized, multicenter, double-blind, placebo-controlled study (NCT05271162) called the *EMPACT* trial (Empagliflozin in the Prevention of Cardiotoxicity in Cancer Patients Undergoing Chemotherapy Based on Anthracyclines) is recruiting patients. The *EMPACT* trial seeks to evaluate whether empagliflozin can effectively prevent left ventricular systolic dysfunction in cancer patients receiving high cumulative doses of anthracycline [7]. The *PROTECTAA* trial (CardioPROTECTion with Dapagliflozin in Breast Cancer Patients Treated with AnthrAcycline) is also currently recruiting. Additionally, many clinical trials (NCT06103279, NCT06427226; NCT06341842, NCT06443645) are also being developed to evaluate SGLT2 inhibitors in protecting against cardiotoxicity. However, these trials are focused on their use in breast neoplasms and breast cancer patients, which may potentially limit generalizability.

Notably, there is ongoing preclinical and clinical work evaluating the cardioprotective role of several phytochemicals [38]. Currently, the cardioprotective effects of the nutritional supplement sulforaphane are being evaluated in a clinical trial (NCT03934905). In some cases, cardiac resynchronization therapy may also be used to improve LVEF in select patients with anticancer therapy-related cardiomyopathy [39].

Current Gaps in Cardio-Oncology

Evidence-Based Guidelines

Given that the field of cardio-oncology is less than 30-years-old, there are several areas that are ripe for improvement and contribution by oncologists, cardiologists, and basic scientists. Notably, there are limited evidence-based guidelines within the field for the management of cardiotoxicity. Existing cardio-oncology guidelines are largely derived from observational studies or small-scale trials, leading to variability in clinical practice [9, 12]. Key challenges to developing robust guidelines center on both a lack of homogenous large RCTs evaluating cardiotoxicity and the heterogeneity of cardiotoxicity mechanisms [40, 41]. In addition, the lack of standardized imaging protocols for risk stratification and monitoring of individual cancer therapies has led to inconsistent management practices [8, 11]. Further, fragmented institutional data hinder the development of comprehensive databases necessary for both population and personalized medical treatment [9].

Cardiac Imaging

While echocardiography remains a first-line cardiotoxic diagnostic imaging tool, it is limited by user variability and low sensitivity for early subclinical toxicity. Advanced imaging such as strain echocardiography and cardiac MRI holds great promise for earlier detection of cardiotoxicity [6, 42, 43]. However, these tools are accompanied by challenges such as prohibitive cost and inconsistent reproducibility. Position emission tomography (PET) may also be valuable in detecting and monitoring cardiometabolic and mitochondrial dysfunction, and vascular alterations in patients, particularly in anthracycline-induced cardiomyopathy [43–46]. For example, [18F]FDG PET/CT showed significantly altered myocardial FDG uptake in patients treated with anthracyclines [44, 47]. Lastly, the use of T1 and T2 mapping for myocardial fibrosis detection using CMR requires further clinical validation [48–50].

Health Disparities in Cardio-Oncology

Largely unexplored areas within cardio-oncology research include health disparities and chronic anticancer-related cardiomyopathy. There is emerging evidence to suggest that there are inequities in cardio-oncology care, particularly in the context of race, ethnicity, and socioeconomic status (SES) [51]. Among patients receiving anthracycline and HER2-targeted therapy, Black women face significantly higher risks of cardiotoxicity [52, 53]. Black women see a *40% higher risk of mortality* after a breast cancer diagnosis, a gap that persists even after accounting for socioeconomic and behavioral factors [54]. Anthracyclines are standard treatment for most advanced breast cancers, but ~10% of all women treated with anthracyclines develop serious or fatal heart failure, also dubbed anthracycline-induced cardiomyopathy (AIC). The development of AIC associates with disparately high mortality, even after adjusting for socioeconomic and traditional risk factors, suggesting AIC may be a key contributor to the higher mortality. Twenty-six percent of Black breast cancer patients treated with anthracyclines develop AIC [54]. Following adjustments for cardiovascular risk factors, SES, and insurance status, Black women treated with HER2-targeted breast cancer therapy still have increased cardiotoxicity compared to non-Hispanic White women, suggesting potential biological or structural contributors [52]. Socioeconomically marginalized populations also experience elevated cardiotoxicity risks due to higher prevalence of cardiovascular risk factors but remain underrepresented in research [9, 10, 52, 55]. Yet, to date, there are no prospective studies addressing the factors predisposing AIC among Black women. Thus, there is a critical need to define mechanisms and potential interventions to reduce these cardiotoxic events. Limited health literacy, inadequate access to cardio-oncology care, and underrepresentation in clinical trials further exacerbate disparities, highlighting the need for targeted interventions and interdisciplinary research [9, 10, 51, 53, 56].

Translation Research

Transgenic mouse models provide insights into disease mechanisms. But considerable differences in gene expression and physiology between humans and mice continue to hamper our ability to translate mechanistic discoveries. Traditional methods to study the human heart have been limited by the challenge of having direct access to *patient-specific* primary cardiomyocytes, as cardiac biopsies are high-risk and not routinely performed.

Human-inducible pluripotent stem cell cardiomyocyte (hiPSC-CM) models have been used to recapitulate clinical toxicities of various anticancer therapies [57, 58] and offer a working model to dissect mechanisms. hiPSC-CMs offer unique advantages over traditional animal models for predicting cardiotoxicity. Since hiPSC-CMs are derived from human cells, this model system provides species-specific insights into cardiac physiology and drug responses, which animal models often fail to replicate due to interspecies differences in cardiac structure, function, and metabolism. Fo example, the rapid heart rate in rats and mice requires different electrophysiological and calcium handling systems compared to humans [59, 60]. In terms of anticancer therapy-induced-arrhythmogenicity, hiPSC-CMs can model electrophysiological abnormalities, contractile dysfunction, and structural toxicity, making them suitable for detecting mechanisms like delayed repolarization or torsade de pointes (TdP) [59].

hiPSCs also offer personalization, allowing for assessment of individual susceptibility to anticancer drugs and reduce the reliance on animal testing [61]. hiPSCs derived from cancer patients allow personalized studies of chemotherapy-induced cardiotoxicity, such as doxorubicin-related heart dysfunction, aiding in the development of targeted interventions [62, 63].

In terms of limitations, evaluations of hiPSC-CMs have demonstrated that they resemble fetal cardiomyocytes in their transcriptome and phenotype, limiting their ability to fully mimic adult cardiac response [64]. Further, owing to their in vitro nature, hiPSC-CMs cannot replicate the complexity of integrated whole organ responses, which animal models provide (e.g., integrated physiological responses such as systemic humoral or hormonal effects). To address these initial challenges with immaturity, tissue engineering and recent advancements in two- and three-dimensional culture systems have made it possible to use hiPSC-CMs as screening platforms to assess the preclinical cardiotoxic effects of anticancer treatments, including anthracyclines, monoclonal antibodies, and tyrosine kinase inhibitors (TKIs) [58, 65].

The Future of Cardio-Oncology

In the coming years, researchers and clinicians envision the development of evidence-based guidelines informed by large-scale RCTs focusing on prevention and management strategies of anticancer therapy-induced cardiotoxicity [9, 12].

Artificial Intelligence in Cardio-Oncology

Modern and developing research on various artificial intelligence models has shown promise for their implementation in the cardio-oncology space. The integration of AI with patient health records could improve risk stratification and enable increased vigilance with serial imaging of high-risk patients [9, 10]. AI and machine learning enabled evaluation of patient health and demographic data could also provide important insight into disparities in onset and treatment of cardiotoxicity in underrepresented groups [51, 66, 67]. Additionally, leveraging machine learning with historical patient health records allows for accurate prediction of short-term cardiotoxicity risk in patients with colorectal cancer [68]. Recent studies have verified the use of AI models with electrocardiogram (ECG) images to evaluate the risk of adverse cardiac events in breast cancer patients, and to assist in diagnosing anticancer therapy-associated cardiac dysfunction [69–71]. AI-ECG has also shown promising results in predicting atrial fibrillation in chronic lymphocytic leukemia patients [72]. In the future, confirmation of these findings in other cancer types treated with various therapies would enable simplified risk assessment for cardiotoxicity in cancer patients. Additionally, because ECG protocols are standardized, AI-ECG screening provides an attractive solution to the current lack of guidelines surrounding detection and management of cardiotoxicity due to anticancer treatment. Further, Papadopoulou et al. showed that nonexpert staff were able to use AI-assisted handheld ultrasound devices to screen cancer patients for altered LVEF with high specificity and sensitivity [73]. These findings are encouraging for interdisciplinary cardio-oncology care teams, indicating that clinicians other than cardiologists can accurately screen for subclinical and early clinical cardiotoxicity symptoms. Future developments in AI-assisted imaging and patient data analysis will be crucial to enabling early screening, detection, and management of anticancer therapy-associated cardiotoxicity.

In addition to the use of AI, genomic data could be leveraged to identify patients at higher risk for cardiotoxicity to guide individualized prevention strategies [9, 74]. Emerging work suggests that specific single nucleotide polymorphisms are associated with increased risk of cardiotoxicity from certain drugs. Equity-focused research will be vital to addressing health disparities with the goal of expanding access to cardio-oncology services to vulnerable and at-risk cancer patients [9, 10].

To address long-term sequelae, comprehensive cardio-oncology programs integrating survivorship care plans can ensure long-term cardiovascular health while maintaining effective cancer treatment outcomes [9]. Further, more preclinical and longitudinal clinical studies are needed to elucidate the chronic cardiovascular impacts of emerging therapies like ICIs [12].

Conclusion

Cardio-oncology represents a new clinical frontier, which has made significant strides in imaging technologies, collaborative care models, and therapeutic innovations. Objective pretreatment assessments of cardiometabolic risks (e.g., blood pressure, blood cholesterol, hemoglobin A1c), behavioral health risk (e.g., exercise, diet), and baseline cardiac biomarker testing should be recorded. Despite this, significant gaps in evidence-based guidelines, equitable care, and long-term monitoring remain unexplored. Toward achieving the best cardiovascular outcomes for cancer patients and survivors, future efforts should focus on large-scale RCT research, health equity, and AI integration. Basic scientists and cardio-oncology researchers should also prioritize evaluation of long-term cardiovascular sequelae of novel cancer therapies, particularly for ICIs and targeted therapies such as immunomodulators, CAR-T therapy, small molecule drugs, and monoclonal antibodies. Furthermore, preclinical and clinical work exploring the role of anti-inflammatory agents in the prevention of ICI- and CAR-T-mediated myocarditis is warranted.

Disclosures The authors have no relevant affiliations or financial involvement to disclose.

Funding This work was supported by National Institute of Health grant, T32HL134616 to SSS and NIH R01 HL160590 to SAS.

References

1. Lefrak EA, Pitha J, Rosenheim S, Gottlieb JA. A clinicopathologic analysis of adriamycin cardiotoxicity. Cancer. 1973;32:302–14.
2. Herrmann J, Lerman A. An update on cardio-oncology. Trends Cardiovasc Med. 2014;24:285–95.
3. Cannizzaro MT, Inserra MC, Passaniti G, Celona A, D'Angelo T, Romeo P, et al. Role of advanced cardiovascular imaging in chemotherapy-induced cardiotoxicity. Heliyon. 2023;9:e15226.
4. Zhang X, Sun Y, Zhang Y, Fang F, Liu J, Xia Y, et al. Cardiac biomarkers for the detection and management of cancer therapy-related cardiovascular toxicity. J Cardiovasc Dev Dis. 2022;9:372.
5. Pudil R, Mueller C, Čelutkienė J, Henriksen PA, Lenihan D, Dent S, et al. Role of serum biomarkers in cancer patients receiving cardiotoxic cancer therapies: a position statement from the Cardio-Oncology Study Group of the Heart Failure Association and the Cardio-Oncology Council of the European Society of Cardiology. Eur J Heart Fail. 2020;22:1966–83.
6. Yu AF, Ky B. Roadmap for biomarkers of cancer therapy cardiotoxicity. Heart. 2016;102:425–30.
7. Dabour MS, George MY, Daniel MR, Blaes AH, Zordoky BN. The cardioprotective and anticancer effects of SGLT2 inhibitors: JACC: CardioOncology state-of-the-art review. JACC: CardioOncology. 2024;6:159–82.
8. Birch S, Otton J. Cardio-oncology and radiation oncology: how collaboration between cardiologists and oncologists can be realised in radiation oncology. J Med Imaging Radiat Oncol. 2024;68:962–73.
9. Pons-Riverola A, Morillas H, Berdejo J, Pernas S, Pomares H, Asiain L, et al. Developing Cardio-Oncology Programs in the new era: beyond ventricular dysfunction due to cancer treatments. Cancers (Basel). 2023;15:5885.
10. Baldassarre LA, Ganatra S, Lopez-Mattei J, Yang EH, Zaha VG, Wong TC, et al. Advances in multimodality imaging in cardio-oncology: JACC state-of-the-art review. J Am Coll Cardiol. 2022;80:1560–78.
11. Addison D, Neilan TG, Barac A, Scherrer-Crosbie M, Okwuosa TM, Plana JC, et al. Cardiovascular imaging in contemporary cardio-oncology: a scientific statement from the American Heart Association. Circulation. 2023;148:1271–86.
12. Moslehi JJ. Cardio-oncology: a new clinical frontier and novel platform for cardiovascular investigation. Circulation. 2024;150:513–5.
13. Lefebvre B, Kang Y, Smith AM, Frey NV, Carver JR, Scherrer-Crosbie M. Cardiovascular effects of CAR T cell therapy: a retrospective study. JACC CardioOncol. 2020;2:193–203.
14. Scott SS, Greenlee AN, Schwendeman EJ, Mohammad SJ, Naughton MT, Matzko A, et al. Intracellular cardiac signaling pathways altered by cancer therapies. In: Parinandi NL, Hund TJ, editors. Cardiovascular signaling in health and disease [Internet]. Cham: Springer

International Publishing; 2022. p. 111–73. [cited 2024 Jun 17]. Available from: https://link.springer.com/10.1007/978-3-031-08309-9_4.
15. Alvi RM, Frigault MJ, Fradley MG, Jain MD, Mahmood SS, Awadalla M, et al. Cardiovascular events among adults treated with Chimeric Antigen Receptor T-cells (CAR-T). J Am Coll Cardiol. 2019;74:3099–108.
16. Zamorano JL, Lancellotti P, Rodriguez Muñoz D, Aboyans V, Asteggiano R, Galderisi M, et al. 2016 ESC Position Paper on cancer treatments and cardiovascular toxicity developed under the auspices of the ESC Committee for practice guidelines: the Task Force for cancer treatments and cardiovascular toxicity of the European Society of Cardiology (ESC). Eur Heart J. 2016;37:2768–801.
17. Biersmith MA, Tong MS, Guha A, Simonetti OP, Addison D. Multimodality cardiac imaging in the era of emerging cancer therapies. J Am Heart Assoc. 2020;9:e013755.
18. Armenian SH, Lacchetti C, Barac A, Carver J, Constine LS, Denduluri N, et al. Prevention and monitoring of cardiac dysfunction in survivors of adult cancers: American Society of Clinical Oncology clinical practice guideline. J Clin Oncol. 2017;35:893–911.
19. Virizuela JA, García AM, De Las PR, Santaballa A, Andrés R, Beato C, et al. SEOM clinical guidelines on cardiovascular toxicity (2018). Clin Transl Oncol. 2019;21:94–105.
20. Liu J, Banchs J, Mousavi N, Plana JC, Scherrer-Crosbie M, Thavendiranathan P, et al. Contemporary role of echocardiography for clinical decision making in patients during and after cancer therapy. JACC Cardiovasc Imaging. 2018;11:1122–31.
21. Clasen SC, Scherrer-Crosbie M. Applications of left ventricular strain measurements to patients undergoing chemotherapy. Curr Opin Cardiol. 2018;33:493–7.
22. Awadalla M, Mahmood SS, Groarke JD, Hassan MZO, Nohria A, Rokicki A, et al. Global longitudinal strain and cardiac events in patients with immune checkpoint inhibitor-related Myocarditis. J Am Coll Cardiol. 2020;75:467–78.
23. Sudasena D, Balanescu DV, Donisan T, Hassan S, Palaskas N, Kim P, et al. Fulminant vascular and cardiac toxicity associated with Tyrosine Kinase inhibitor Sorafenib. Cardiovasc Toxicol. 2019;19:382–7.
24. Agha A, Zarifa A, Kim P, Iliescu C, Gladish G, Hassan S, et al. The role of cardiovascular imaging and serum biomarkers in identifying cardiotoxicity related to cancer therapeutics. Methodist Debakey Cardiovasc J. 2019;15:258–66.
25. Kopp LM, Womer RB, Schwartz CL, Ebb DH, Franco VI, Hall D, et al. Effects of dexrazoxane on doxorubicin-related cardiotoxicity and second malignant neoplasms in children with osteosarcoma: a report from the Children's Oncology Group. Cardio-Oncology. 2019;5:15.
26. Hlatky MA, Greenland P, Arnett DK, Ballantyne CM, Criqui MH, Elkind MSV, et al. Criteria for evaluation of novel markers of cardiovascular risk: a scientific statement from the American Heart Association. Circulation. 2009;119:2408–16.
27. Armenian S, Bhatia S. Predicting and preventing Anthracycline-related cardiotoxicity. Am Soc Clin Oncol Educ Book. 2018;38:3–12.
28. Kim Y-H, Park S-M, Kim M, Kim SH, Lim S-Y, Ahn J-C, et al. Cardioprotective effects of rosuvastatin and carvedilol on delayed cardiotoxicity of doxorubicin in rats. Toxicol Mech Methods. 2012;22:488–98.
29. Gulati G, Heck SL, Ree AH, Hoffmann P, Schulz-Menger J, Fagerland MW, et al. Prevention of cardiac dysfunction during adjuvant breast cancer therapy (PRADA): a 2 × 2 factorial, randomized, placebo-controlled, double-blind clinical trial of candesartan and metoprolol. Eur Heart J. 2016;37:1671–80.
30. Georgakopoulos P, Roussou P, Matsakas E, Karavidas A, Anagnostopoulos N, Marinakis T, et al. Cardioprotective effect of metoprolol and enalapril in doxorubicin-treated lymphoma patients: a prospective, parallel-group, randomized, controlled study with 36-month follow-up. Am J Hematol. 2010;85:894–6.
31. Lipshultz SE, Lipsitz SR, Sallan SE, Simbre VC, Shaikh SL, Mone SM, et al. Long-term enalapril therapy for left ventricular dysfunction in doxorubicin-treated survivors of childhood cancer. J Clin Oncol. 2002;20:4517–22.

32. Guglin M, Krischer J, Tamura R, Fink A, Bello-Matricaria L, McCaskill-Stevens W, et al. Randomized trial of Lisinopril versus Carvedilol to prevent Trastuzumab cardiotoxicity in patients with breast cancer. J Am Coll Cardiol. 2019;73:2859–68.
33. Boekhout AH, Gietema JA, Milojkovic Kerklaan B, van Werkhoven ED, Altena R, Honkoop A, et al. Angiotensin II-receptor inhibition with Candesartan to prevent Trastuzumab-related Cardiotoxic effects in patients with early breast cancer: a randomized clinical trial. JAMA Oncol. 2016;2:1030–7.
34. Nabati M, Janbabai G, Esmailian J, Yazdani J. Effect of Rosuvastatin in preventing chemotherapy-induced cardiotoxicity in women with breast cancer: a randomized, single-blind, placebo-controlled trial. J Cardiovasc Pharmacol Ther. 2019;24:233–41.
35. Neilan TG, Quinaglia T, Onoue T, Mahmood SS, Drobni ZD, Gilman HK, et al. Atorvastatin for Anthracycline-associated cardiac dysfunction: the STOP-CA randomized clinical trial. JAMA. 2023;330:528–36.
36. Vaziri Z, Saleki K, Aram C, Alijanizadeh P, Pourahmad R, Azadmehr A, et al. Empagliflozin treatment of cardiotoxicity: a comprehensive review of clinical, immunobiological, neuroimmune, and therapeutic implications. Biomed Pharmacother. 2023;168:115686.
37. Shaaban A, Scott SS, Greenlee AN, Binda N, Noor A, Webb A, et al. Atrial fibrillation in cancer, anticancer therapies, and underlying mechanisms. J Mol Cell Cardiol. 2024;194:118–32.
38. Abushouk AI, Ismail A, Salem AMA, Afifi AM, Abdel-Daim MM. Cardioprotective mechanisms of phytochemicals against doxorubicin-induced cardiotoxicity. Biomed Pharmacother. 2017;90:935–46.
39. Anker MS, Hadzibegovic S, Lena A, Belenkov Y, Bergler-Klein J, de Boer RA, et al. Recent advances in cardio-oncology: a report from the "Heart Failure Association 2019 and World Congress on Acute Heart Failure 2019". ESC Heart Fail. 2019;6:1140–8.
40. Avila MS, Siqueira SRR, Ferreira SMA, Bocchi EA. Prevention and treatment of chemotherapy-induced cardiotoxicity. Methodist Debakey Cardiovasc J. 2019;15:267–73.
41. Scott SS, Greenlee AN, Matzko A, Stein M, Naughton MT, Zaramo TZ, et al. Intracellular signaling pathways mediating Tyrosine Kinase inhibitor cardiotoxicity. Heart Fail Clin. 2022;18:425–42.
42. Awadalla M, Hassan MZO, Alvi RM, Neilan TG. Advanced imaging modalities to detect cardiotoxicity. Curr Probl Cancer. 2018;42:386–96.
43. Seraphim A, Westwood M, Bhuva AN, Crake T, Moon JC, Menezes LJ, et al. Advanced imaging modalities to monitor for cardiotoxicity. Curr Treat Options in Oncol. 2019;20:73.
44. Kalnina M, Nikolajevs A, Vikentjeva A, Vagalis E. [18F]FDG PET/CT as a novel imaging biomarker in the diagnosis of anthracycline cardiotoxicity in hematologic patients. European Heart Journal - Cardiovascular Imaging. 2024;25:jeae142.116.
45. Becker MMC, Arruda GFA, Berenguer DRF, Buril RO, Cardinale D, Brandão SCS. Anthracycline cardiotoxicity: current methods of diagnosis and possible role of 18F-FDG PET/CT as a new biomarker. Cardiooncology. 2023;9:17.
46. Suthar PP, Singh JS, Gupta K. 18F-FDG PET/CT imaging features of cardiac Arrhythmia in a patient treated with Panitumumab. J Nucl Med Technol. 2021;49:360–1.
47. Palomino-Fernández D, Bueno H, Jiménez-López-Guarch C, Moreno G, Seiffert AP, Gómez EJ, et al. Reduced homogeneous myocardial [18F]FDG uptake in routine PET/CT studies as an early indicator of chemotherapy-induced cardiotoxicity. Appl Sci. 2024;14:11653.
48. Tahir E, Azar M, Shihada S, Seiffert K, Goy Y, Beitzen-Heineke A, et al. Myocardial injury detected by T1 and T2 mapping on CMR predicts subsequent cancer therapy-related cardiac dysfunction in patients with breast cancer treated by epirubicin-based chemotherapy or left-sided RT. Eur Radiol. 2022;32:1853–65.
49. Park HS, Hong YJ, Han K, Kim PK, An E, Lee JY, et al. Ultrahigh-field cardiovascular magnetic resonance T1 and T2 mapping for the assessment of anthracycline-induced cardiotoxicity in rat models: validation against histopathologic changes. J Cardiovasc Magn Reson. 2021;23:76.

50. Yu AF, Chan AT, Steingart RM. Cardiac magnetic resonance and cardio-oncology: does T2 signal the end of Anthracycline cardiotoxicity? J Am Coll Cardiol. 2019;73:792–4.
51. Fazal M, Malisa J, Rhee J-W, Witteles RM, Rodriguez F. Racial and ethnic disparities in cardio-oncology: a Call to action. JACC: CardioOncology. 2021;3:201–4.
52. Al-Sadawi M, Hussain Y, Copeland-Halperin RS, Tobin JN, Moskowitz CS, Dang CT, et al. Racial and socioeconomic disparities in cardiotoxicity among women with HER2-positive breast cancer. Am J Cardiol. 2021;147:116–21.
53. Ohman RE, Yang EH, Abel ML. Inequity in cardio-oncology: identifying disparities in cardiotoxicity and links to cardiac and cancer outcomes. J Am Heart Assoc. 2021;10:e023852.
54. Jatoi I, Sung H, Jemal A. The emergence of the racial disparity in U.S. breast-cancer mortality. N Engl J Med. 2022;386:2349–52.
55. Lu Y, Gehr AW, Anikpo I, Meadows RJ, Craten KJ, Narra K, et al. Cardiotoxicity among socioeconomically marginalized breast cancer patients. Breast Cancer Res Treat. 2022;195:401–11.
56. Taylor LL, Hong AS, Hahm K, Kim D, Smith-Morris C, Zaha VG. Health literacy, individual and community engagement, and cardiovascular risks and disparities: JACC: CardioOncology state-of-the-art review. JACC CardioOncol. 2024;6:363–80.
57. Stella Stoter AM, Hirt MN, Stenzig J, Weinberger F. Assessment of cardiotoxicity with stem cell-based strategies. Clin Ther. 2020;42:1892–910.
58. Lyra-Leite DM, Burridge PW. Pluripotent stem cell modeling of anticancer therapy-induced cardiotoxicity. Curr Cardiol Rep. 2020;22:56.
59. Gintant G, Burridge P, Gepstein L, Harding S, Herron T, Hong C, et al. Use of human induced pluripotent stem cell-derived Cardiomyocytes in preclinical cancer drug cardiotoxicity testing: a scientific statement from the American Heart Association. Circ Res. 2019;125:e75–92.
60. Correia CD, Ferreira A, Fernandes MT, Silva BM, Esteves F, Leitão HS, et al. Human stem cells for cardiac disease modeling and preclinical and clinical applications—are we on the road to success? Cells. 2023;12:1727.
61. Schwach V, Cofino-Fabres C, Slaats RH, Rivera Arbelaez J, Ten Den SA, Ribeiro MC, et al. Innovative models for cardiotoxicity assessment in human pluripotent stem cell-derived cardiomyocytes that enable clinically relevant treatment regimen. Cardiovasc Res. 2024;120:cvae088.068.
62. Haupt LP, Rebs S, Maurer W, Hübscher D, Tiburcy M, Pabel S, et al. Doxorubicin induces cardiotoxicity in a pluripotent stem cell model of aggressive B cell lymphoma cancer patients. Basic Res Cardiol. 2022;117:13.
63. Huang M-F, Pang LK, Chen Y-H, Zhao R, Lee D-F. Cardiotoxicity of antineoplastic therapies and applications of induced pluripotent stem cell-derived Cardiomyocytes. Cells. 2021;10:2823.
64. Zhou B, Shi X, Tang X, Zhao Q, Wang L, Yao F, et al. Functional isolation, culture and cryopreservation of adult human primary cardiomyocytes. Sig Transduct Target Ther. 2022;7:254.
65. Schwach V, Slaats RH, Cofino Fabres C, Ten Den SA, Rivera-Arbelaez J, Van Der Zanden S, et al. Abstract 13272: Modeling cardiotoxicity utilizing human Pluripotent stem cell-derived Cardiomyocytes: towards protective therapy for the heart. Circulation [Internet]. 2021 [cited 2025 Apr 10];144. Available from: https://www.ahajournals.org/doi/10.1161/circ.144.suppl_1.13272.
66. Zheng Y, Chen Z, Huang S, Zhang N, Wang Y, Hong S, et al. Machine learning in cardio-oncology: new insights from an emerging discipline. Rev Cardiovasc Med. 2023;24:296.
67. Kim J wan, Monlezun D, Alsheikh-Kassim M, Javaid A, Honan K, Park J kun, et al. Sex, racial, income, and regional disparities in PCI Among cardio-oncology patients: propensity score and machine learning augmented nationally representative case-control study of over 30 million hospitalizations [Internet]. Research Square; 2022 [cited 2025 Jan 25]. Available from: https://www.researchsquare.com/article/rs-1705689/v1.

68. Li C, Chen L, Chou C, Ngorsuraches S, Qian J. Using machine learning approaches to predict short-term risk of cardiotoxicity among patients with colorectal cancer after starting Fluoropyrimidine-based chemotherapy. Cardiovasc Toxicol. 2022;22:130–40.
69. Oikonomou EK, Sangha V, Dhingra LS, Aminorroaya A, Coppi A, Krumholz HM, et al. Artificial intelligence–enhanced risk stratification of cancer therapeutics–related cardiac dysfunction using electrocardiographic images. Circ Cardiovasc Qual Outcomes. 2025;18:e011504.
70. Jacobs JEJ, Greason G, Mangold KE, Wildiers H, Willems R, Janssens S, et al. Artificial intelligence electrocardiogram as a novel screening tool to detect a newly abnormal left ventricular ejection fraction after anthracycline-based cancer therapy. Eur J Prev Cardiol. 2024;31:560–6.
71. Halasz G, Mazzacuva P, Mistrulli R. Artificial intelligence–assisted electrocardiography: a new and easily accessible approach for diagnosing cancer therapy–related cardiac dysfunction. Eur J Prev Cardiol. 2024;31:557–9.
72. Christopoulos G, Attia ZI, Noseworthy PA, Call TG, Ding W, Leis JF, et al. Use of artificial intelligence electrocardiography to predict atrial fibrillation (AF) in patients with chronic lymphocytic leukemia (CLL). Blood. 2020;136:50–1.
73. Papadopoulou S-L, Dionysopoulos D, Mentesidou V, Loga K, Michalopoulou S, Koukoutzeli C, et al. Artificial intelligence-assisted evaluation of cardiac function by oncology staff in chemotherapy patients. Eur Heart J Digit Health. 2024;5:278–87.
74. Campia U, Moslehi JJ, Amiri-Kordestani L, Barac A, Beckman JA, Chism DD, et al. Cardio-oncology: vascular and metabolic perspectives: a scientific statement from the American Heart Association. Circulation. 2019;139:e579–602.

Index

A

ACE-inhibitors, 95, 99, 101, 118, 119, 170, 172, 196
Alkylating agents, 108–131
Anthracycline-induced cardiotoxicity (AIC), 4–15, 35, 36, 39, 40, 42, 92–101, 109, 116, 123, 199
Anthracyclines, 4, 24, 34, 55, 92, 108, 160, 194
Anticancer therapies, 12, 22, 24, 28, 45, 125, 159–178, 195, 197, 199
Artificial intelligence (AI), 195, 200–201
Autophagy, 14, 34, 35, 39, 44, 67, 117, 160, 168, 177

B

Bakuchiol, 66, 74, 75
Berberine, 53, 58, 65, 73–74, 163, 164, 166
Beta-blockers, 43, 95, 99, 101
Bevacizumab, 34, 167–169, 173
Biomarkers, 13, 15, 38, 65, 78, 92, 94, 95, 97, 99–101, 125, 127, 128, 131, 161, 178, 194, 196, 201
Breast cancer, 4, 7, 9–14, 23–25, 27, 43, 47, 69, 78, 109, 110, 121, 129, 160, 167, 170, 172, 194, 197–199, 201

C

Cardiac magnetic resonance imaging (CMRI), 23, 100, 194
Cardiac signaling, 48
Cardio-oncology, 7, 22–28, 92, 98–101, 131, 194–202
Cardioprotection, 40, 41, 53–78, 92, 93, 97, 98, 121, 168, 172–174, 196–198
Cardiotoxicity, 4, 23, 34, 53, 92, 108, 161, 194
Carfilzomib, 176–178
Chemotherapy, 5, 9–13, 23, 24, 34, 35, 38, 47, 48, 64, 108, 117, 131, 159–161, 174, 178, 194, 197
Coenzyme Q (CoQ), 53, 61, 68, 76, 77
Cryptotanshinone (CTS), 68, 76, 77
Curcumin, 11, 62, 64, 71, 128, 129, 168, 169, 173

D

Disparities, 4, 6, 7, 22–28, 94, 199–201
Doxorubicin, 8, 27, 34, 61, 92, 109, 170, 194

F

5-fluorouracil (5-FU), 159–178
Flavonoids, 53, 56, 57, 64, 70, 78, 97, 100, 115, 165, 173, 176

H

Heart failure, 6–8, 11, 24, 25, 37, 53, 65, 68, 92, 95, 96, 98, 99, 108, 118, 167, 170–172, 194, 196, 197, 199
HiPSC-CMs, 45, 46, 199, 200

S. A. Smith, S. S. Scott (eds.), *Myocardial Oxidative Stress in Anti-Cancer Therapies*, Oxidative Stress in Applied Basic Research and Clinical Practice, https://doi.org/10.1007/978-3-031-97517-2

I
Imaging, 38, 92, 94, 99–101, 194–196, 198, 200, 201
Immunotherapy, 11, 34

L
Left ventricular dysfunction, 4, 8, 92, 108, 167, 196, 197

M
Mitochondrial dysfunction, 9, 34, 43, 53, 54, 63, 68, 70, 93, 117, 128, 129, 170, 198

O
Oxidative stress, 5, 34, 53, 92, 108, 159
Oxygen, 54, 58, 60, 61, 109–111, 115, 121, 127, 159, 170, 176

P
Pharmacotherapies, 55, 108–131, 160, 165, 167
Phytochemicals, 53–78, 112, 114–116, 128, 163, 168, 173, 176, 178, 198
Polyphenols, 53, 56–63, 78, 97–98, 100, 114, 124, 168, 175
Proteasome inhibitors (PIs), 176

Q
Quinones, 53, 61, 68, 76–78, 109

R
Race, 4, 5, 7, 24–27, 199
Radiotherapy, 5, 23–25, 94
Reactive oxygen species (ROS), 9, 11, 34, 35, 43–45, 53–56, 58, 60, 61, 63, 64, 66–72, 74, 76–78, 93, 95, 99, 100, 109–116, 119–121, 123, 125–130, 159–170, 173–176, 178
Risk factors, 4–15, 26, 37, 48, 94, 98, 100, 199

S
Sex, 4, 5, 7, 23–26
Sodium glucose transporter 2 (SGLT2), 96, 99, 100, 109, 116, 127, 161, 195, 197
Sorafenib, 34, 161, 163–165
Spironolactone, 35

T
Taxane, 108–131
Terpenes, 53, 59–61, 66, 74–78
Trastuzumab, 4, 9, 10, 14, 25, 34, 43–45, 167, 169–173, 194, 197
Tyrosine kinase inhibitors (TKIs), 45, 46, 55, 159–178, 194, 200

GPSR Compliance

The European Union's (EU) General Product Safety Regulation (GPSR) is a set of rules that requires consumer products to be safe and our obligations to ensure this.

If you have any concerns about our products, you can contact us on ProductSafety@springernature.com

In case Publisher is established outside the EU, the EU authorized representative is:

Springer Nature Customer Service Center GmbH
Europaplatz 3
69115 Heidelberg, Germany

Batch number: 10755332

Printed by Printforce, the Netherlands